Land Surface Monitoring Based on Satellite Imagery

Land Surface Monitoring Based on Satellite Imagery

Editors

Sara Venafra
Carmine Serio
Guido Masiello

MDPI • Basel • Beijing • Wuhan • Barcelona • Belgrade • Manchester • Tokyo • Cluj • Tianjin

Editors

Sara Venafra
School of Engineering,
University of Basilicata,
Potenza, Italy

Carmine Serio
School of Engineering,
University of Basilicata,
Potenza, Italy

Guido Masiello
School of Engineering,
University of Basilicata,
Potenza, Italy

Editorial Office
MDPI
St. Alban-Anlage 66
4052 Basel, Switzerland

This is a reprint of articles from the Special Issue published online in the open access journal *Land* (ISSN 2073-445X) (available at: https://www.mdpi.com/journal/land/special_issues/land_surface_monitoring_based_on_satellite_imagery).

For citation purposes, cite each article independently as indicated on the article page online and as indicated below:

LastName, A.A.; LastName, B.B.; LastName, C.C. Article Title. *Journal Name* **Year**, *Volume Number*, Page Range.

ISBN 978-3-0365-5929-2 (Hbk)
ISBN 978-3-0365-5930-8 (PDF)

Cover image courtesy of Guido Masiello.

Contents

About the Editors

Sara Venafra

Sara Venafra, (Ph.D.) has been a technologist at the Italian Space Agency since 2022. During her academic studies and working experiences, Sara Venafra has worked on several Topics, all of which were mainly concerned with remote sensing. Sara Venafra received his master's degree in electronic engineering at the Telecommunications Program of Study from the Politecnico of Bari in 2010. At the beginning of her career, she focused her research activities on sea surface wind speed retrieval using SAR (Synthetic Aperture Radar) satellite data and its applications in the fields of renewable energies with the development of algorithms. She applied this software to estimate the sea surface wind field from the SAR satellite data in the X (Cosmo-SkyMed) and C (ERS and ENVISAT) bands, and to evaluate the energy production of an offshore wind farm on the basis of wind model retrieved from SAR satellite data analysis. During her Ph.D. (2011–2015) and post-doc positions at Università degli Studi della Basilicata she worked on *Applied Spectroscopy* Topics, developing Level 2 processors for the inversion of the thermodynamic state of the Earth's atmosphere via the use of physical radiative transfer models. She currently works on the development of a fast radiative transfer model and relative inverse module for the retrieval of state vectors in all-sky conditions (cloudy and clear sky) in the Far-InfraRed spectral band (FIR, 100-667 cm-1), also giving information about the microphysical properties of clouds by exploiting the next ESA mission, FORUM (Far-infrared Outgoing Radiation Understanding and Monitoring). She is a co-investigator in various projects in the area of high spectral resolution infrared sounders from satellites (far and thermal, from 100 to 12 micron). She was a visiting and associate scientist at EUMETSAT headquarters. She has published her research in more than 40 publications, acting as a reviewer for several International Journals of remote sensing, physics, and atmospheric science.

Carmine Serio

Prof. Dr Carmine Serio received his doctorate in Physics with honor in 1978 from University of Napoli (Italy). He was appointed assistant professor in 1984 by University of Napoli and after 16 years of distinguished academic career in 2000 he became Full Professor of Experimental Physics. At present he teaches and works at the School of Engineering, University of Basilicata (Italy), where he moved in 1992. In 1998 he founded the Applied Spectroscopy group, which he is still leading. Since 1990 Carmine Serio has been performing research in the field of Fourier Spectroscopy, methods and experimental techniques, applied to the study of Environment, Earth Atmosphere, land processes and remote sensing of atmospheric and surface parameters. His experience of Fourier transform infrared spectrometers or FTIR instruments includes ground-based, airborne and satellite platforms. His most important achievements include the first ground-based field observations of atmospheric emitted radiance in the far infrared, which led to a complete validation and revision of state-of-the-art water–vapor continuum absorption models. The observations were performed with a novel detector concept and beam splitter technology for Fourier Transform spectrometers (FTS) which led to the fabrication of two instruments: REFIR and I-BEST.

In the last 20 years, he has contributed to the development of radiative transfer models for application to high spectral resolution infrared observations from satellite (such as IASI: Infrared Atmospheric Sounding Interferometer) readings and has developed the technique of correlation interferometry for the retrieval of atmospheric trace gas molecules from Fourier transform spectrometer observations. He is currently a member of the ESA/CNES/EUMETSAT teams for the design, development and remote sensing applications of FTIR from polar and geostationary Space platforms. These include the ISSWG-2 (IASI Next-Generation Sounding Science Working Group) team and the MTG-IRS team.

Guido Masiello

Guido Masiello teaches at the School of Engineering, University of Basilicata, Potenza, Italy, where he is currently an Associate Professor of Earth and Atmospheric Physics. He works on applied spectroscopy Topics, developing methods and instruments for the retrieval of the thermodynamic state of the Earth's atmosphere. He is a co-investigator in various projects in the area of high spectral resolution infrared sounders from satellites (far and thermal, from 100 to 3.5 micron): IASI, the Infrared Atmospheric Sounder Interferometer of the French Space Agency, CNES, and EUMETSAT, REFIR, the Radiation Explorer in the Far Infrared (a three-years EU project supported within the 4th framework program), MTG-IRS, the Meteosat Third Generation Infrared Sounder (a joint program ESA/EUMETSAT), IASI-NG, the next generation of IASI and the European Space Agency next Earth explorer mission FORUM (Far-infrared Outgoing Radiation Understanding and Monitoring).

In the period 2009-2016, he was elected as member (2009 to 2016) of the IRC (International Radiation Commission) and in 2013 and 2014 he worked as a visiting scientist in the Remote Sensing and Product Division of EUMETSAT.

Mr. Masiello was a reviewer for several International Journal of Physics, Optics, and Atmospheric Sciences and for Italian and international research funding agencies.

Preface to "Land Surface Monitoring Based on Satellite Imagery"

This book offers readers an overview of the different means of exploiting remote sensing to monitor land surfaces, a phenomenon which plays a significant role in the study of climate change and global warming. This overview highlights how novel approaches employ satellite measurements to improve the detection of land surface changes at the level of global coverage by compensating the in situ measurements lack in terms of spatial and temporal resolution. Land surface parameters from remote sensing are incredibly attractive for applications in different environmental fields, such as land use/change, monitoring of vegetation and soil water stress, and early warning and detection of forest fires and drought. Typically, the practice of monitoring land cover changes is based on the definition of vegetation indices, employing methods to exploit the surface information provided by the channels in the visible and the infrared spectra.

Acknowledgements

We thank all contributing authors for choosing our Special Issue to share their expertise. We gratefully acknowledge the work of MDPI in this publication, in particular editor Ms. Carol Ma. We thank all reviewers for their valuable insight into the content, which helped to improve the quality of all published manuscripts.

Sara Venafra, Carmine Serio, and Guido Masiello
Editors

 land

Article

The IASI Water Deficit Index to Monitor Vegetation Stress and Early Drying in Summer Heatwaves: An Application to Southern Italy

Guido Masiello [1,*], Francesco Ripullone [2], Italia De Feis [3], Angelo Rita [4], Luigi Saulino [4], Pamela Pasquariello [1], Angela Cersosimo [1], Sara Venafra [1,*] and Carmine Serio [1]

1 Scuola di Ingegneria, Università della Basilicata, 85100 Potenza, Italy
2 Scuola di Scienze Agrarie, Forestali, Alimentari ed Ambientali, Università della Basilicata, 85100 Potenza, Italy
3 Istituto per le Applicazioni del Calcolo "Mauro Picone", IAC-CNR, 80131 Napoli, Italy
4 Dipartimento di Agraria, Università di Napoli Federico II, 80055 Portici, Italy
* Correspondence: guido.masiello@unibas.it (G.M.); sara.venafra@unibas.it (S.V.)

Abstract: The boreal hemisphere has been experiencing increasing extreme hot and dry conditions over the past few decades, consistent with anthropogenic climate change. The continental extension of this phenomenon calls for tools and techniques capable of monitoring the global to regional scales. In this context, satellite data can satisfy the need for global coverage. The main objective we have addressed in the present paper is the capability of infrared satellite observations to monitor the vegetation stress due to increasing drought and heatwaves in summer. We have designed and implemented a new water deficit index (*wdi*) that exploits satellite observations in the infrared to retrieve humidity, air temperature, and surface temperature simultaneously. These three parameters are combined to provide the water deficit index. The index has been developed based on the Infrared Atmospheric Sounder Interferometer or IASI, which covers the infrared spectral range 645 to 2760 cm^{-1} with a sampling of 0.25 cm^{-1}. The index has been used to study the 2017 heatwave, which hit continental Europe from May to October. In particular, we have examined southern Italy, where Mediterranean forests suffer from climate change. We have computed the index's time series and show that it can be used to indicate the atmospheric background conditions associated with meteorological drought. We have also found a good agreement with soil moisture, which suggests that the persistence of an anomalously high water deficit index was an essential driver of the rapid development and evolution of the exceptionally severe 2017 droughts.

Keywords: climate change; drought; water deficit index; infrared observations; satellite; remote sensing; surface temperature; air temperature; humidity; dew point temperature

Citation: Masiello, G.; Ripullone, F.; De Feis, I.; Rita, A.; Saulino, L.; Pasquariello, P.; Cersosimo, A.; Venafra, S.; Serio, C. The IASI Water Deficit Index to Monitor Vegetation Stress and Early Drying in Summer Heatwaves: An Application to Southern Italy. *Land* **2022**, *11*, 1366. https://doi.org/10.3390/land11081366

Academic Editor: Dailiang Peng

Received: 27 July 2022
Accepted: 19 August 2022
Published: 21 August 2022

1. Introduction

The ECMWF (European Centre for Medium-Range Weather Forecasts) has determined that the winter of 2020 was the hottest winter season ever recorded in Europe (e.g., see https://climate.copernicus.eu/boreal-winter-season-1920-was-far-warmest-winter-season-ever-recorded-europe-0 (accessed on 15 August 2022)). This is an event that is now repeated year after year [1,2], as evidenced by the Copernicus Climate Change Service (C3S) dataset (e.g., see https://climate.copernicus.eu/esotc/2021/globe-in-2021 (accessed on 15 August 2022)), which shows that the last seven years have been the warmest on record, with 2021 varying from the fifth to the seventh warmest.

The present analysis is most relevant to temperate regions and the Mediterranean vegetation. In this respect, ref. [3] discussed the risks of climate change altering sustainable development in the Mediterranean area. Furthermore, in [4], it has been shown that long-lasting droughts induce dieback phenomena in temperate and Mediterranean climate regions, an issue that has also been addressed in [5], which analysed the effect of the 2017 summer heatwave in Europe.

The continental extension of the phenomenon calls for tools and techniques capable of monitoring the global to regional scales. For this reason, we have set up a methodology based on satellite data with the objective of using infrared satellite observations to monitor early drying in summer because of drought and heatwaves.

Vegetation stress due to water deficit is widespread in many countries due to climate change (e.g., see [4,5]). Drought is an extreme natural event typical of semi-arid areas and much of the Mediterranean, especially regions located at middle latitudes. The lack of rain for long periods increases the danger and risk of forest fires in lands rich in vegetation and wooded areas [6]. Furthermore, the lack of rain in semi-arid regions causes water stress (e.g., [7,8]). Therefore, the deficit of rainfall and/or water, in general, requires specific actions to monitor and detect drought conditions aiming to mitigate its adverse impacts on human health, wildlife, and plant communities.

Water deficit can be estimated using (1) meteorological data (e.g., [9–11]); and (2) remote sensing (e.g., [12–15]).

The present study aims at a synergetic use of these two different methods to develop new vegetation dryness indices based on the surface temperature, complemented with atmospheric temperature and the water vapor mixing ratio or parameters depending on it, such as dew point temperature.

In general, the problem has been studied through the use of indexes such as the vegetation dryness index (or VDI), the temperature vegetation dryness index (TVDI), and the improved TVDI (or iTVDI) (among many others, see [16–18]). These indices are based on the NDVI (normalized differential vegetation index), the surface temperature, T_s, and the air temperature close to the surface, or T_a. The problem with NDVI is that it is a greenness index and cannot distinguish bare soil from senescent vegetation (e.g., see [19]). In addition, neither T_s nor T_a are directly linked with soil moisture. It should be observed that the use of T_s-NDVI relationships has been long investigated for application to drought assessment, and it has been found to produce inconsistent results in some specific situations (e.g., [20]).

Conversely, we propose to follow the strategy of using surface temperature (T_s), and the dew point temperature (T_d), which are more closely related to surface type and coverage, and soil moisture. The water deficit index is then defined according to the linear difference $T_s - T_d$.

The water deficit index is meant for analysis at the regional scale; therefore, we need the use of satellite data to ensure the correct spatial coverage and time sampling. Toward this objective, we have used the hyper-spectral satellite infrared sounder (Infrared Atmospheric Sounder Interfemoter or IASI, e.g., [21]) flying on board the European Meteorological Platforms (MetOp). By adequately exploiting IASI observations, we can simultaneously retrieve T_s and T_d, which limit problems of time-space colocation. However, satellite data are available at uneven grid points, making it challenging to check spatial patterns. In this respect, our objective is two-fold: first, we want to define and compute a suitable water deficit index based on direct satellite soundings; and second, we want to define a strategy to resample the sparse satellite retrievals on a regular grid for the better understanding of spatial patterns.

We acknowledge that water deficit indices are common in-field analyses related to horticulture, e.g., irrigation management, evaluation of crop water stress, and so on (e.g., see [6] and references therein). However, in these cases, we are generally in the presence of temporary water deficit anomalies. In contrast, our approach is meant to account for the background atmospheric humidity and temperature related to drought onset and development (e.g., see [22]). For satellite-based analysis, a similar approach has been proposed in [23], using the concept of vapour pressure deficit (VPD), the difference between the saturation and actual vapour pressure for a given time. In contrast, our approach uses

T_d, which is related to VPD, and T_s to build the difference $T_s - T_d$, allowing us to better separate the hot-dry from humid-warm weather conditions.

The paper is organized as follows. Section 2 deals with data and methods; in particular, the section illustrates the IASI retrieval system we have developed and used for the present analysis. Results are shown in Section 3 and discussed in Section 4. Finally, conclusions are drawn in Section 5.

2. Materials and Methods

2.1. Material and Data

The retrieval from space observations of T_s and T_d have been performed using the Infrared Atmospheric Sounder Interferometer (or IASI) [21]. IASI has been developed in France by CNES and is flying on board the Metop platforms, which are satellites of the EUMETSAT European Polar System (EPS). IASI has been primarily designed as a meteorological mission; hence, its main objective is to provide relevant information on temperature and water vapour profiles. The spectral coverage of the instrument extends from 645 to 2760 cm^{-1}, and its sampling interval is $\Delta\sigma = 0.25$ cm^{-1}; therefore, the instrument provides 8461 channels, i.e., spectral observations for every spectrum.

IASI is a cross-track scanner with 30 adjacent fields of regard (FOR) per scan, spanning an angular range of $\pm 48.33°$ on either side of the nadir. The FOR viewing geometry consists of a 2×2 matrix of instantaneous fields of views (IFOVs). In turn, the single IFOV has a diameter of $0.8394°$, corresponding to a ground resolution of 12 km per nadir for a satellite altitude of 819 km. The 2×2 IFOV matrix is centered on the viewing direction. At nadir, a FOR of 4 IASI IFOVs (or pixels) covers the ground a square area of $\approx 50 \times 50$ km^2. The corresponding FORs (among the 30 views) are $\pm 1.67°$ on each side from the nadir direction. Further details about IASI and its mission objectives are referred to in [21].

Figure 1 shows the target area we have focused on in the paper. The site corresponds to southern Italy, with the Apennine chains covered by forest, as exemplified by the 2018 CORINE land cover (https://land.copernicus.eu/pan-european/corine-land-cover (accessed on 15 August 2022)). The black dots identify two dieback forest areas, where forest monitoring, by ecophysiological and dendrochronological approaches, has been running since 2013 [24].

The two locations circled in the maps of Figure 1 correspond to the forest stands of San Paolo Albanese (40.02° N, 16.34° E, 950–1050 m.a.s.l.) and Gorgoglione (40.40° N, 16.14° E, 800–850 m.a.s.l.), which are suffering from long-lasting drought-induced tree mortality (e.g., [4]). In the San Paolo Albanese site, the vegetation is formed by a pure high forest of *Quercus frainetto Ten.* for a stand density of 348 trees ha^{-1}. As far as the most affected stands are concerned, recent studies observed that more than 50% of the mature specimens showed symptoms of death, while about 15% died recently [25]. On the other hand, the Gorgoglione woodland is a highly mixed forest, with an average density of about 600 stems ha^{-1}. The vegetation is dominated by *Quercus cerris* L. (71%), followed by *Quercus pubescens* L. (25%) and, at a lower density (4%), other species of deciduous trees [25].

The two main studied tree species (i.e., *Quercus cerris* L. and *Quercus pubescens* L.) have shown recent drought-induced decline symptoms since the early 2000s (shoot dieback, summer leaf loss, withering, growth decline, and high mortality). According to local reports about the study area, the yearly oak mortality affected ca. 450 ha. The incidence of the decline syndrome raised mortality from 5 to 10%, from 2002 to 2004 [24].

IASI soundings have been acquired for the whole year of 2017 when an intense heat wave hit Europe and the Mediterranean area in summer (e.g., see [5]). For comparison, we have also acquired IASI data for 2020 and 2021.

For a proper comparison with our IASI $T_s - T_d$ index, for the same target area and the year 2017, the Copernicus Global Land Service (https://land.copernicus.eu/global/products (accessed on 15 August 2022)) was used to obtain data about the surface soil moisture (*ssm*) and the leaf area index (LAI).

Figure 1. Target region for which IASI data have been selected for the present analysis. The figure also shows the CORINE land cover for 2018 to help identify forest regions, which are primarily of interest for this study. The two upper panels help to determine the target area (magenta square) on the globe and Italy.

The surface soil moisture was derived by observing the band C SAR onboard the satellite Sentinel-1. Data were provided with a timeliness of one day at a spatial resolution of ~1 km. For details about the *ssm* product, we refer the interested reader to [26].

The leaf area index was globally estimated at a spatial resolution of about 300 m through a neural net approach. The input to the net was obtained from instantaneous top-of-canopy reflectances from the OLCI (Ocean and Land Colour Imager) instrument onboard the Sentinel-3 satellite, or daily top-of-aerosol reflectances from the PROBA-V satellite. We refer the interested reader to [27] for further details about the LAI data.

Finally, data about the ecophysiological responses of trees for the forest stands of San Paolo Albanese and Gorgoglione were measured and used in the present analysis during two field campaigns performed from July–September in 2020 and 2021.

2.2. Methods

IASI will add unique capabilities to the present study because we were able to simultaneously retrieve T_s and T_d (e.g., [28–32]) from this instrument. To this end, we developed two retrieval prototypes: one for simultaneous inversion of infrared observations (level 2 or L2 prototype), and the second for remapping L2 products on a regular grid (L3 prototype).

The layout of the overall scheme we developed is sketched in Figure 2. The procedure consists of three main steps identified in Figure 2 with grey boxes.

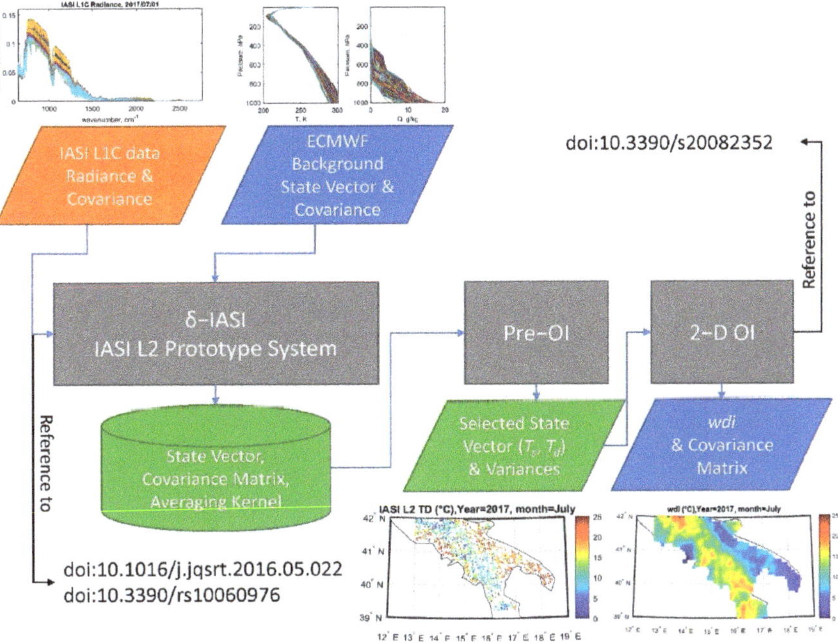

Figure 2. Schematic flow chart of the methodology developed in the present study to yield Level 3 monthly maps of the water deficit index [30,33,34].

The IASI Level 2 and 3 prototypes have been developed in previous studies. The most up-to-date versions of both schemes can be found in [30,33] for the L2 package and [34] for Level 3 Optimal Interpolation. The digital object identifier (doi) shown in Figure 2 allows the interested reader to open online references where the two schemes are analytically presented. For this reason, they are just summarized in the present paper. In contrast, the Pre-OI scheme is described in more detail, as it implements the equations and formulas needed for calculating the state vector and associated covariance matrix, which are passed to the OI scheme to compute the maps of the water deficit index.

2.2.1. The L2 Retrieval System

The L2 prototype, which we also call δ-IASI, consists of an optimal estimation scheme (e.g., [35]), which simultaneously inverts the full IASI spectrum to retrieve the state vector, which is made up of the surface emissivity (ε), the surface temperature (T_s), the atmospheric profiles of temperature (T), water vapour (Q), ozone (O), HDO (D), carbonyl sulfide or OCS, and scalar scaling factors for the column amount of CO_2, CO, N_2O, CH_4, SO_2, HNO_3, NH_3, and CF_4. However, the parameters relevant to the present analysis are T_s, and the atmospheric profiles for T and Q. Our L2 prototype for IASI has been variously validated as far as the surface parameters and T and Q profiles are concerned. Validation for surface parameters can be found, e.g., in [32,33], whereas for T and Q they can be found in [30,36].

2.2.2. The L2 Pre-OI and the Definition of the Water Deficit Index

Regarding Figure 2, the Pre-OI acts on the IASI Level 2 data to extract the geophysical parameters close to the surface and compute the water deficit index and its variance to input the final optimal interpolation scheme. From the profiles of T and Q, we considered only the elements, which correspond to the lowermost atmospheric layer, say T_1 (in units of K) and Q_1 (in units of gr/Kg). The corresponding layer pressure was denoted with P_1 (in units of hPa). From the L2 products, we also extracted the surface temperature, T_s. The three parameters (T_s, T_1, Q_1) were piled up in a vector, $\boldsymbol{x}_1$ of size $n = 3$, whose covariance matrix was denoted with $\boldsymbol{S}_1$, whose size was $n \times n$. Both $\boldsymbol{x}_1$, $\boldsymbol{S}_1$ are outputs of the IASI L2 system.

The computation of the dew point temperature, T_d, involves the calculation of the actual and saturation water vapour pressures. These are referred to as P_w and P_{ws} , respectively. From Q_1, we can compute P_w according to:

$$P_w = 10^{-3} P_1 Q_1 \frac{R_w}{R_{air}} = \beta P_1 Q_1; \beta = 10^{-3} \frac{R_w}{R_{air}} \tag{1}$$

with P_w in hPa and where $R_w = 461.5$ J K^{-1} Kg^{-1} and $R_{air} = 286.9$ J K^{-1}Kg^{-1} are the specific gas constants of water vapour and air, respectively. According to [37], P_{ws} is computed with the formula:

$$P_{ws} = 10^{-2} \frac{\exp\left(a_1 - \frac{a_2}{t_1 + a_3}\right)}{(t_1 + a_4)^{a_5}} \tag{2}$$

with $t_1 = T_1 - 273.15$ (temperature in degrees Celsius) and P_{ws} in hPa. Equation (2) is valid for $t_1 > 0$ (vapor pressure of water), and where $a_1 = 34.494$, $a_2 = 4924.99$, $a_3 = 237.1$, $a_4 = 105$, $a_5 = 1.57$ are fit parameters that in case t_1 are expressed in degrees Celsius. From (1) and (2), we obtain the fractional relative humidity:

$$rh = \frac{P_w}{P_{ws}} \tag{3}$$

From (1) and (2), we can also compute the vapour pressure deficit or $VPD = P_{ws} - P_w$. Finally, the dew point temperature, T_d can be calculated by using the well-known Magnus formula (e.g., [38]):

$$t_d = \frac{cx}{b - x}, \; x = \ln(rh) + \frac{bt_1}{c + t_1} \tag{4}$$

where t_d is in degrees Celsius (we will use T_d when referring to degrees Kelvin units), and $b = 17.62$ (dimensionless), $c = 243.12$ C. Finally, the IASI-based water deficit index, *wdi*, is defined according to:

$$wdi = T_s - T_d = t_s - t_d \tag{5}$$

Equation (5) stresses that the index can be computed indifferently with both temperatures in K or C degrees, although the computation of the dew point temperature has to be performed in C, according to Equation (4), before converting it to K.

For the application of the optimal interpolation to the mapping of the water deficit index, we also need the variance of the index, σ^2_{wdi}. Considering the chain of equations from (2) to (5), we can formally write wdi as a function $wdi = f(T_s, T_1, Q_1)$, from which, using the usual rule of variance propagation (see, e.g., [39]), we obtain:

$$\sigma^2_{wdi} = \boldsymbol{g}^t \boldsymbol{S}_1 \boldsymbol{g} \tag{6}$$

with the superscript t indicating the transpose operation, and

$$\boldsymbol{g} = \left(\frac{\partial f}{\partial T_s}, \frac{\partial f}{\partial T_1}, \frac{\partial f}{\partial Q_1}\right)^t \tag{7}$$

We stress that the parameters defined by Equations (5) and (6) have to be computed for the IASI retrievals and the ECMWF background, as shown in the diagram of Figure 2. For the background, the covariance matrix is assumed to be diagonal, as we use background derived from climatology (see [29]) for which we do not consider correlation among air temperature, humidity, and surface temperature.

Considering that

$$f(T_s,\ T_1, Q_1) = T_s - T_d = t_s - t_d = t_s - \frac{cx}{b-x} \tag{8}$$

we have

$$\begin{aligned} \frac{\partial f}{\partial T_s} &= \frac{\partial f}{\partial t_s} = 1 \\ \frac{\partial f}{\partial T_1} = \frac{\partial f}{\partial t_1} = \frac{\partial f}{\partial x}\frac{\partial x}{\partial t_1} &= -\frac{cb}{(b-x)^2}\left(-\frac{a_2}{(t_1+a_3)^2} + \frac{a_5}{(t_1+a_4)} + \frac{bc}{(t_1+c)^2}\right) \\ \frac{\partial f}{\partial Q_1} = \frac{\partial f}{\partial x}\frac{\partial x}{\partial Q_1} &= -\frac{cb}{(b-x)^2}\frac{1}{Q_1} \end{aligned} \tag{9}$$

The parameter *wdi*, when referring to a surface covered by vegetation or crops, can help to understand the water stress or deficit during long-lasting droughts or heatwaves. This is because vegetation releases water into the atmosphere through transpiration. The process involves the vaporization of liquid water in plant tissues and the consequent release of vapour into the atmosphere (for example, see [40])). Similar to direct evaporation, transpiration depends on the amount of energy available: solar radiation, wind, and vapour pressure gradient at the surface–atmosphere interface. Consequently, solar radiation, air temperature and humidity, and wind velocity must be considered when evaluating and assessing a satellite-based index to quantify water deficit.

In Equation (5), the role of energy supply is modelled with T_s. The sun's radiation will cause a rapid increase in the surface temperature of the land. On the other hand, T_d will take into account both air temperature and air humidity. The effect of wind is more difficult to introduce. However, drought and heatwave conditions minimize the spatial gradient and wind intensity. The air subsidence and low intense pressure gradients characterize meteorological conditions that favour summer heatwaves.

It is also important to stress that evaporation and transpiration co-occur, and it is not easy to distinguish between the two processes. For this reason, we mention evapotranspiration when referring to the water exchange between vegetation and the air. In addition to water availability in the topsoil, evaporation from the cultivated terrain depends, as already mentioned, on the amount of impinging solar radiation. The solar energy at the surface decreases during crop growth because its foliage or canopy shadows the area below from the sun's rays as the crop develops. Therefore, water is predominately lost by soil evaporation when the crop is small, or when the leaves are not well developed. However, transpiration becomes the main process once the crop and leaves are well developed and completely cover the soil.

With this in mind, the parameter *wdi* can help to identify different regimes of water deficit:

1. $wdi \gg 0$; this regime characterizes very hot and dry conditions that favour evapotranspiration. Furthermore, in this regime, the evapotranspiration increases almost linearly with the wind speed (e.g., [40]);
2. $wdi \geq 0$; this regime characterizes warm and humid conditions when the air is already close to saturation; therefore, less additional water can be stored, so the evapotranspiration rate is even lower than for arid land;
3. $wdi \leq 0$; this is the regime $T_s < T_d$, and therefore the vapour condenses in liquid water at the surface.

2.2.3. The 2-D OI scheme

It is worth noting that when *wdi* is determined by L2 satellite observations, as in our case, we obtain data that are sparse and not homogeneously covering a given spatial region. Therefore, to better compare with other data sources and perform a correct collocation with stations at the ground, we used a resampling tool, which can remap *wdi* data to a regular grid. To this end, we used the tool developed in [34]. The technique is based on a 2-dimensional (2D) optimal interpolation (OI) scheme, and is derived from the broad class of Kalman filter or Bayesian estimation theory. For further details, we refer the interested reader to [34].

The steps involved in the mapping on a regular grid are exemplified in Figure 3 using the IASI retrieval for *wdi* for July 2017. Figure 3a,c show the IASI data points for *wdi*, and its square root of the variance (standard deviation) as estimated by the L2 retrieval scheme and the Pre-OI step (see Figure 2). These values are accumulated considering all the IASI overpasses for July 2017. As said before, the IASI scan pattern is made up of footprints with circular diameters of about 12 km at nadir, and the scanning lines are 50 km apart along the flight direction of the satellite. The IASI scan pattern over the target area for a single overpass is shown in Figure 4 for the benefit of the reader. Comparing Figure 3 with Figure 3a,c, it can be seen that after one month, the IASI clear sky footprints (we stress that we use only observations in a clear sky, which is diagnosed based on a stand-alone algorithm for cloud detection, e.g., [41]) are densely distributed over the area much more than the single IASI scan pattern overpass. The monthly ensemble of satellite overpasses improves the sampling of spatial data, and therefore allows, for example, a better comparison with in situ observations. We use the ensemble of multiple observations to build a map with a better spatial sampling. Towards this objective, we use the 2-D OI method, which remaps the data into a grid with a finer mesh than the original data.

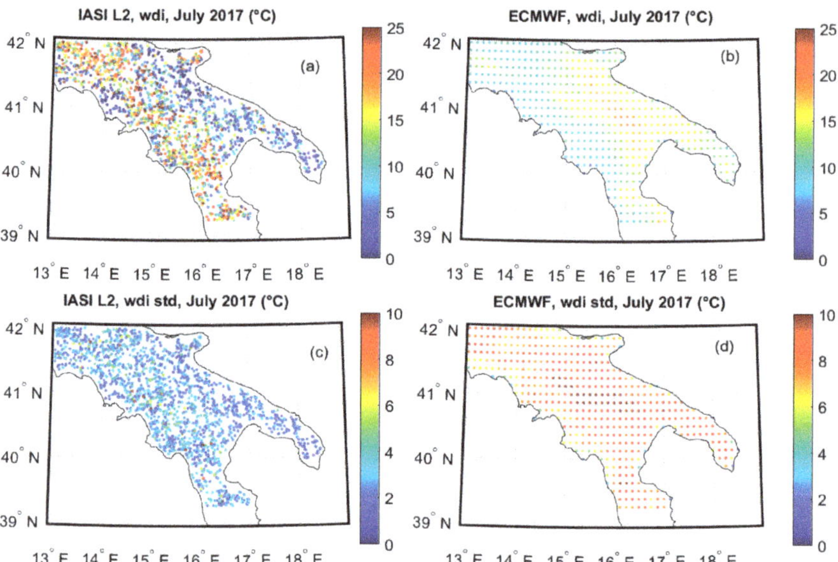

Figure 3. July 2017. IASI L2 products for *wdi* (panel (**a**)) and its standard deviation (panel (**c**)) over the target area. The figure also shows the ECMWF background field (both mean (panel (**b**) and standard deviation (**d**)) at its native spatial resolution of $0.125° \times 0.125°$.

Figure 4. Target region showing the IASI footprint scan pattern (red ovals) for one single overpass. The IASI morning overpass for May, the first of 2020 is shown in the figure.

The final mesh we use has a spatial sampling of $0.05° \times 0.05°$. Another important aspect of OI remapping is the use of background fields. These fields are built up by using the time and space co-located ECMWF (European Centre for Medium-Range Weather Forecasts) analysis. The ECMWF fields are available on a grid-mesh of $0.125° \times 0.125°$, and, for the case at hand, the values for *wdi* and its square root of the variance, i.e., standard deviation, are exemplified in Figure 3b,d, respectively. Based on the coarse ECMWF background, the un-gridded L2 IASI observations and the 2-D OI yields the results are shown in Figure 5; that is, the maps of *wdi* (panel (a)) and its standard deviation (panel (b)) at a sampling of $0.05° \times 0.05°$. In this process, we lose temporal resolution, but we obtain a map with improved spatial sampling and precision, as shown by the standard deviation map, which, apart from boundary effects, is one °C or less.

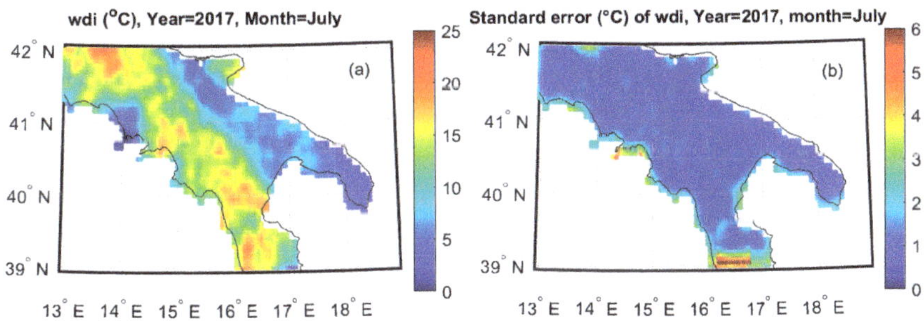

Figure 5. Level 3 map at a grid step of 0.05° for the index *wdi* obtained from the source data shown in Figure 3 (panel (**a**)) and its standard deviation (panel (**b**)). The map is exemplified for July 2017.

3. Results

The rise and fall of the exceptionally hot and dry summer are well captured by the monthly time series of *wdi* maps shown in Figure 6. Of particular interest for us is the Apennine chain, which is covered by broad-leaved, deciduous forests. If we compare Figure 5 to the land cover map shown in Figure 1, we see that the *wdi* closely follows the forested area in the summer season. In July and August 2017, the index was above ~10 °C in the regions covered by forests, which shows that the vegetation ecosystem was suffering from a water deficit.

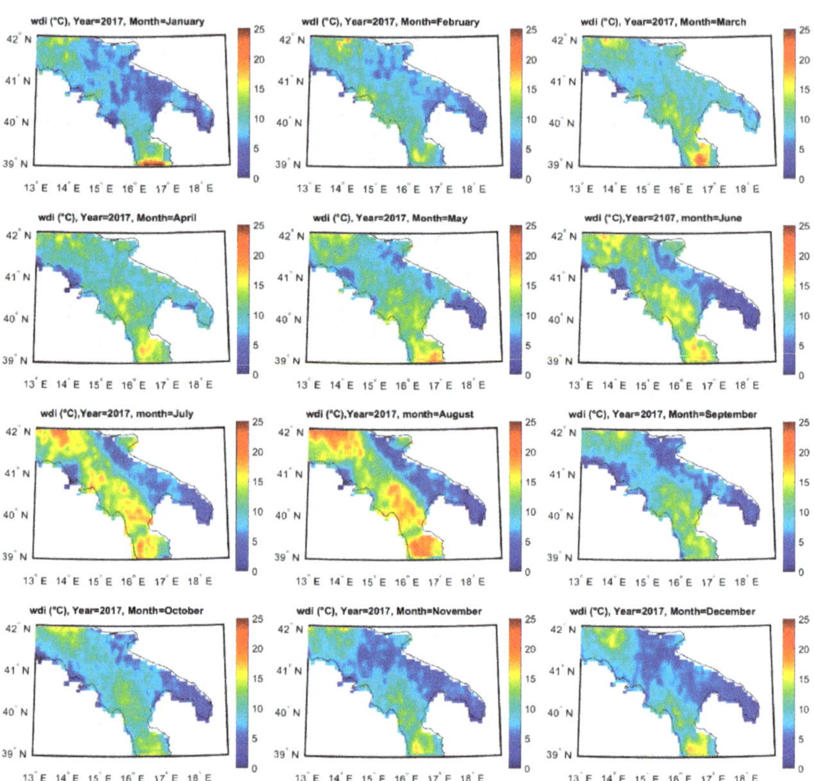

Figure 6. Level 3 map at a grid step of 0.05 degrees for the index *wdi* for 2017.

To understand the index's sensitivity to heat waves, we have compared the *wdi* parameter over three consecutive years, 2017, 2020, and 2021, for July. We know that July 2020 has been relatively wetter than 2017 and 2021 (e.g., see https://climate.copernicus.eu/esotc/2021 (accessed on 15 August 2022)). The comparison is shown in Figure 7, and we see that *wdi* is able to indicate that the year 2020 was less warm than the other two. This situation is reflected in the soil moisture maps shown for the same target area and year and month. When we focus on the forested area, especially in the southern part of the map, we see that the soil moisture follows the same spatial-time evolution as *wdi* and, in particular, the soil moisture is lower in 2017 and 2021 than in 2020. This is a significant result because it shows that the *wdi* is capable of capturing processes at the surface–atmosphere interface. A large *wdi* means a high rate of evapotranspiration; that is, trees lose water in the atmosphere. The fact that the soil moisture is getting lower means that the vegetation can catch less water from the surface.

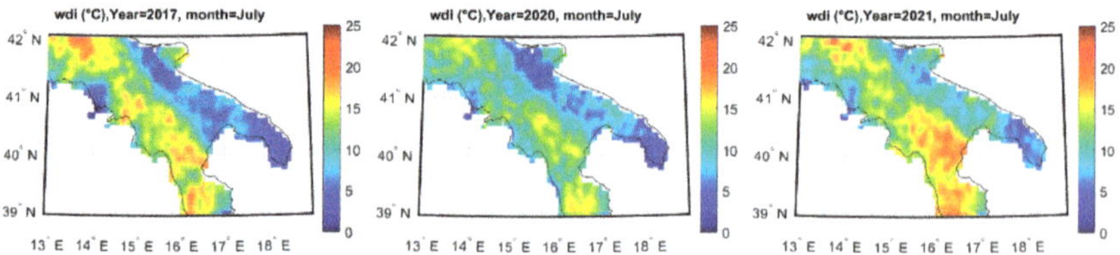

Figure 7. Exemplifying the *wdi* evolution through the years. From left to right: July 2017, 2020, and 2021.

The anti-correlation between *wdi* and soil moisture proves that *wdi* is a good metric for monitoring water deficit during intense heatwaves. The more significant values we saw in summer are not merely a consequence of the hotter weather, but also reflects the decrease in water vapour exchange between the surface and the atmosphere. We stress that, unlike other indices, *wdi* considers the surface-air temperature and humidity fields simultaneously.

A further comparison with other parameters sensitive to vegetation stress is shown in Figures 8 and 9. Concerning the 2017 heatwave, Figure 8 compares the surface soil moisture (*ssm*) against *wdi* for the period June to August. It is seen that while *wdi* tends to increase with time, *ssm* does the opposite. The leaf area index (LAI) is another crucial parameter to be monitored for investigating vegetation stress. Indeed, under the action of an intense heat wave, trees tend to lose leaves to protect from the fierce evapotranspiration. Trees use this mechanism, e.g., in winter, when the light is not enough to sustain the photosynthesis activity. The comparison with LAI is shown in Figure 9, and we see that consistently with the increasing *wdi* behaviour, LAI is decreasing from June to July. In normal situations, the LAI. decrease is not expected in the summer when there is a more significant availability of light to sustain photosynthesis.

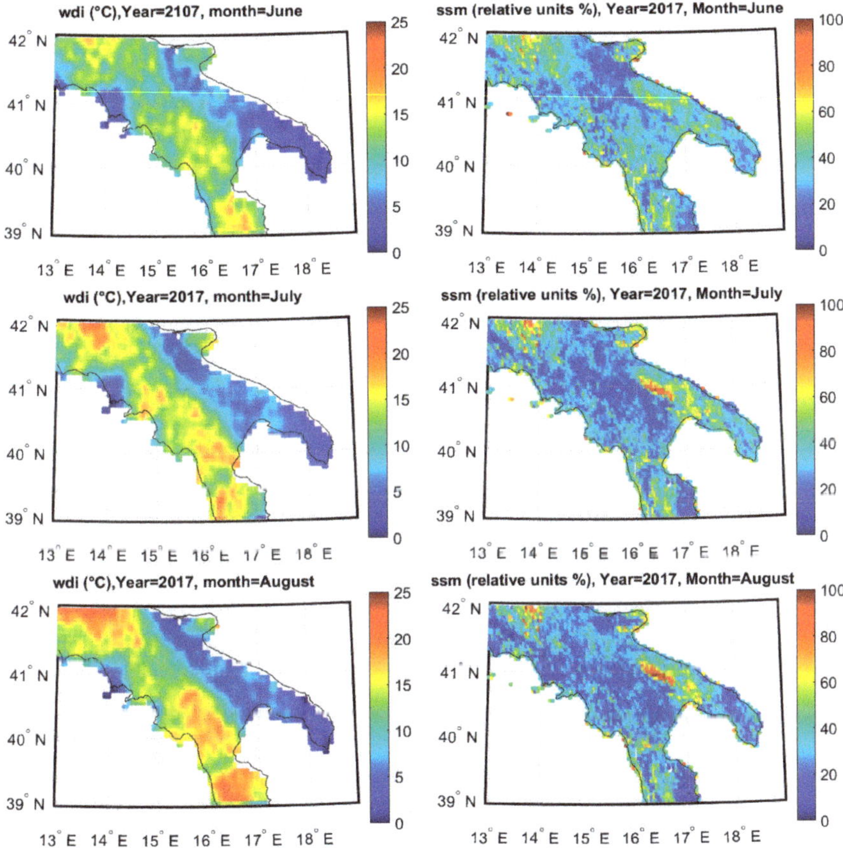

Figure 8. Comparison of *ssm* vs. *wdi* for the period of June to August in 2017. Top to bottom, June to August.

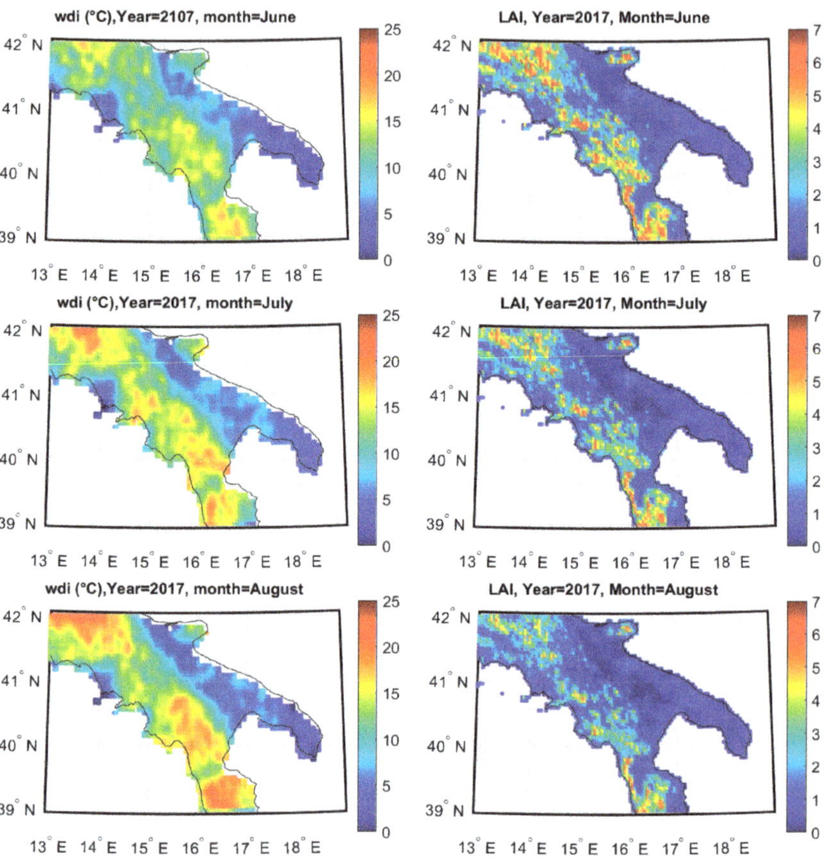

Figure 9. Comparison of LAI (m^2/m^2) vs. *wdi* for the period of June to August in 2017. Top to bottom, June to August.

We have checked that the good consistency among *ssm*, LAI, and *wdi* also persists at the local scale. In fact, for the two stations of S. Paolo Albanese and Gorgoglione, shown in Figure 1, we have computed the monthly time series of *ssm* and *wdi* for 2017. The time series are shown in Figure 10, and we see that starting from May until September, *wdi* goes up, whereas *ssm* has the opposite behaviour. Again, this is an important result because it shows that the *wdi* is capturing a water deficit condition for the vegetation, especially in the area where we know there are declining trees [6,24].

The most striking agreement is seen when comparing in situ observations for the flux exchange of CO_2 and H_2O from trees to the *wdi* parameter. In the summers of 2020 and 2021, CO_2 exchange measurements were performed at the leaf scale on declining and non-declining *Q. frainetto* trees growing at the S. Paolo Albanese study site. In each tree, net photosynthesis rate (An, $\mu molCO_2\ m^{-2}\ s^{-1}$), stomatal conductance (gsw, $mmolH_2O\ m^{-2}\ s^{-1}$), and intrinsic water use efficiency (WUEi, $\mu molCO_2\ mmol^{-1}\ H_2O$) were measured by using a portable Photosynthesis System LiCOR 6400xt equipped with a 6400-40 Leaf Chamber Fluorometer.

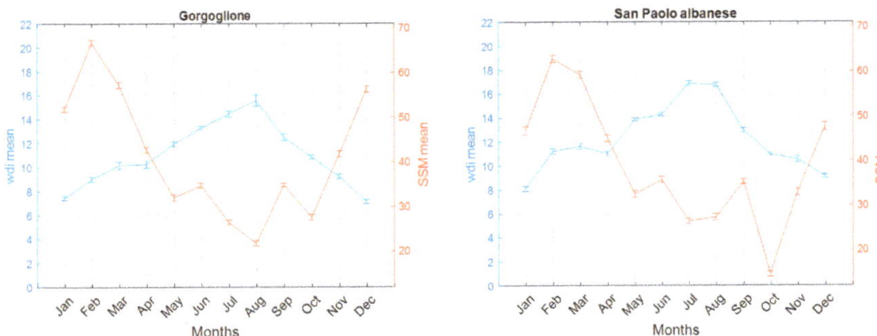

Figure 10. Monthly time series of *ssm* and *wdi* in 2017 for the two-tower stations of Gorgoglione (**left**) and S. Paolo Albanese (**right**). Data points are the mean values from a circle of diameter 0.1° around the stations. The error bars represent the variability (standard deviation) of the samples.

In the summers of 2020 and 2021, the ecophysiological response of *Q. frainetto* trees exhibiting decline and non-decline symptoms is shown in Figure 11. In 2020, when no heatwave occurred, *Q. frainetto* ecophysiological responses were similar for declining and non-declining trees, suggesting that there was no evident sign of water stress in the summer of 2020. From Figure 7, we see that *wdi* is in fact below 10 °C in July. In contrast, in 2021, not only is the water vapour exchange more than doubled, showing that the evapotranspiration has increased because of the larger difference $T_s - T_d$, but also the declining trees behave differently with respect to the non-declining vegetation, showing that the non-declining trees are suffering from the water deficit much more than the healthy vegetation.

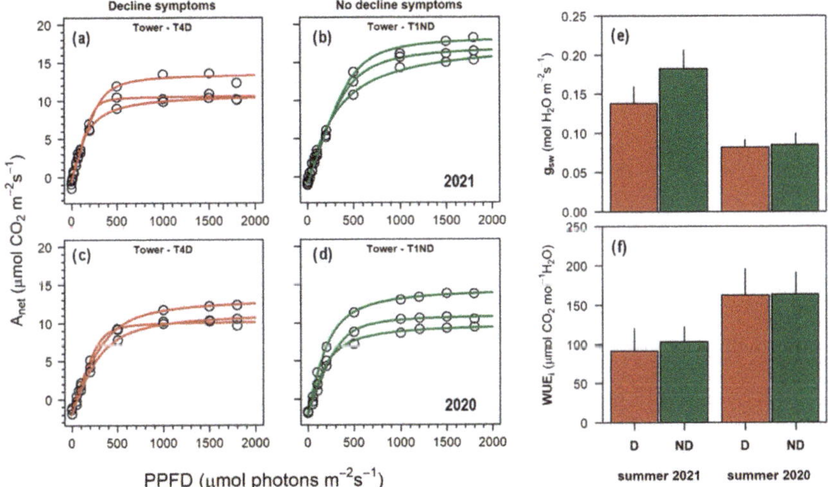

Figure 11. Ecophysiological responses of declining (D) and non-declining (ND) *Q. frainetto* trees of the San Paolo Albanese forest stand site. Panels (**a–d**) present the net photosynthesis curve (An, μmolCO_2 m^{-2} s^{-1}), while panels (**e**,**f**) show the average values of stomatal conductance (gsw, mmolH_2O m^{-2} s^{-1}) and intrinsic water use efficiency (WUEi, μmolCO_2 $mmol^{-1}$ H_2O) measured in the summers of 2020 and 2021. PPFD represents the photosynthetic photon flux density (μmol photons m^{-2} s^{-1}). The black vertical bar represents the 1st deviation standard.

It is also interesting to note that CO_2, flux exchange exhibited the opposite behaviour to H_2O. In the summer of 2020, when there were good climatic conditions, we observed an exchange larger than in 2021. In 2021, the results showed that the vegetation had reduced photosynthesis activity because of stress conditions.

4. Discussion

In the summer of 2017, southern Europe and the Euro-Mediterranean were hit by an exceptional heat wave [6,42]. After an outstandingly warm June in western Europe, the heat returned to southern Italy in July. It contributed to more than 400 wildfires, which destroyed approximately 800 km^2 of forest and vegetated areas. The number of fires has been unprecedented in the last 20 years. Furthermore, early August saw a particularly intense heat wave described as the "worst heat wave since 2003", with the air temperature above 40 °C in many parts of Italy [42].

In the quest for possible satellite indices to assess and possibly mitigate the effect of long-lasting drought on vegetation, we have devised an index, *wdi*, which takes advantage of the IASI capability to retrieve surface data simultaneously with atmospheric parameters. Other methods that use satellite data exploit the visible region of the electromagnetic spectrum (e.g., NDVI, NDMI, and related indices) or the microwave band (e.g., *ssm* and LAI). The normalized difference moisture index (NDMI) (e.g., [43]) is mainly intended to detect humidity in vegetation using a combination of near-infrared (NIR) and short-wave infrared (SWIR) spectral bands. The index NDMI and the original greenness index, NDVI, with the same *ssm*, have also been used coupled to surface temperature and air temperature (e.g., see [8,11–14,18]). Other tools have tried to couple surface temperature and the humidity field, e.g., [6].

In contrast, our *wdi* exploits the thermal band of the Earth's emission spectrum and simultaneously uses the surface temperature, air temperature, and humidity. To our knowledge, this combination is unique. In effect, the water deficit index we have defined can monitor water deficit and assess vegetation stress, as the comparison with in situ measurements has demonstrated. It can be used complementary to *ssm*, LAI, and the set of NDVI-related indices to better understand the intensity and danger of heatwaves for the vegetation. Sequences of increasing *wdi* can help to identify the onset of water deficit for the vegetation, hence the increased risk of fire, especially in forests.

The *wdi* index is meant to identify regions where particular weather conditions can produce water deficits. The index is not intended as an estimate or an estimator of evapotranspiration. This process is also affected by vegetation/crop characteristics, environmental conditions, and cultivation types. Therefore, there is too much variability, which cannot be condensed into a single index. The *wdi* parameter is a bulk index, which can help to monitor forest and wood regions suffering from long-lasting droughts because of adverse weather conditions. It can be mapped on a regional and even global scale, allowing us to monitor drought processes at a glance. The *wdi* maps could be important to monitor and evaluate the risk of fires in the large forested area, which is otherwise inaccessible. In addition, we have shown that in regions where the vegetal ecosystem has a particular fragility to water deficit, the index can soon quantify the possible danger and require more accurate in situ observations.

In this respect, *wdi* is most effective in the case of a heatwave. In the wintertime, for example, large values of *wdi* could be linked to a dry atmosphere and low air temperature. In effect, this is the case in January 2017 for the more southern area on the map of Figure 5, which belongs to the high mountains of the Sila chain. Additionally, in summer, very humid and warm conditions could lead to $wdi \sim 0$. For example, this is the case for the coastal regions in July–August 2017, as seen again in the map in Figure 5. For these cases, it is better to look separately at the maps of T_s and T_d. In this respect, we observe that the dew point temperature has been individuated as a key parameter to compute sophisticated indicators of health stress for human beings during heatwaves [44].

Some words of caution should also be said about the temporal sampling of *wdi*. The occasional occurrence of a high *wdi* for one day should be of no concern. Drought is a process that takes several days or months. The severity of the process depends on its time continuity and persistence. Therefore, it is crucial to assess the persistence of the process, which can be done by looking at the time series. Averaging over several days can help to understand the persistence of the phenomenon. Another important point concerns the capability of the retrieval system to solve the daily cycle, which cannot be done with the present polar satellite IASI instrument. During the night, the surface temperature could go below the dew point temperature and cause water vapour to condense at the surface. Therefore, it could be interesting to examine day and night separately. Hopefully, this could be the case when the MTG-IRS (https://www.eumetsat.int/mtg-infrared-sounder (accessed on 15 August 2022)) is put in orbit.

5. Conclusions

Exploiting the capability of the IASI instrument to perform simultaneous retrievals of surface and thermodynamical parameters, we have developed an index called the water deficit index, or *wdi*. The index is intended to be used in the case of evident droughts, as it can assess the severe water deficit of vegetation, and in particular, forests.

The tool has been exemplified in a target area in the south of Italy, which suffered from an intense drought and heatwave in 2017. When the heatwave is developing, we have shown, with the help of correlative observations of surface soil moisture and the leaf area index, that *wdi* can assess the severity of the water deficit. Of particular interest is the anti-correlation with the surface soil moisture. The soil water content and the ability of the soil to transport water to the roots govern the transpiration rate of vegetation. In cases where the *wdi* becomes large, we have found that *ssm* gets smaller, which shows how *wdi* is capable of capturing processes occurring at the surface–atmosphere interface.

The possible usage of *wdi* includes monitoring large forested areas for the increased risks of wildfire and assessing mitigation measures for regions whose green ecosystems are more fragile and in danger because of climate change.

Author Contributions: G.M. contributed to the overall conceptualization, organized the field campaigns and revised the paper; F.R. revised the paper, contributed to funding acquisition, and organized the field campaigns; I.D.F. developed the 2D-OI scheme; A.R. was involved in the formal analysis of the field campaign data and contributed to the final editing; L.S. contributed to the editing of the paper and was involved in the field campaigns; P.P. contributed to the analysis of IASI data; A.C. contributed to the implementation of the 2D-OI scheme; S.V. contributed to the software development and data acquisition; C.S. wrote the paper, and contributed to the funding acquisition and conceptualization of the *wdi* index. All authors have read and agreed to the published version of the manuscript.

Funding: The Italian Ministry of University has supported this research in the framework of the project ARS01_00405, "OT4CLIMA" (D.D. 2261 del 6.9.2018, PON R & I 2014-2020 and FSC).

Data Availability Statement: The IASI L1C data used in this study are directly available from EUMETSAT. They are received through the EUMETCast near real-time data distribution service. Surface soil moisture data were downloaded from the site https://land.copernicus.eu/global/products/ssm (accessed on 15 August 2022). Leaf area index data were downloaded from the site https://land.copernicus.eu/global/products/lai (accessed on 15 August 2022). All L2 IASI data computed in the paper are available on request from the authors. Data used to build up Figure 10 are available on request from the authors.

Acknowledgments: IASI has been developed and built under the responsibility of the Centre National d'Etudes Spatiales (CNES, France). Instrumentation flies onboard the Metop satellites as part of the EUMETSAT Polar System.

Conflicts of Interest: The authors declare no conflict of interest.

Abbreviations

LAI	leaf area index (m^2/m^2)
NDMI	normalized difference moisture index (dimensionless)
NDVI	normalized difference vegetation index (dimensionless)
iTVDI	improved temperature vegetation dryness index (dimensionless)
P	pressure (hPa)
P_w	water vapour pressure (hPa)
P_{ws}	saturation water vapour pressure (hPa)
Q	water vapour mixing ratio profile (g/kg)
Q_1	water vapour mixing ratio at the surface level (g/kg)
$rh = \frac{P_w}{P_{ws}}$	relative humidity (dimensionless)
$\boldsymbol{x_1} = (T_s,\ T_1, Q_1)$	vector of size $n = 3$
$\mathbf{S_1}$	covariance matrix of $\boldsymbol{x_1}$ size (3×3)
ssm	surface soil moisture (dimensionless)
$\boldsymbol{T}$	temperature profile (K)
$T_1 = T_a$	air temperature at the surface level (K)
$t_1 = t_a$	air temperature at the surface level (C)
T_d	dew point temperature at the surface level (K)
t_d	dew point temperature at the surface level (C)
T_s	surface temperature at the surface level (K)
t_s	surface temperature at the surface level (C)
TVDI	temperature vegetation dryness index (dimensionless)
VDI	vegetation dryness index (dimensionless)
VPD	vapour pressure deficit (hPa)
$wdi = T_s - T_d = t_s - t_d$	water deficit index (difference temperature, in units of K or C)
σ^2_{wdi}	variance of wdi (K^2 or C^2)

References

1. Voosen, P. Global Temperatures in 2020 Tied Record Highs. *Science* **2021**, *371*, 334–335. [CrossRef] [PubMed]
2. Cheng, L.; Abraham, J.; Trenberth, K.E.; Fasullo, J.; Boyer, T.; Mann, M.E.; Zhu, J.; Wang, F.; Locarnini, R.; Li, Y.; et al. Another Record: Ocean Warming Continues through 2021 despite La Niña Conditions. *Adv. Atmos. Sci.* **2022**, *39*, 373–385. [CrossRef] [PubMed]
3. Cramer, W.; Guiot, J.; Fader, M.; Garrabou, J.; Gattuso, J.-P.; Iglesias, A.; Lange, M.A.; Lionello, P.; Llasat, M.C.; Paz, S.; et al. Climate Change and Interconnected Risks to Sustainable Development in the Mediterranean. *Nat. Clim. Chang.* **2018**, *8*, 972–980. [CrossRef]
4. Rita, A.; Camarero, J.J.; Nolè, A.; Borghetti, M.; Brunetti, M.; Pergola, N.; Serio, C.; Vicente-Serrano, S.M.; Tramutoli, V.; Ripullone, F. The Impact of Drought Spells on Forests Depends on Site Conditions: The Case of 2017 Summer Heat Wave in Southern Europe. *Glob. Chang. Biol.* **2020**, *26*, 851–863. [CrossRef]
5. Anderegg, W.R.L.; Trugman, A.T.; Badgley, G.; Anderson, C.M.; Bartuska, A.; Ciais, P.; Cullenward, D.; Field, C.B.; Freeman, J.; Goetz, S.J.; et al. Climate-Driven Risks to the Climate Mitigation Potential of Forests. *Science* **2020**, *368*, eaaz7005. [CrossRef]
6. Mishra, A.K.; Singh, V.P. A Review of Drought Concepts. *J. Hydrol.* **2010**, *391*, 202–216. [CrossRef]
7. Feiziasl, V.; Jafarzadeh, J.; Sadeghzadeh, B.; Mousavi Shalmani, M.A. Water Deficit Index to Evaluate Water Stress Status and Drought Tolerance of Rainfed Barley Genotypes in Cold Semi-Arid Area of Iran. *Agric. Water Manag.* **2022**, *262*, 107395. [CrossRef]
8. Spinoni, J.; Barbosa, P.; De Jager, A.; McCormick, N.; Naumann, G.; Vogt, J.V.; Magni, D.; Masante, D.; Mazzeschi, M. A New Global Database of Meteorological Drought Events from 1951 to 2016. *J. Hydrol. Reg. Stud.* **2019**, *22*, 100593. [CrossRef]
9. Sutanto, S.J.; van der Weert, M.; Wanders, N.; Blauhut, V.; Van Lanen, H.A.J. Moving from Drought Hazard to Impact Forecasts. *Nat. Commun.* **2019**, *10*, 4945. [CrossRef]
10. Stoyanova, J.S.; Georgiev, C.G. SVAT Modelling in Support to Flood Risk Assessment in Bulgaria. *Atmos. Res.* **2013**, *123*, 384–399. [CrossRef]
11. Gouveia, C.; Trigo, R.M.; DaCamara, C.C. Drought and Vegetation Stress Monitoring in Portugal Using Satellite Data. *Nat. Hazards Earth Syst. Sci.* **2009**, *9*, 185–195. [CrossRef]
12. Bento, V.; Trigo, I.; Gouveia, C.; DaCamara, C. Contribution of Land Surface Temperature (TCI) to Vegetation Health Index: A Comparative Study Using Clear Sky and All-Weather Climate Data Records. *Remote Sens.* **2018**, *10*, 1324. [CrossRef]
13. Stoyanova, J.; Georgiev, C.; Neytchev, P.; Kulishev, A. Spatial-Temporal Variability of Land Surface Dry Anomalies in Climatic Aspect: Biogeophysical Insight by Meteosat Observations and SVAT Modeling. *Atmosphere* **2019**, *10*, 636. [CrossRef]
14. Feldman, A.F.; Short Gianotti, D.J.; Trigo, I.F.; Salvucci, G.D.; Entekhabi, D. Land-Atmosphere Drivers of Landscape-Scale Plant Water Content Loss. *Geophys. Res. Lett.* **2020**, *47*, e2020GL090331. [CrossRef]

15. Vicente-Serrano, S.M.; Pons-Fernández, X.; Cuadrat-Prats, J.M. Mapping Soil Moisture in the Central Ebro River Valley (Northeast Spain) with Landsat and NOAA Satellite Imagery: A Comparison with Meteorological Data. *Int. J. Remote Sens.* **2004**, *25*, 4325–4350. [CrossRef]
16. Vicente-Serrano, S.M.; Cuadrat-Prats, J.M.; Romo, A. Aridity Influence on Vegetation Patterns in the Middle Ebro Valley (Spain): Evaluation by Means of AVHRR Images and Climate Interpolation Techniques. *J. Arid Environ.* **2006**, *66*, 353–375. [CrossRef]
17. Chen, S.; Wen, Z.; Jiang, H.; Zhao, Q.; Zhang, X.; Chen, Y. Temperature Vegetation Dryness Index Estimation of Soil Moisture under Different Tree Species. *Sustainability* **2015**, *7*, 11401–11417. [CrossRef]
18. Masiello, G.; Cersosimo, A.; Mastro, P.; Serio, C.; Venafra, S.; Pasquariello, P. Emissivity-Based Vegetation Indices to Monitor Deforestation and Forest Degradation in the Congo Basin Rainforest. In Proceedings of the Remote Sensing for Agriculture, Ecosystems, and Hydrology XXII, Online, 20 September 2020; Neale, C.M., Maltese, A., Eds.; p. 17.
19. Karnieli, A.; Agam, N.; Pinker, R.T.; Anderson, M.; Imhoff, M.L.; Gutman, G.G.; Panov, N.; Goldberg, A. Use of NDVI and Land Surface Temperature for Drought Assessment: Merits and Limitations. *J. Clim.* **2010**, *23*, 618–633. [CrossRef]
20. Ru, C.; Hu, X.; Wang, W.; Ran, H.; Song, T.; Guo, Y. Evaluation of the Crop Water Stress Index as an Indicator for the Diagnosis of Grapevine Water Deficiency in Greenhouses. *Horticulturae* **2020**, *6*, 86. [CrossRef]
21. Hilton, F.; Armante, R.; August, T.; Barnet, C.; Bouchard, A.; Camy-Peyret, C.; Capelle, V.; Clarisse, L.; Clerbaux, C.; Coheur, P.-F.; et al. Hyperspectral Earth Observation from IASI: Five Years of Accomplishments. *Bull. Am. Meteorol. Soc.* **2012**, *93*, 347–370. [CrossRef]
22. Behrangi, A.; Loikith, P.; Fetzer, E.; Nguyen, H.; Granger, S. Utilizing Humidity and Temperature Data to Advance Monitoring and Prediction of Meteorological Drought. *Climate* **2015**, *3*, 999–1017. [CrossRef]
23. Gentilesca, T.; Camarero, J.; Colangelo, M.; Nolè, A.; Ripullone, F. Drought-Induced Oak Decline in the Western Mediterranean Region: An Overview on Current Evidences, Mechanisms and Management Options to Improve Forest Resilience. *iForest* **2017**, *10*, 796–806. [CrossRef]
24. Ripullone, F.; Camarero, J.J.; Colangelo, M.; Voltas, J. Variation in the Access to Deep Soil Water Pools Explains Tree-to-Tree Differences in Drought-Triggered Dieback of Mediterranean Oaks. *Tree Physiol.* **2020**, *40*, 591–604. [CrossRef] [PubMed]
25. Colangelo, M.; Camarero, J.J.; Borghetti, M.; Gentilesca, T.; Oliva, J.; Redondo, M.-A.; Ripullone, F. Drought and Phytophthora Are Associated with the Decline of Oak Species in Southern Italy. *Front. Plant Sci.* **2018**, *9*, 1595. [CrossRef] [PubMed]
26. Bauer-Marschallinger, B.; Freeman, V.; Cao, S.; Paulik, C.; Schaufler, S.; Stachl, T.; Modanesi, S.; Massari, C.; Ciabatta, L.; Brocca, L.; et al. Toward Global Soil Moisture Monitoring with Sentinel-1: Harnessing Assets and Overcoming Obstacles. *IEEE Trans. Geosci. Remote Sens.* **2019**, *57*, 520–539. [CrossRef]
27. Fuster, B.; Sánchez-Zapero, J.; Camacho, F.; García-Santos, V.; Verger, A.; Lacaze, R.; Weiss, M.; Baret, F.; Smets, B. Quality Assessment of PROBA-V LAI, FAPAR and FCOVER Collection 300 m Products of Copernicus Global Land Service. *Remote Sens.* **2020**, *12*, 1017. [CrossRef]
28. Amato, U.; Masiello, G.; Serio, C.; Viggiano, M. The σ-IASI Code for the Calculation of Infrared Atmospheric Radiance and Its Derivatives. *Environ. Model. Softw.* **2002**, *17*, 651–667. [CrossRef]
29. Carissimo, A.; De Feis, I.; Serio, C. The Physical Retrieval Methodology for IASI: The δ-IASI Code. *Environ. Model. Softw.* **2005**, *20*, 1111–1126. [CrossRef]
30. Liuzzi, G.; Masiello, G.; Serio, C.; Venafra, S.; Camy-Peyret, C. Physical Inversion of the Full IASI Spectra: Assessment of Atmospheric Parameters Retrievals, Consistency of Spectroscopy and Forward Modelling. *J. Quant. Spectrosc. Radiat. Transf.* **2016**, *182*, 128–157. [CrossRef]
31. Serio, C.; Masiello, G.; Liuzzi, G. Demonstration of Random Projections Applied to the Retrieval Problem of Geophysical Parameters from Hyper-Spectral Infrared Observations. *Appl. Opt.* **2016**, *55*, 6576–6587. [CrossRef]
32. Masiello, G.; Serio, C.; Venafra, S.; DeFeis, I.; Borbas, E.E. Diurnal Variation in Sahara Desert Sand Emissivity during the Dry Season from IASI Observations: Diurnal Emissivity Variation. *J. Geophys. Res. Atmos.* **2014**, *119*, 1626–1638. [CrossRef]
33. Masiello, G.; Serio, C.; Venafra, S.; Liuzzi, G.; Poutier, L.; Göttsche, F.-M. Physical Retrieval of Land Surface Emissivity Spectra from Hyper-Spectral Infrared Observations and Validation with In Situ Measurements. *Remote Sens.* **2018**, *10*, 976. [CrossRef]
34. De Feis, I.; Masiello, G.; Cersosimo, A. Optimal Interpolation for Infrared Products from Hyperspectral Satellite Imagers and Sounders. *Sensors* **2020**, *20*, 2352. [CrossRef] [PubMed]
35. Rodgers, C.D. *Inverse Methods for Atmospheric Sounding: Theory and Practice*; Series on Atmospheric, Oceanic and Planetary Physics; World Scientific: Singapore, 2000; ISBN 978-981-02-2740-1.
36. Masiello, G.; Serio, C.; Deleporte, T.; Herbin, H.; Di Girolamo, P.; Champollion, C.; Behrendt, A.; Bosser, P.; Bock, O.; Wulfmeyer, V.; et al. Comparison of IASI Water Vapour Products over Complex Terrain with COPS Campaign Data. *Meteorol. Z.* **2013**, *22*, 471–487. [CrossRef]
37. Huang, J. A Simple Accurate Formula for Calculating Saturation Vapor Pressure of Water and Ice. *J. Appl. Meteorol. Climatol.* **2018**, *57*, 1265–1272. [CrossRef]
38. Sonntag, D. Important New Values of the Physical Constants of 1986, Vapour Pressure Formulations Based on the ITS-90, and Psychrometer Formulae. *Z. Meteorol.* **1990**, *40*, 340–344.
39. Tellinghuisen, J. Statistical Error Propagation. *J. Phys. Chem. A* **2001**, *105*, 3917–3921. [CrossRef]
40. Allen, R.G.; Pereira, L.S.; Raes, D.; Smith, M. *Crop Evapotranspiration: Guidelines for Computing Crop Water Requirements*; FAO Irrigation and Drainage Paper; Food and Agriculture Organization of the United Nations: Rome, Italy, 1998. ISBN 978-92-5-104219-9.

41. Amato, U.; Lavanant, L.; Liuzzi, G.; Masiello, G.; Serio, C.; Stuhlmann, R.; Tjemkes, S.A. Cloud Mask via Cumulative Discriminant Analysis Applied to Satellite Infrared Observations: Scientific Basis and Initial Evaluation. *Atmos. Meas. Tech.* **2014**, *7*, 3355–3372. [CrossRef]
42. Kew, S.F.; Philip, S.Y.; Jan van Oldenborgh, G.; van der Schrier, G.; Otto, F.E.L.; Vautard, R. The Exceptional Summer Heat Wave in Southern Europe 2017. *Bull. Am. Meteorol. Soc.* **2019**, *100*, S49–S53. [CrossRef]
43. 4Khanmohammadi, F.; Homaee, M.; Noroozi, A.A. Soil Moisture Estimating with NDVI and Land Surface Temperature and Normalized Moisture Index Using MODIS Images. *J. Water Soil Resour. Conserv.* **2015**, *4*, 37–45.
44. Lee, S.-Y.; Lung, S.-C.C.; Chiu, P.-G.; Wang, W.-C.; Tsai, I.-C.; Lin, T.-H. Northern Hemisphere Urban Heat Stress and Associated Labor Hour Hazard from ERA5 Reanalysis. *IJERPH* **2022**, *19*, 8163. [CrossRef] [PubMed]

 land

Article

Correlation Analysis of Evapotranspiration, Emissivity Contrast and Water Deficit Indices: A Case Study in Four Eddy Covariance Sites in Italy with Different Environmental Habitats

Michele Torresani [1,2,*], Guido Masiello [3], Nadia Vendrame [4], Giacomo Gerosa [5], Marco Falocchi [6], Enrico Tomelleri [1], Carmine Serio [3], Duccio Rocchini [7,8] and Dino Zardi [2,4]

1 Faculty of Science and Technology, Free University of Bolzano/Bozen, Piazza Università/Universitätsplatz 1, 39100 Bolzano/Bozen, Italy
2 Department of Civil, Environmental and Mechanical Engineering, University of Trento, 38123 Trento, Italy
3 School of Engineering, University of Basilicata, Via Ateneo Lucano 10, 85100 Potenza, Italy
4 Center Agriculture Food Environment (C3A), University of Trento, 38010 San Michele All'Adige, Italy
5 Department of Mathematics and Physics, Università Cattolica del Sacro Cuore, 25121 Brescia, Italy
6 Cisma S.r.l., Via Ipazia 2 c/o NOI Techpark, 39100 Bolzano, Italy
7 BIOME Lab., Department of Biological, Geological and Environmental Sciences, Alma Mater Studiorum University of Bologna, Via Irnerio 42, 40126 Bologna, Italy
8 Department of Spatial Sciences, Faculty of Environmental Sciences, Czech University of Life Sciences Prague, Kamýcka 129, 16500 Praha, Czech Republic
* Correspondence: michele.torresani@unibz.it

Abstract: Evapotranspiration (ET) represents one of the essential processes controlling the exchange of energy by terrestrial vegetation, providing a strong connection between energy and water fluxes. Different methodologies have been developed in order to measure it at different spatial scales, ranging from individual plants to an entire watershed. In the last few years, several methods and approaches based on remotely sensed data have been developed over different ecosystems for the estimation of ET. In the present work, we outline the correlation between ET measured at four eddy covariance (EC) sites in Italy (situated either in forest or in grassland ecosystems) and (1) the emissivity contrast index (ECI) based on emissivity data from thermal infrared spectral channels of the MODIS and ASTER satellite sensors (CAMEL data-set); (2) the water deficit index (WDI), defined as the difference between the surface and dew point temperature modeled by the ECMWF (European Centre for Medium-Range Weather Forecasts) data. The analysis covers a time-series of 1 to 7 years depending on the site. The results showed that both the ECI and WDI correlate to the ET calculated through EC. In the relationship WDI-ET, the coefficient of determination ranges, depending on the study area, between 0.5 and 0.9, whereas it ranges between 0.5 and 0.7 when ET was correlated to the ECI. The slope and the sign of the latter relationship is influenced by the vegetation habitat, the snow cover (particularly in winter months) and the environmental heterogeneity of the area (calculated in this study through the concept of the spectral variation hypothesis using Rao's Q heterogeneity index).

Keywords: emissivity; evapotranspiration; heterogeneity; Rao's Q index; spectral variation hypothesis; thermal infrared

Citation: Torresani, M.; Masiello, G.; Vendrame, N.; Gerosa, G.; Falocchi, M.; Tomelleri, E.; Serio, C.; Rocchini, D.; Zardi, D. Correlation Analysis of Evapotranspiration, Emissivity Contrast, and Water Deficit Indices: A Case Study in Four Eddy Covariance Sites in Italy with Different Environmental Habitats. *Land* **2022**, *11*, 1903. https://doi.org/10.3390/land11111903

Academic Editor: Jin Wu

Received: 5 September 2022
Accepted: 23 October 2022
Published: 26 October 2022

1. Introduction

Evapotranspiration (ET) is an important component of the forest hydrological budget, and influences the flow of water to downstream users, including aquatic habitats and human populations. Furthermore, it represents a considerable water loss in the landscape [1,2]. As an example, ET has been reported to inject into the atmosphere approximately 70% of annual precipitation in a loblolly pine (*Pinus taeda*) plantation in south-eastern USA [3], more than 85% in a Canadian black spruce (*Picea mariana*) forest [4] and more than 85% in a ponderosa pine (*Pinus ponderosa*) forest in Arizona [5]. Consequently, the magnitude and seasonality of forest ET are important regulators of water resources available to humans and ecosystems. ET represents a crucial process within a broad range of systems, including ecology, hydrology and meteorology. For this reason, different methodologies have been developed in order to measure it at different spatial scales, ranging from individual plants to entire watersheds [6]. Various techniques have been developed to measure ET [6], including sap flow analysis [7], by weighing lysimeters [8], plant chambers, stable isotope [7,9], soil water budgets [10], land surface models [10] and eddy covariance (EC) [7,11]. More recently, remote sensing data, offering large area coverage, frequent updates and consistent quality, have been used in different studies to collect a quantitative information of ET over different ecosystems world-wide [12,13].

ET cannot be measured directly from remote sensing data. Indirect approaches [14], such as the energy balance approach [15], the Priestley–Talor approach [16,17] and through the use of spectral indices [18], are commonly applied. In general, process-based models that couple remote sensing information and ET have been widely used in science in the last several years, at both local and global scale. The models reproduce physical and plant physiological mechanisms that regulate ET, such as stomata processes, radiation absorption and water interception [14]. Different remote sensing approaches use land surface characteristics such as the leaf area index (LAI) and the albedo to estimate ET via surface energy balance or within-scene scaling [19,20]. Remote sensing thermal infrared measurements have also been largely used for the retrieval of ET information [21,22]. As an example, Hamberg et al.[23] illustrated the potential of thermal information derived from the ECOSTRESS satellite sensor for inferring land surface temperature and ET in different forest sites in Southern Ontario, Canada. Carlson et al. [24], again using the HCMM satellite, introduced a method for inferring different variables, including the distribution of evaporative fluxes and surface heat, in the cities of Los Angeles and St. Louis (USA). We refer to the following articles for an exhaustive overview of the use of infrared thermal radiation for ET retrievals [25–28]. For more general information about ET estimation techniques based on remote sensing data, Zhang et al. [12] provided an exhaustive review.

The thermal infrared (TIR) spectral region is also susceptible to soil moisture, allowing for the retrieval of the atmosphere's thermodynamic state along with the hydrometeorological conditions near the surface. The thermodynamic state close to the surface and the surface itself can be related straightforwardly to surface ET. A recent study by Masiello et al. [29] made use of the remote-sensed emissivity contrast index (ECI) based on TIR emissivity data derived from infrared atmospheric sounding interferometer (IASI) measurements [30,31] and demonstrated that it correlates with the water deficit index, or WDI, defined as the difference between the surface and dew point temperature close to it [32]. In [29], both the ECI and WDI have been obtained with a technique that enables the simultaneous retrieval of spectral emissivity and the vertical distribution of temperature (T), water vapor (Q) and other trace gases [33]. The WDI can be computed using in situ measurements or using modeled information, such as that of the European Centre for Medium Range Weather Forecasts (ECMWF) .

The ECI, firstly introduced by French et al. [34], in Masiello et al. [35] has been computed as the difference between the CAMEL emissivity channels (derived from the CAMEL database CAM5K30EM v002 [36–39]) at 8.6, 10.8 and 12.1 μm. The index was developed with an NDVI synergy to better classify vegetation cover and to overcome the limitations of the vegetation index, particularly in the discrimination of bare soil and senescent vegetation. It showed promising results in the classification of changes in land use when, for example, a vegetation regeneration follows the deforestation or forest degradation events [35]. The CAMEL dataset, where the emissivity information is stored, is produced by the combination of two distinct databases to take advantage of each product's characteristics. The first is the ASTER Global Emissivity (ASTER GEDv4), developed at the Jet Propulsion Laboratory (JPL): it has a temporal resolution of 1 month, a spatial resolution of 5 km and a spectral range from 8 to 12.0 μm. The MODIS baseline-t emissivity (MODBF) represents the second database: it is provided by the University of Wisconsin-Madison and it has a spectral emissivity range from 3.6–12.0 µm. The resulting dataset is available globally in mean monthly time-steps with a spatial resolution of 5 km, with several layers providing information of emissivity (13 bands ranging from 3.6–14.3 µm), NDVI, snow fraction and related quality flags. The CAMEL dataset has been produced to design a uniform, long-term and calibrated emissivity database in order to advance the analysis of different applications, such as atmospheric retrievals and radiative transfer simulations. Within such a coarse spatial resolution, the environmental heterogeneity could be very high. For this reason, there is need for a sub-pixel heterogeneity assessment. The concept behind the spectral variation hypothesis (SVH) [40] could be used to assess the environmental heterogeneity within each pixel. This concept hypothesizes that the spectral response of a remotely sensed image could be used as a proxy to assess habitat heterogeneity and species diversity. Areas with a high spectral heterogeneity (SH) in a remotely sensed image have a high environmental heterogeneity with a higher number of available ecological niches. This concept was established firstly by Palmer et al. [40] and later developed by other authors [41]. The SVH has been tested in different ecosystems using various remote sensing data through the use of different SH indices. In the last few years, Rao's Q index (developed by Rao [42] for ecological purposes) has been proposed as an original SH measure [43] and has gained popularity due to the positive results obtained in various studies [44,45]. As stated by Rocchini et al. [43], *"given an image of N pixels, the Rao's Q is related to the sum of all the pixel values pairwise distances, each of which is multiplied by the relative abundance of each pair of pixels in the analyzed image"*. Hence, Rao's Q index, in comparison to other heterogeneity indices, has the advantages of considering both the values (through the distance/difference between the pixel) and the abundance of the pixels in a considered image [46].

The main aim of this paper is to analyze the relationship between ET, derived from ground-based eddy-covariance (EC) surface measurements at four different sites in Italy, and both the ECI (based on emissivity data from the CAMEL database) and the WDI (based on the difference in the surface and dew-point temperature modeled by ECMWF data). In the first relationship, the effects of the snow cover, the different vegetations and the environmental heterogeneity (calculated through the concept of the SVH using Rao's Q index) were analyzed. The paper is organized as follows. Section 2 deals with data and methods. Results are shown in Section 3 and discussed in Section 4. Conclusions are drawn in Section 5.

2. Materials and Method

2.1. Study Areas

Four EC sites were used to assess the relationship between both ECI and WDI with ET.

The Renon site [47–49] is located in the province of Bolzano/Bozen in the Alps, in the municipality of Renon/Ritten at an elevation of 1740 m asl. The EC tower is located in a *Picea abies*-dominated forest (around 85 %), but also including *Pinus cembra* L., (12%) and *Larix decidua* Mill., (3%). The forest canopy is irregular, with maximal height of around 30 m. The annual average temperature is around 4.6 °C, and the average annual precipitation is approximately 900 mm.

The Monte Bondone site [50] is located on a mountain plateau (called "Viote del Monte Bondone") near the city of Trento at 1550 m asl. The mean annual air temperature is 5.5 °C and the mean annual rainfall is 1190 mm. The site is managed as productivity-extensive meadow, typical of the alpine regions, characterized by the presence of *Festuca rubra* (basal cover of 25%), *Nardus stricta* (13%) and *Trifolium* sp. (14.5%).

The Lavarone EC tower [47] is situated near the town of Lavarone in the province of Trento at an elevation of around 1350 m asl. The tower is located in uneven-aged mixed forest dominated by *Abies alba* (around 70%), *Fagus sylvatica* (15%) and *Picea abies* (15%). The forest canopy reaches an elevation of approximately 35 m. The average annual precipitation is approximately 1290 mm and the mean annual temperature is around 7.8 °C.

Finally, the Bosco della Fontana site [51,52] is located near the city of Mantova (at an elevation of 19 m asl.) in the middle of the Po valley, within a forest nature reserve of around 235 ha. The wood canopy is 26 m high and dominated by *Carpinus betulus* L. and *Quercus robur* L. (57%), with a minor presence of *Acer campestre* L., *Prunus avium* L., *Fraxinus ornus* L. and *Ulmus minor* Mill., with *Alnus glutinosa* L. along the little rivers. The average annual precipitation is approximately 930 mm and the mean annual temperature is around 13.9 °C. The EC tower data are measured from a 42 m tall tower Figure 1.

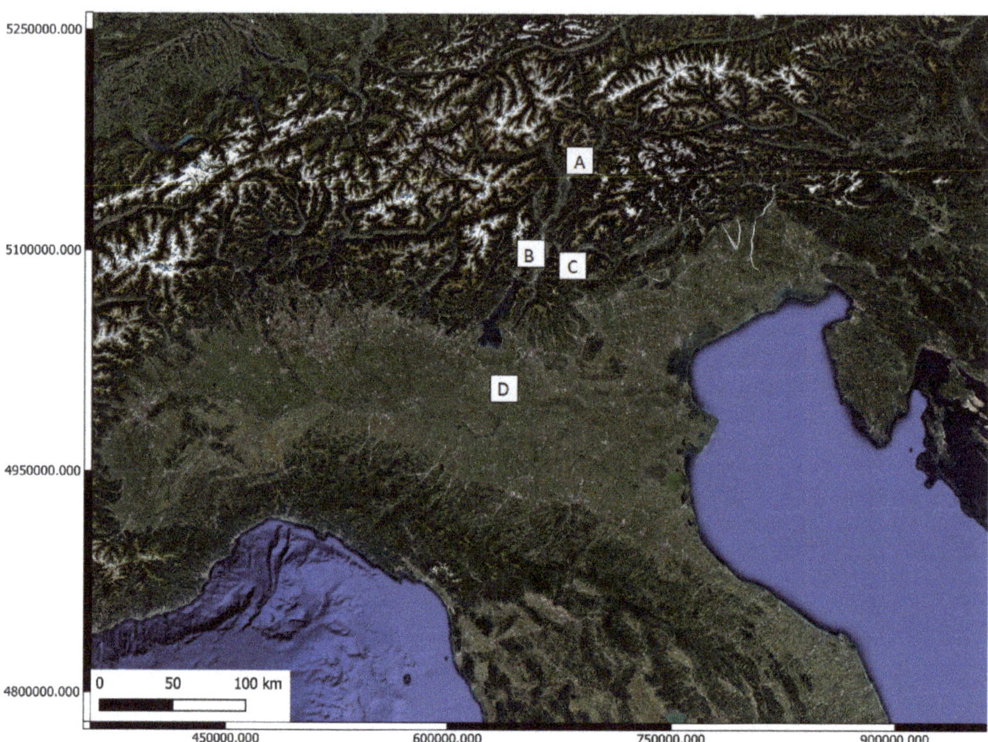

Figure 1. Google Earth image with the location of the four study sites in Italy. (A) Renon (Fluxnet code: IT-Ren); (B) Monte Bondone (Fluxnet code: IT-MBo); (C) Lavarone (Fluxnet code: IT-La2); (D) Bosco della Fontana (ICOS code: IT-BFt). CRS map: WGS84 /UTM zone 32N.

2.2. Eddy Covariance Data

Renon, Lavarone and Monte Bondone eddy covariance sites are part of the FLUXNET Network, where CO_2 fluxes, water vapour and other ancillary meteorological variables are measured at half-hourly intervals. Data are processed and quality controlled following the Fluxnet methodology [53]. Within the Fluxnet network, the data availability for the Renon site ranges from 1998 to 2013, for Lavarone, from 2003 to 2014 and, for Monte Bondone, from 2003 to 2013.

Bosco della Fontana is part of the ICOS Ecosystem Network. The eddy covariance tower measures half-hourly turbulent fluxes of CO_2, water vapor and different meteorological data following the ICOS protocols [54].

Despite the corrections applied in the calculation of EC fluxes, sensible and latent heat fluxes are usually underestimated at most EC sites with respect to the available energy at the surface [55], resulting in some uncertainty in the quantification of water lost by ecosystems through ET. For this reason, in our analysis, we used the latent heat flux adjusted by a correction factor based on the ratio between available energy and the sum of turbulent energy fluxes for each half hour [53]. The half-hourly latent heat flux data from the eddy covariance sites obtained from the FLUXNET and ICOS datasets were converted to ET using "LE.to.ET" function of the "bigleaf" R package, applying the correlation parameter between depth units and energy of ET [56] (Formula (1)). The conversion was corrected using the half-hourly air temperature.

$$\text{ET} = \text{LE}/\lambda \tag{1}$$

where:

- ET is the evapotranspiration (kg m^{-2} s^{-1});
- LE is the latent heat flux (W m^{-2});
- λ is the latent heat of vaporization 2.45 MJ kg^{-1}.

Successively, daily ET values were then accumulated and converted into monthly ET in order to be correlated to the ECI and WDI that were assessed on a monthly basis.

2.3. Emissivity Data and ECI Estimation

The ECI (that ranges in the interval [0, 1]) has been developed to discriminate between bare soil and vegetation [34] and to better classify vegetation cover [35]. For this present study, the methods introduced by Masiello et al. [35] were used to calculate the ECI from the CAMEL dataset. We used only the CAMEL pixels that had a "good" emissivity quality flag (value 1) in order to have an adequate overall accuracy.

The ECI is based on the channels at 8.6, 10.8 and 12.1 µm for the CAMEL dataset. According to different studies [34,35], these channels are indeed the most sensitive to bare, green and senescent vegetation. As a consequence, ECI is calculated as:

$$\text{ECI} = 1 - \delta\epsilon \tag{2}$$

where $\delta\epsilon$ represents the difference between the maximum and the minimum value of emissivity (ϵ) among the three CAMEL spectral channels.

For each study area, monthly ECI was successively correlated to the monthly ET by a time series analysis and, successively, by linear regression. R^2 and p values were used to assess the strength and significance of the correlations. Due to the different temporal range data availability of ET and ECI, the correlations were tested differently for each study area. For the Renon study area, the time-series range from 2008 to 2013, for Lavarone, from 2008 to 2014, for Monte Bondone, from 2010 to 2013 and, for Bosco della Fontana, only the data from 2013 were available.

Furthermore, snow cover information, derived from the "snow fraction" layer of the CAMEL dataset, was included in the time-series correlation ET–ECI data. This layer provides information of snow cover on the basis of the normalized difference snow index (NDSI) [57], which ranges from 0 (no snow cover) to 100 (full snow cover). It is used to identify possible anomalies in the ET-ECI index correlation, particularly in the mountain sites (Renon, Lavarone and Monte Bondone), where snow remains on the ground for several winter months.

2.4. Meteorological Data and WDI Calculation

Monthly modeled data of surface temperature (Ts_{ECMWF}) and dew point temperature (Td_{ECMWF}) derived from the ECMWF [58] were used to compute the water deficit index, or WDI. ECMWF data were from the "Operational Analysis", and were released over a regular grid of 0.125° × 0.125°. For each EC site, the closest point of the ECMWF grid was chosen. Surface temperature (Ts_{ECMWF}) and dew point temperature (Td_{ECMWF}) were, respectively, the skin temperature and the 2 m dew point temperature from surface analyses.

WDI was then computed according to [32]:

$$\text{WDI} = Ts_{ECMWF} - Td_{ECMWF} \tag{3}$$

For this reason, WDI values depend on surface and dew point temperatures. High WDI values are expected in summer, especially in dry conditions, when the surface temperature becomes significantly higher than the dew temperature near it, whereas lower values are expected in winter.

Because of its definition and calculation, the WDI has a coarser spatial resolution than the ECI. However, the temperature and humidity fields are expected to be more homogeneous than the surface emissivity, which can have space scales of variability of a few meters or less.

As for the correlation ECI-ET, for each study area, the monthly WDI was successively correlated to the monthly ET by a time series analysis and, successively, by linear regression. R^2 and p values were used to assess the strength and significance of the correlations. Due to the different temporal range data availability of WDI and ET, the correlations were tested differently for each study area. In the Renon study area ,the correlation range was from 2010 to 2013, for Lavarone, from 2010 to 2014, for Monte Bondone, from 2010 to 2013 and, for Bosco della Fontana, only for 2013.

2.5. Assessment of the Environmental Heterogeneity

In order to assess the effect of the environmental heterogeneity within the ECI pixel (Figure 2), the SVH was assessed through Rao's Q index (Formula (4)) using an NDVI MODIS image (resolution of 500 m) captured on 8 June 2014. The choice of this date is related to the work of Torresani et al. [59], where they stated that the NDVI at this time of the year (summer), when it reaches the highest seasonal values, is more able to capture small variations in reflectance of different vegetation and, thus, of different ecosystems. For this purpose, the R-package function "Rao" of the R package *rasterdiv* [60] was implemented to retrieve a Rao's Q value for the single ECI pixel.

$$Q_{rs} = \sum_{i=1}^{F-1} \sum_{j=i+1}^{F} d_{ij} * p_i * p_j, \tag{4}$$

where:

- Q_{rs} is the Rao's Q applied to remote sensing data;
- p is the relative abundance of a pixel value in a selected study area (F). In our case, it is the CAMEL pixel;
- d_{ij} is the distance between the i-th and j-th pixel value ($d_{ij} = d_{ji}$ and $d_{ii} = 0$);
- i and j identify two pixels within the area F.

The relative abundance p was calculated as the ratio between the considered pixel (p_i and p_j) and the total number of pixels in F. The distance matrix d_{ij} can be built in different dimensions, allowing for the consideration of more than one band or raster at a time. In our case, the d_{ij} was calculated as a simple Euclidean distance based on the NDVI image.

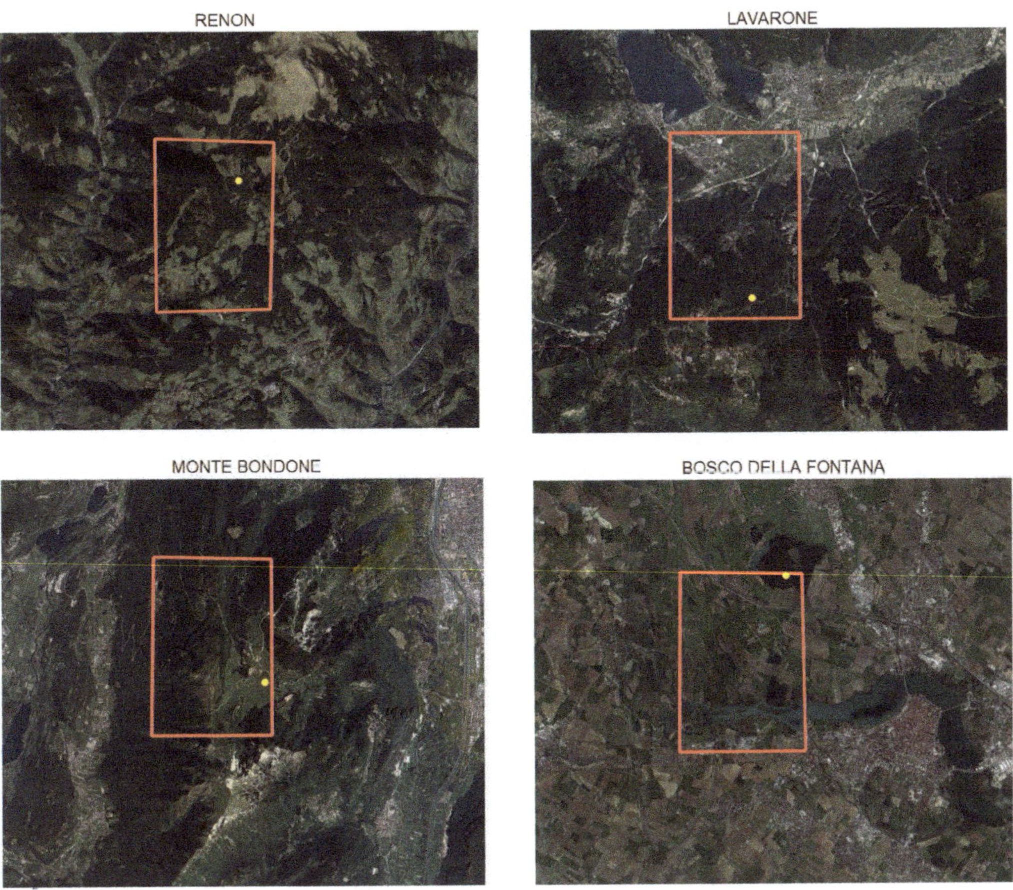

Figure 2. Google earth image with the location of the four eddy covariance towers (yellow points) for each site. In red, the CAMEL pixel used to derive the ECI. Date picture: Lavarone and Monte Bondone 6 December 2017; Renon 19 October 2017; Bosco delle Fontana 18 June 2013.

3. Results

3.1. Seasonal Evolution and Correlations

The temporal evolution of the ET and the monthly ECI with reference to the snow cover information is shown in Figure 3. For the Renon, Lavarone and Monte Bondone study area, both the ET and ECI seem to follow the same seasonal evolution, particularly in the summer months, when both the curves reaches the peak. In several situations, generally in the cold months, the ECI behaves differently, creating a "second peak" that does not follow the normal ET trend. The snow cover information explains the "winter peak" of the ECI, where the amplitude and the size of both the blue curve and black histogram are indeed similar. On the other hand, in the Bosco della Fontana study area, ET and the ECI have an opposite seasonal evolution. ET has a normal seasonal trend with the highest values in summer, decreasing in winter, whereas the ECI has its low values in summer.

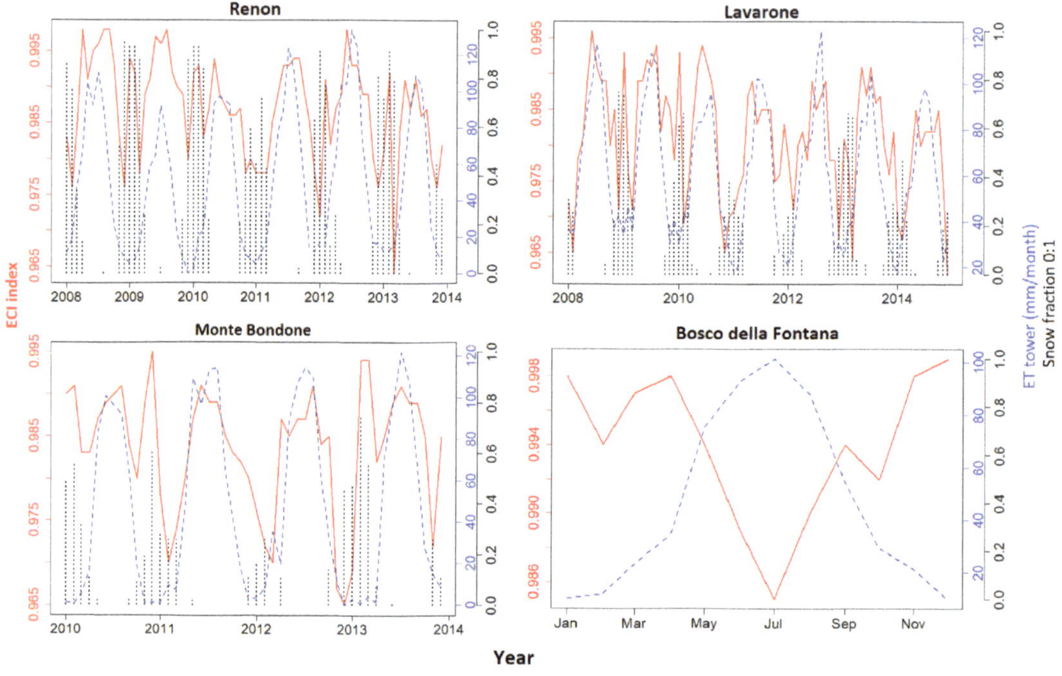

Figure 3. Relationship between the ET (ET tower—blue line) and the monthly ECI (red line).

Figure 4 shows the linear regression between ET-ECI considering the months with low snow cover (lower than 50%). This was carried out to assess the strength of the correlation without the "anomalous peaks" created by the ECI in the cold months. In the Renon, Lavarone and Monte Bondone study areas, the correlations are all positive, reaching an R^2 between 0.49 (Lavarone) to 0.68 (Renon). In the Bosco della Fontana study area, the correlation is negative, reaching an R^2 of 0.72.

The monthly time series of ET and the WDI are shown in Figure 5. In the four study areas, both the curves show the same trend, with the peak in the summer months and a lower value in winter.

Figure 6 shows the linear regression between ET and the WDI for the four considered study areas. R^2 ranges from 0.48 (Lavarone) to 0.89 (Bosco della Fontana). Unlike the correlation ET-ECI, the slope remains positive for all of the considered areas.

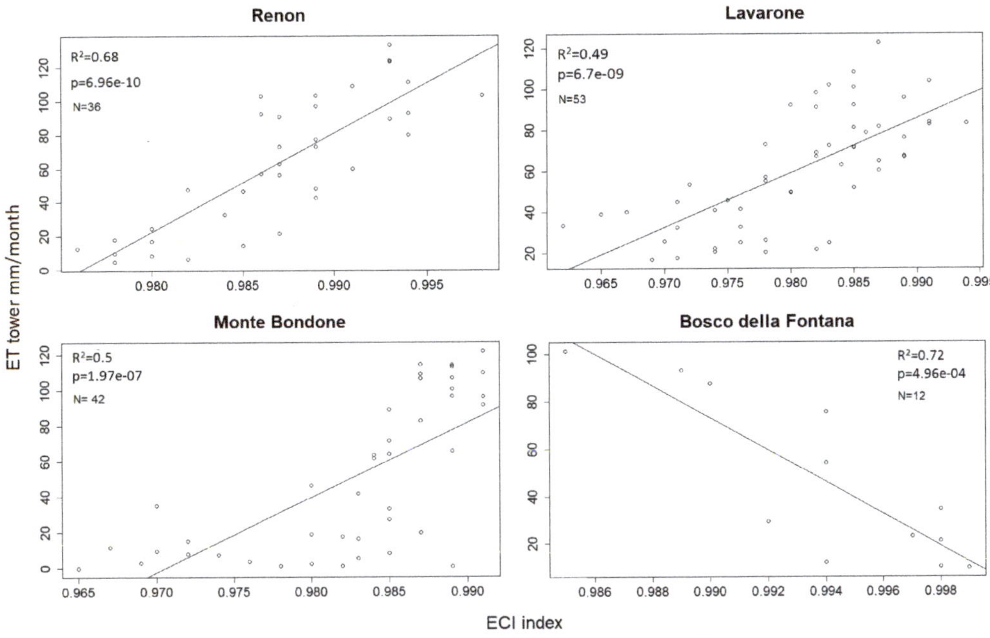

Figure 4. Linear regression between the ET (ET tower) and the monthly ECI when the snow cover is lower than 50%. *N* is the number of data points.

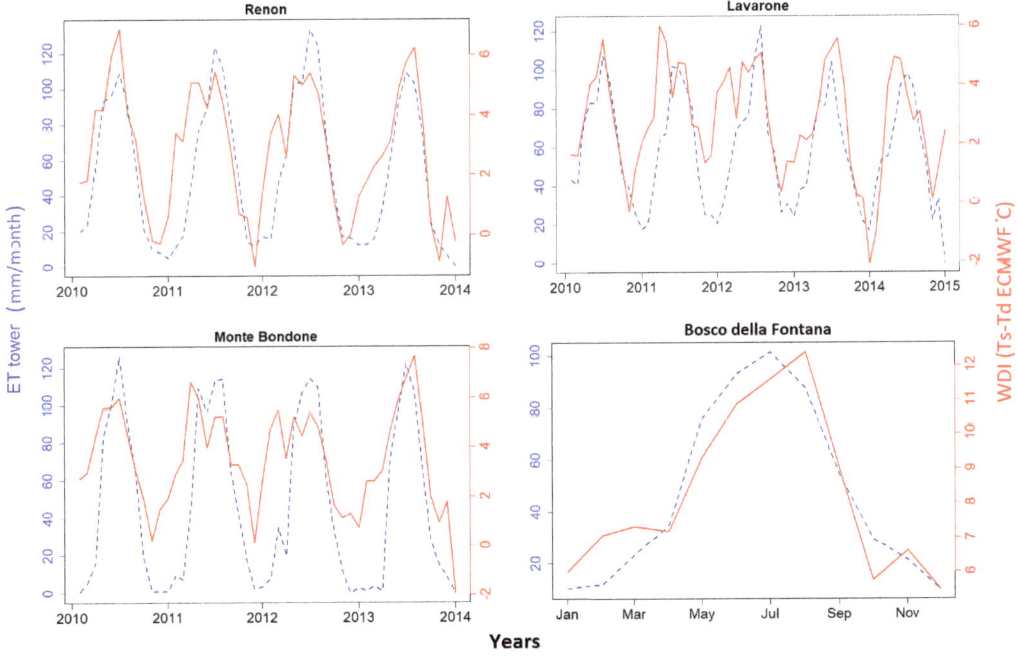

Figure 5. Time series of ET (blue dashed line) and WDI (red line)

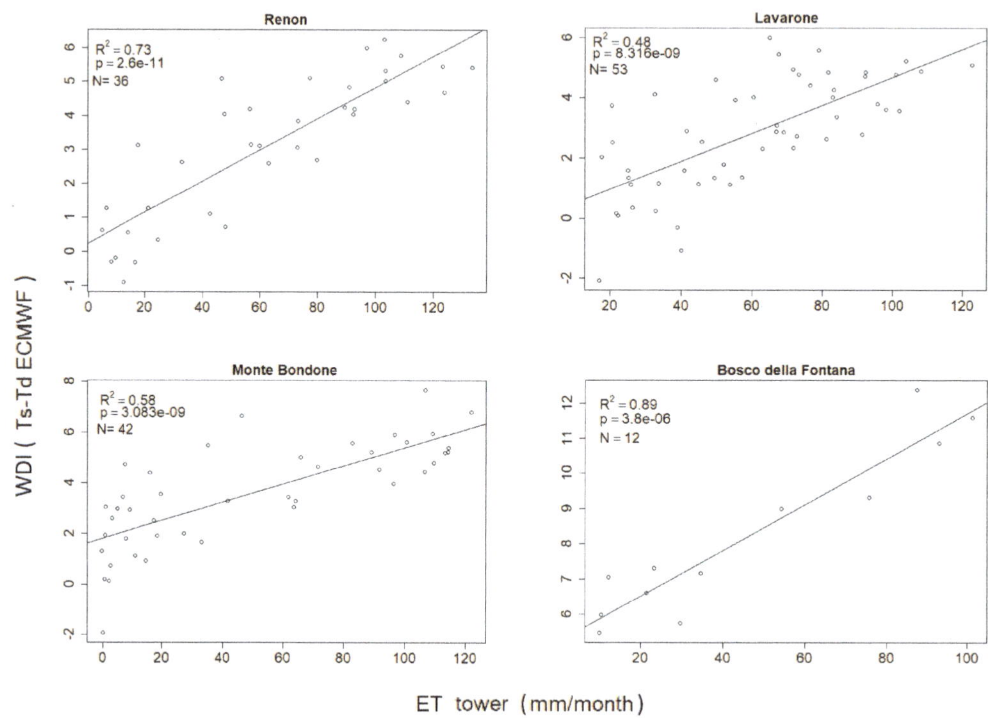

Figure 6. Linear regression between ET and WDI for the four considered study areas.

3.2. Environmental Heterogeneity

In order to understand the impact of the environmental heterogeneity, especially within the ECI pixel, the SH, as an indicator of habitat heterogeneity, was calculated through Rao's Q index over the four study areas. Table 1 summarizes Rao's Q index values for each site.

Table 1. Rao's Q value over the four ECI pixels of each study area.

Area	Rao's Q Index
Monte Bondone	0.036
Renon	0.061
Lavarone	0.075
Bosco della Fontana	0.082

The highest Rao's Q value is in the area of Bosco della Fonata (higher environmental heterogeneity), whereas the lowest is in the area of Monte Bondone (lower heterogeneity).

Figure 7 shows both the above mentioned study areas with the exact position of the eddy covariance tower (yellow dot) and the pixel size of the CAMEL database (red rectangle). Also from a first view, it is possible to notice the difference in habitat heterogeneity between the two sites. In the area of Monte Bondone (Figure 7A), the forested and grassland area are predominant. In the area of Bosco della Fontana (Figure 7B), the environmental heterogeneity is higher. Different habitats fall within the CAMEL pixel: the forested area (in the upper right corner), grassland, farmland, uncultivated areas and the aquatic ecosystems of the Mincio river and of the Mantova's lakes.

Figure 7. (**A**) shows the area of Monte Bondone and (**B**) shows the area of Bosco della Fontana; the yellow dot shows the position of the eddy covariance tower. The red rectangle shows the size of the CAMEL pixel.

4. Discussion

In this paper, we presented the correlation between the ECI (based on emissivity data from infrared spectral channels of the CAMEL dataset) and the WDI (based on the data from ECMWF analyses) against ET measured in four EC sites in Italy (Renon, Monte Bondone, Lavarone and Bosco della Fontana). Furthermore, the behavior of the above mentioned correlations in relation to the vegetation cover was discussed. Additionally, the environmental heterogeneity was assessed (using NDVI MODIS images and analyzed through the SH Rao's Q index) in order to evaluate the performance, particularly of the ECI in relation to the habitat fragmentation over the four study areas.

The results show that the correlation between ECI and ET is statistically significant for the four study areas. In the areas of Renon, Lavarone and Monte Bondone, the relationships are positive and they are influenced by the snow cover (estimated with the "snow fraction" layer of the CAMEL dataset), which interferes with the emissivity values. In the winter months, when the snow covers the earth surface, the ECI yields a second peak that does not follow the ET trend. As described by Masiello et al. [29,35], the ECI is highly influenced by the dryness and by the land cover. As an example, the ECI has a value of around 0.98 for water-covered surfaces (e.g., snow), whereas, for dry and senescent vegetation, the index reaches lower values. The ECI was initially developed to build synergy with the NDVI to overcome the drawback of the vegetation index that, in certain cases (as in the case of Masiello et al. [35] over the Congo basin area for a temporal range of seventeen years), is not able to discriminate senescent vegetation and bare soil; in particular, when the vegetation regeneration start after a deforestation or degradation event [35]. On the other hand, in the area of Bosco della Fonatana, the correlation ECI–ET was negative: in the considered year, the curves indeed had an opposite trend. Several reasons can explain this different trend: since the ECI is influenced by the dryness and by the land cover, its values could be distorted by the humidity of the soil and by the presence of surface water, which is very high in this area, located near the aquatic ecosystems of the Mincio river and of the Mantova's lakes. The other reason is related to the high habitat fragmentation within the CAMEL pixel that alters the ECI value (as shown in in Figure 7) and summarized in Table 1. The results of the table indeed show that the Rao's Q index, used to estimate the environmental heterogeneity (by the assessment of the SH), has the highest value in the area of Bosco della Fontana.

The concept of the SVH and the related environmental heterogeneity is therefore crucial not only in the estimation of biodiversity and the assessment of species diversity [41,61,62], but also in all studies where the ground information (e.g., energy fluxes) is correlated with the remote sensing data. This aspect is particularly common when using remote sensing information. An image, through the diversity of the various pixels, provides knowledge about the remote sensing response caused by the physical interaction of the measured information (emissivity in our case) with the Earth's surface. Thus, the raster grid of pixels that build up an image, represents just an average response of a real information. The question regarding the size of the pixels, in order to describe a certain area or to characterize a real situation or an event is, as previously stated, still debated in the research community [63,64]. Like in our case, images with a coarse spatial resolution tend to integrate the information of various subjects (e.g., vegetation, human artefacts, rivers.), homogenizing the signal and causing difficulties toward clearly identifying boundaries between spatial entities (individuals, vegetation types, ecosystem types) [65]. On the other hand, a fine spatial resolution may lead to a level of details within spatial entities that may cause a strong heterogeneity, leading to strong noises and uncertainties [66].

As far as the relationship of ET vs. WDI is concerned (Figures 5 and 6), we have found that the correlation is less sensitive to the vegetation changes and environmental heterogeneity. The slope of the linear regression remains positive, although we have explored sites with a very large heterogeneity. The difference in temperature (Ts_{ECMWF}—Td_{ECMWF}) is correlated with the ET because the surface temperature is strongly dependent on the impinging solar radiation, whereas the dew point temperature is dependent on both the air temperature and the humidity field. The partial mismatch between the two variables might be due to the lack of an additional meteorological parameter, such as the wind. The detail of the regression still seems to be site-dependent; however, in this case, there is no ambiguity that ET and the WDI are positive correlated. Since the WDI is based on temperatures, it is less influenced by the habitat fragmentation and by the environmental heterogeneity compared to the ECI, which is based on the emissivity of the surface. For this reason, the correlation holds true also in the area of Bosco della Fontana, which, as previously stated, showed an opposite result in the correlation ET-ECI.

Finally, we believe that the synergistic use of the ECI and WDI might increase the accuracy of ET estimation because of their different sensitivities to different aspects of the vegetation. The ECI is better suited to detecting changes in the vegetation state, green to senescent or transition to bare soil. These states can influence ET, but in a way that can be highly nonlinear. Conversely, the WDI is more linearly related to ET. The synergistic use of the two could be, e.g., of some interest during an intense heatwave, which has become common in temperate regions because of climate change (e.g., see https://climate.copernicus.eu/esotc/2021/globe-in-2021 (accessed on 8 May 2012)). In the event of heatwaves, we expect the WDI > 0 (e.g., see [32]), and a decreasing value of the ECI could show an early decay of the foliage to the senescent state, and hence vegetation stress.

5. Conclusions and Outlook

In this study, the correlation between data of ET derived from four eddy covariance sites in Italy (Renon in the Province of Bolzano, Monte Bondone and Lavarone in the Province of Trento and Bosco della Fontana in the Province of Mantova) and two indices—(1) the emissivity contrast index, or ECI; (2) the water stress index, or WDI—was assessed. Both indices were shown to correlate with in situ observations, which is good from the perspective of using remote-sensed data to monitor the state of vegetation from satellite. The correlation ECI-ET is influenced by the habitat heterogeneity and by the presence of snow/water in the surface. This could be critical, especially in areas covered by snow (e.g., mountain regions in winter), with surface water or with high environmental heterogeneity. The WDI showed generally fewer uncertainties in detecting the correct evolution of ET, in that the index is directly related to the thermodynamic parameters that govern ET and to the intensity of solar radiation. Furthermore, we believe that the synergistic use of the ECI and WDI could lead to a more accurate ET estimation, bringing the benefits of both indices. Improvements on this side would also be greatly beneficial for providing a more accurate input to numerical weather and climate prediction models, for which, reliable estimates of fluxes over snow-covered terrain are still a challenging situation [67,68].

Further refinements can also be obtained from a more precise evaluation of ET from EC, taking into account timescales associated with different atmospheric conditions [69]. This is particularly applicable to the mountainous sites, where daily periodic flows, such as thermally driven slope wind and valley winds, are well established, and documented meteorological features of the mountain boundary layer [70–72]. This goal is among the scopes of the ongoing international cooperation effort TEAMx (Multi-Scale Transport and Exchange Processes in the Atmosphere over Mountains–Program and Experiment) through intensive field campaigns performed at selected target areas in the Alps, combining ground-based, airborne and remote sensing observations [73].

Author Contributions: Data curation, M.F.; Investigation, E.T. ; Methodology, G.G.; Supervision, D.R. and D.Z.; Writing—original draft, M.T.; Writing—review & editing, G.M., N.V. and C.S. All authors have read and agreed to the published version of the manuscript.

Funding: This research was carried out in the framework of the project 'OT4CLIMA', which was funded by the Italian Ministry of Education, University and Research (D.D. 2261 del 6.9.2018, PON R&I 2014-2020 and FSC). M.T. and D.R. were partially supported by the European Union's Horizon 2020 research and innovation program under grant agreement No 862480 (SHOWCASE). DR was partially supported by the Horizon Europe project EarthBridge and by the H2020 COST Action CA17134 'Optical synergies for spatiotemporal sensing of scalable ecophysiological traits (SENSECO).

Data Availability Statement: Not applicable.

Conflicts of Interest: The authors declare no conflict of interest.

References

1. Oishi, A.C.; Oren, R.; Novick, K.A.; Palmroth, S.; Katul, G.G. Interannual invariability of forest evapotranspiration and its consequence to water flow downstream. *Ecosystems* **2010**, *13*, 421–436. [CrossRef]
2. Irmak, S.; Haman, D.Z. Evapotranspiration: Potential or reference? *EDIS* **2003**, *2003*. [CrossRef]
3. Sun, G.; McNulty, S.; Amatya, D.; Skaggs, R.; Swift, L., Jr.; Shepard, J.; Riekerk, H. A comparison of the watershed hydrology of coastal forested wetlands and the mountainous uplands in the Southern US. *J. Hydrol.* **2002**, *263*, 92–104. [CrossRef]
4. Arain, M.; Black, T.; Barr, A.G.; Griffis, T.; Morgenstern, K.; Nesic, Z. Year-round observations of the energy and water vapour fluxes above a boreal black spruce forest. *Hydrol. Process.* **2003**, *17*, 3581–3600. [CrossRef]
5. Dore, S.; Montes-Helu, M.; Hart, S.C.; Hungate, B.A.; Koch, G.W.; Moon, J.B.; Finkral, A.J.; Kolb, T.E. Recovery of ponderosa pine ecosystem carbon and water fluxes from thinning and stand-replacing fire. *Glob. Chang. Biol.* **2012**, *18*, 3171–3185. [CrossRef] [PubMed]
6. Wilson, K.B.; Hanson, P.J.; Mulholland, P.J.; Baldocchi, D.D.; Wullschleger, S.D. A comparison of methods for determining forest evapotranspiration and its components: Sap-flow, soil water budget, eddy covariance and catchment water balance. *Agric. For. Meteorol.* **2001**, *106*, 153–168. [CrossRef]
7. Williams, D.; Cable, W.; Hultine, K.; Hoedjes, J.; Yepez, E.; Simonneaux, V.; Er-Raki, S.; Boulet, G.; De Bruin, H.; Chehbouni, A.; et al. Evapotranspiration components determined by stable isotope, sap flow and eddy covariance techniques. *Agric. For. Meteorol.* **2004**, *125*, 241–258. [CrossRef]
8. Agam, N.; Evett, S.R.; Tolk, J.A.; Kustas, W.P.; Colaizzi, P.D.; Alfieri, J.G.; McKee, L.G.; Copeland, K.S.; Howell, T.A.; Chávez, J.L. Evaporative loss from irrigated interrows in a highly advective semi-arid agricultural area. *Adv. Water Resour.* **2012**, *50*, 20–30. [CrossRef]
9. Cienciala, E.; Lindroth, A. Gas-exchange and sap flow measurements of Salix viminalis trees in short-rotation forest. *Trees* **1995**, *9*, 295–301. [CrossRef]
10. Wang, S.; Pan, M.; Mu, Q.; Shi, X.; Mao, J.; Brümmer, C.; Jassal, R.S.; Krishnan, P.; Li, J.; Black, T.A. Comparing evapotranspiration from eddy covariance measurements, water budgets, remote sensing, and land surface models over Canada. *J. Hydrometeorol.* **2015**, *16*, 1540–1560. [CrossRef]
11. Kosugi, Y.; Katsuyama, M. Evapotranspiration over a Japanese cypress forest. II. Comparison of the eddy covariance and water budget methods. *J. Hydrol.* **2007**, *334*, 305–311. [CrossRef]
12. Zhang, K.; Kimball, J.S.; Running, S.W. A review of remote sensing based actual evapotranspiration estimation. *Wiley Interdiscip. Rev. Water* **2016**, *3*, 834–853. [CrossRef]

13. Chen, J.M.; Liu, J. Evolution of evapotranspiration models using thermal and shortwave remote sensing data. *Remote Sens. Environ.* **2020**, *237*, 111594. [CrossRef]
14. Liu, J.; Chen, J.; Cihlar, J. Mapping evapotranspiration based on remote sensing: An application to Canada's landmass. *Water Resour. Res.* **2003**, *39*. [CrossRef]
15. Choudhury, B.J. Estimating areal evaporation using multispectral satellite observations. In *Land Surface Processes in Hydrology*; Springer: Berlin/Heidelberg, Germany, 1997; pp. 347–382.
16. Jiang, L.; Islam, S. Estimation of surface evaporation map over southern Great Plains using remote sensing data. *Water Resour. Res.* **2001**, *37*, 329–340. [CrossRef]
17. Agam, N.; Kustas, W.P.; Anderson, M.C.; Norman, J.M.; Colaizzi, P.D.; Howell, T.A.; Prueger, J.H.; Meyers, T.P.; Wilson, T.B. Application of the Priestley–Taylor approach in a two-source surface energy balance model. *J. Hydrometeorol.* **2010**, *11*, 185–198. [CrossRef]
18. Mokhtari, A.; Noory, H.; Pourshakouri, F.; Haghighatmehr, P.; Afrasiabian, Y.; Razavi, M.; Fereydooni, F.; Naeni, A.S. Calculating potential evapotranspiration and single crop coefficient based on energy balance equation using Landsat 8 and Sentinel-2. *ISPRS J. Photogramm. Remote Sens.* **2019**, *154*, 231–245. [CrossRef]
19. Xue, J.; Anderson, M.C.; Gao, F.; Hain, C.; Yang, Y.; Knipper, K.R.; Kustas, W.P.; Yang, Y. Mapping Daily Evapotranspiration at Field Scale Using the Harmonized Landsat and Sentinel-2 Dataset, with Sharpened VIIRS as a Sentinel-2 Thermal Proxy. *Remote Sens.* **2021**, *13*, 3420. [CrossRef]
20. Chen, H.; Zhu, G.; Zhang, K.; Bi, J.; Jia, X.; Ding, B.; Zhang, Y.; Shang, S.; Zhao, N.; Qin, W. Evaluation of evapotranspiration models using different LAI and meteorological forcing data from 1982 to 2017. *Remote Sens.* **2020**, *12*, 2473. [CrossRef]
21. Carlson, T.N.; Petropoulos, G.P. A new method for estimating of evapotranspiration and surface soil moisture from optical and thermal infrared measurements: The simplified triangle. *Int. J. Remote Sens.* **2019**, *40*, 7716–7729. [CrossRef]
22. García-Santos, V.; Sánchez, J.M.; Cuxart, J. Evapotranspiration Acquired with Remote Sensing Thermal-Based Algorithms: A State-of-the-Art Review. *Remote Sens.* **2022**, *14*, 3440. [CrossRef]
23. Hamberg, L.J.; Fisher, J.B.; Ruppert, J.L.; Tureček, J.; Rosen, D.H.; James, P.M. Assessing and modeling diurnal temperature buffering and evapotranspiration dynamics in forest restoration using ECOSTRESS thermal imaging. *Remote Sens. Environ.* **2022**, *280*, 113178. [CrossRef]
24. Carlson, T.N.; Dodd, J.K.; Benjamin, S.G.; Cooper, J.N. Satellite estimation of the surface energy balance, moisture availability and thermal inertia. *J. Appl. Meteorol. Climatol.* **1981**, *20*, 67–87. [CrossRef]
25. Fisher, J.B.; Tu, K.P.; Baldocchi, D.D. Global estimates of the land–atmosphere water flux based on monthly AVHRR and ISLSCP-II data, validated at 16 FLUXNET sites. *Remote Sens. Environ.* **2008**, *112*, 901–919. [CrossRef]
26. Anderson, M.; Norman, J.; Diak, G.; Kustas, W.; Mecikalski, J. A two-source time-integrated model for estimating surface fluxes using thermal infrared remote sensing. *Remote Sens. Environ.* **1997**, *60*, 195–216. [CrossRef]
27. Allen, R.G.; Tasumi, M.; Trezza, R. Satellite-based energy balance for mapping evapotranspiration with internalized calibration (METRIC)—Model. *J. Irrig. Drain. Eng.* **2007**, *133*, 380–394. [CrossRef]
28. Bastiaanssen, W.G.; Menenti, M.; Feddes, R.; Holtslag, A. A remote sensing surface energy balance algorithm for land (SEBAL). 1. Formulation. *J. Hydrol.* **1998**, *212*, 198–212. [CrossRef]
29. Masiello, G.; Serio, C.; Venafra, S.; Cersosimo, A.; Mastro, P.; Falabella, F.; Pasquariello, P. Emissivity Based Indices for Drought and Forest Fire. In Proceedings of the 2021 IEEE International Geoscience and Remote Sensing Symposium IGARSS, Brussels, Belgium, 11–16 July 2021; pp. 930–933.
30. Hilton, F.; Armante, R.; August, T.; Barnet, C.; Bouchard, A.; Camy-Peyret, C.; Capelle, V.; Clarisse, L.; Clerbaux, C.; Coheur, P.F.; et al. Hyperspectral Earth Observation from IASI: Four years of accomplishments. *Bull. Am. Meteorol. Soc.* **2012**, *93*, 347–370. [CrossRef]
31. Masiello, G.; Serio, C.; Venafra, S.; DeFeis, I.; Borbas, E.E. Diurnal variation in Sahara desert sand emissivity during the dry season from IASI observations. *J. Geophys. Res. Atmos.* **2014**, *119*, 1626–1638. [CrossRef]
32. Masiello, G.; Ripullone, F.; De Feis, I.; Rita, A.; Saulino, L.; Pasquariello, P.; Cersosimo, A.; Venafra, S.; Serio, C. The IASI Water Deficit Index to Monitor Vegetation Stress and Early Drying in Summer Heatwaves: An Application to Southern Italy. *Land* **2022**, *11*, 1366. . [CrossRef]
33. Liuzzi, G.; Masiello, G.; Serio, C.; Venafra, S.; Camy-Peyret, C. Physical inversion of the full IASI spectra: Assessment of atmospheric parameters retrievals, consistency of spectroscopy and forward modelling. *J. Quant. Spectrosc. Radiat. Transf.* **2016**, *182*, 128–157. [CrossRef]
34. French, A.; Schmugge, T.; Kustas, W. Discrimination of senescent vegetation using thermal emissivity contrast. *Remote Sens. Environ.* **2000**, *74*, 249–254. [CrossRef]
35. Masiello, G.; Cersosimo, A.; Mastro, P.; Serio, C.; Venafra, S.; Pasquariello, P. Emissivity-based vegetation indices to monitor deforestation and forest degradation in the Congo basin rainforest. In *Remote Sensing for Agriculture, Ecosystems, and Hydrology XXII*. International Society for Optics and Photonics: Bellingham, WA, USA, 2020; Volume 11528, p. 115280L.
36. Borbas, E.; Hulley, G.; Feltz, M.; Knuteson, R.; Hook, S. The Combined ASTER MODIS Emissivity over Land (CAMEL) Part 1: Methodology and High Spectral Resolution Application. *Remote Sens.* **2018**, *10*, 643. . [CrossRef]
37. Feltz, M.; Borbas, E.; Knuteson, R.; Hulley, G.; Hook, S. The combined ASTER MODIS emissivity over land (CAMEL) part 2: Uncertainty and validation. *Remote Sens.* **2018**, *10*, 664. [CrossRef]

38. Feltz, M.; Borbas, E.; Knuteson, R.; Hulley, G.; Hook, S. The combined ASTER and MODIS emissivity over land (CAMEL) global broadband infrared emissivity product. *Remote Sens.* **2018**, *10*, 1027. [CrossRef]
39. Loveless, M.; Borbas, E.E.; Knuteson, R.; Cawse-Nicholson, K.; Hulley, G.; Hook, S. Climatology of the Combined ASTER MODIS Emissivity over Land (CAMEL) version 2. *Remote Sens.* **2020**, *13*, 111. [CrossRef]
40. Palmer, M.W.; Earls, P.G.; Hoagland, B.W.; White, P.S.; Wohlgemuth, T. Quantitative tools for perfecting species lists. *Environ. Off. J. Int. Environ. Soc.* **2002**, *13*, 121–137. [CrossRef]
41. Torresani, M.; Feilhauer, H.; Rocchini, D.; Féret, J.B.; Zebisch, M.; Tonon, G. Which optical traits enable an estimation of tree species diversity based on the Spectral Variation Hypothesis? *Appl. Veg. Sci.* **2021**, *24*, e12586. [CrossRef]
42. Rao, C.R. Diversity and dissimilarity coefficients: A unified approach. *Theor. Popul. Biol.* **1982**, *21*, 24–43. [CrossRef]
43. Rocchini, D.; Marcantonio, M.; Ricotta, C. Measuring Rao's Q diversity index from remote sensing: An open source solution. *Ecol. Indic.* **2017**, *72*, 234–238. [CrossRef]
44. Torresani, M.; Rocchini, D.; Sonnenschein, R.; Zebisch, M.; Hauffe, H.C.; Heym, M.; Pretzsch, H.; Tonon, G. Height variation hypothesis: A new approach for estimating forest species diversity with CHM LiDAR data. *Ecol. Indic.* **2020**, *117*, 106520. [CrossRef]
45. Tamburlin, D.; Torresani, M.; Tomelleri, E.; Tonon, G.; Rocchini, D. Testing the Height Variation Hypothesis with the R rasterdiv Package for Tree Species Diversity Estimation. *Remote Sens.* **2021**, *13*, 3569. [CrossRef]
46. Rocchini, D.; Salvatori, N.; Beierkuhnlein, C.; Chiarucci, A.; de Boissieu, F.; Förster, M.; Garzon-Lopez, C.X.; Gillespie, T.W.; Hauffe, H.C.; He, K.S.; et al. From local spectral species to global spectral communities: A benchmark for ecosystem diversity estimate by remote sensing. *Ecol. Inform.* **2021**, *61*, 101195. [CrossRef]
47. Cescatti, A.; Marcolla, B. Drag coefficient and turbulence intensity in conifer canopies. *Agric. For. Meteorol.* **2004**, *121*, 197–206. [CrossRef]
48. Nicolini, G.; Aubinet, M.; Feigenwinter, C.; Heinesch, B.; Lindroth, A.; Mamadou, O.; Moderow, U.; Mölder, M.; Montagnani, L.; Rebmann, C.; et al. Impact of CO_2 storage flux sampling uncertainty on net ecosystem exchange measured by eddy covariance. *Agric. For. Meteorol.* **2018**, *248*, 228–239. [CrossRef]
49. Marcolla, B.; Cescatti, A.; Montagnani, L.; Manca, G.; Kerschbaumer, G.; Minerbi, S. Importance of advection in the atmospheric CO_2 exchanges of an alpine forest. *Agric. For. Meteorol.* **2005**, *130*, 193–206. [CrossRef]
50. Gianelle, D.; Vescovo, L.; Marcolla, B.; Manca, G.; Cescatti, A. Ecosystem carbon fluxes and canopy spectral reflectance of a mountain meadow. *Int. J. Remote Sens.* **2009**, *30*, 435–449. [CrossRef]
51. Schallhart, S.; Rantala, P.; Taipale, R.; Nemitz, E.; Tillmann, R.; Mentel, T.; Ruuskanen, T.; Rinne, J. Ecosystem scale VOC exchange measurements at Bosco Fontana (IT) and Hyytiälä (FI). In *AGU Fall Meeting Abstracts;* American Geophysical Union: Washington, DC, USA, 2013; Volome 2013, pp. A31C–0070.
52. Gerosa, G.; Marzuoli, R.; Monteleone, B.; Chiesa, M.; Finco, A. Vertical ozone gradients above forests. Comparison of different calculation options with direct ozone measurements above a mature forest and consequences for ozone risk assessment. *Forests* **2017**, *8*, 337. [CrossRef]
53. Pastorello, G.; Trotta, C.; Canfora, E.; Chu, H.; Christianson, D.; Cheah, Y.W.; Poindexter, C.; Chen, J.; Elbashandy, A.; Humphrey, M.; et al. The FLUXNET2015 dataset and the ONEFlux processing pipeline for eddy covariance data. *Sci. Data* **2020**, *7*, 1–27. [CrossRef]
54. Rebmann, C.; Aubinet, M.; Schmid, H.; Arriga, N.; Aurela, M.; Burba, G.; Clement, R.; De Ligne, A.; Fratini, G.; Gielen, B.; et al. ICOS eddy covariance flux-station site setup: A review. *Int. Agrophysics* **2018**, *32*, 471–494. [CrossRef]
55. Foken, T. The energy balance closure problem: An overview. *Ecol. Appl.* **2008**, *18*, 1351–1367. [CrossRef]
56. Velpuri, N.M.; Senay, G.B.; Singh, R.K.; Bohms, S.; Verdin, J.P. A comprehensive evaluation of two MODIS evapotranspiration products over the conterminous United States: Using point and gridded FLUXNET and water balance ET. *Remote Sens. Environ.* **2013**, *139*, 35–49. [CrossRef]
57. Hall, D.K.; Riggs, G.A. Normalized-difference snow index (NDSI). In *Encyclopedia of Snow, Ice and Glaciers;* Springer Science & Business Media: Berlin/Heidelberg, Germany, 2010.
58. Georgiev, C.G. Quantitative relationship between Meteosat WV data and positive potential vorticity anomalies: A case study over the Mediterranean. *Meteorol. Appl.* **1999**, *6*, 97–109. [CrossRef]
59. Torresani, M.; Rocchini, D.; Sonnenschein, R.; Zebisch, M.; Marcantonio, M.; Ricotta, C.; Tonon, G. Estimating tree species diversity from space in an alpine conifer forest: The Rao's Q diversity index meets the spectral variation hypothesis. *Ecol. Inform.* **2019**, *52*, 26–34. [CrossRef]
60. Rocchini, D.; Thouverai, E.; Marcantonio, M.; Iannacito, M.; Da Re, D.; Torresani, M.; Bacaro, G.; Bazzichetto, M.; Bernardi, A.; Foody, G.M.; et al. rasterdiv-an Information Theory tailored R package for measuring ecosystem heterogeneity from space: To the origin and back. *Methods Ecol. Evol.* **2021**, *12*, 1093–1102. [CrossRef] [PubMed]
61. Rocchini, D.; Balkenhol, N.; Carter, G.A.; Foody, G.M.; Gillespie, T.W.; He, K.S.; Kark, S.; Levin, N.; Lucas, K.; Luoto, M.; et al. Remotely sensed spectral heterogeneity as a proxy of species diversity: Recent advances and open challenges. *Ecol. Inform.* **2010**, *5*, 318–329. [CrossRef]
62. Rocchini, D.; Santos, M.J.; Ustin, S.L.; Féret, J.B.; Asner, G.P.; Beierkuhnlein, C.; Dalponte, M.; Feilhauer, H.; Foody, G.M.; Geller, G.N.; et al. The spectral species concept in living color. *J. Geophys. Res. Biogeosci.* **2022**, *127*, e2022JG007026. [CrossRef]

63. Nagendra, H.; Rocchini, D.; Ghate, R.; Sharma, B.; Pareeth, S. Assessing plant diversity in a dry tropical forest: Comparing the utility of Landsat and IKONOS satellite images. *Remote Sens.* **2010**, *2*, 478–496. [CrossRef]
64. Michele, T.; Duccio, R.; Marc, Z.; Ruth, S.; Giustino, T. Testing the spectral variation hypothesis by using the RAO-Q index to estimate forest biodiversity: Effect of spatial resolution. In Proceedings of the IGARSS 2018–2018 IEEE International Geoscience and Remote Sensing Symposium, Valencia, Spain, 22–27 July 2018; pp. 1183–1186.
65. Feilhauer, H.; Zlinszky, A.; Kania, A.; Foody, G.M.; Doktor, D.; Lausch, A.; Schmidtlein, S. Let your maps be fuzzy!—Class probabilities and floristic gradients as alternatives to crisp mapping for remote sensing of vegetation. *Remote Sens. Ecol. Conserv.* **2021**, *7*, 292–305. [CrossRef]
66. Nagendra, H.; Rocchini, D. High resolution satellite imagery for tropical biodiversity studies: the devil is in the detail. *Biodivers. Conserv.* **2008**, *17*, 3431–3442. [CrossRef]
67. Tomasi, E.; Giovannini, L.; Zardi, D.; de Franceschi, M. Optimization of Noah and Noah_MP WRF Land Surface Schemes in Snow-Melting Conditions over Complex Terrain. *Mon. Weather. Rev.* **2017**, *145*, 4727–4745. [CrossRef]
68. De Wekker, S.F.J.; Kossmann, M.; Knievel, J.C.; Giovannini, L.; Gutmann, E.D.; Zardi, D. Meteorological Applications Benefiting from an Improved Understanding of Atmospheric Exchange Processes over Mountains. *Atmosphere* **2018**, *9*, 371. [CrossRef]
69. Falocchi, M.; Giovannini, L.; de Franceschi, M.; Zardi, D. A method to determine the characteristic time-scales of quasi-isotropic surface-layer turbulence over complex terrain: AÂ case-study in the Adige Valley (Italian Alps). *Q. J. R. Meteorol. Soc.* **2019**, *145*, 495–512. [CrossRef]
70. Montagnani, L.; Manca, G.; Canepa, E.; Georgieva, E.; Acosta, M.; Feigenwinter, C.; Janous, D.; Kerschbaumer, G.; Lindroth, A.; Minach, L.; et al. A new mass conservation approach to the study of CO_2 advection in an alpine forest. *J. Geophys. Res. Atmos.* **2009**, *114*, D07306. [CrossRef]
71. Laiti, L.; Zardi, D.; de Franceschi, M.; Rampanelli, G.; Giovannini, L. Analysis of the diurnal development of a lake-valley circulation in the Alps based on airborne and surface measurements. *Atmos. Chem. Phys.* **2014**, *14*, 9771–9786. [CrossRef]
72. Giovannini, L.; Laiti, L.; Zardi, D.; de Franceschi, M. Climatological characteristics of the Ora del Garda wind in the Alps. *Int. J. Climatol.* **2015**, *35*, 4103–4115. [CrossRef]
73. Rotach, M.W.; Serafin, S.; Ward, H.C.; Arpagaus, M.; Colfescu, I.; Cuxart, J.; Wekker, S.F.J.D.; GrubiÅ¡ic, V.; Kalthoff, N.; Karl, T.; et al. A Collaborative Effort to Better Understand, Measure, and Model Atmospheric Exchange Processes over Mountains. *Bull. Am. Meteorol. Soc.* **2022**, *103*, E1282–E1295. [CrossRef]

 land

Article

Characterization of Land-Cover Changes and Forest-Cover Dynamics in Togo between 1985 and 2020 from Landsat Images Using Google Earth Engine

Arifou Kombate [1,*], Fousseni Folega [2], Wouyo Atakpama [3], Marra Dourma [2], Kperkouma Wala [2] and Kalifa Goïta [1]

[1] Centre d'Applications et de Recherches en Télédétection (CARTEL), Département de Géomatique Appliquée, Université de Sherbrooke, Sherbrooke, QC J1K 2R1, Canada
[2] Géomatique et Modélisation des Écosystèmes, Laboratoire de Botanique et Écologie Végétale (LBEV), Département de botanique, Faculté des sciences, Université de Lomé, 01 BP 1515 Lomé, Togo
[3] Unité de Recherche en Systématique et Conservation de la Biodiversité, Laboratoire de Botanique et Écologie Végétale (LBEV), Département de botanique, Faculté des sciences, Université de Lomé, 01 BP 1515 Lomé, Togo
* Correspondence: arifou.kombate@usherbrooke.ca

Abstract: Carbon stocks in forest ecosystems, when released as a result of forest degradation, contribute to greenhouse gas (GHG) emissions. To quantify and assess the rates of these changes, the Intergovernmental Panel on Climate Change (IPCC) recommends that the REDD+ mechanism use a combination of Earth observational data and field inventories. To this end, our study characterized land-cover changes and forest-cover dynamics in Togo between 1985 and 2020, using the supervised classification of Landsat 5, 7, and 8 images on the Google Earth Engine platform with the Random Forest (RF) algorithm. Overall image classification accuracies for all target years ranged from 0.91 to 0.98, with Kappa coefficients ranging between 0.86 and 0.96. Analysis indicated that all land cover classes, which were identified at the beginning of the study period, have undergone changes at several levels, with a reduction in forest area from 49.9% of the national territory in 1985, to 23.8% in 2020. These losses of forest cover have mainly been to agriculture, savannahs, and urbanization. The annual change in forest cover was estimated at −2.11% per year, with annual deforestation at 422.15 km^2 per year, which corresponds to a contraction in forest cover of 0.74% per year over the 35-year period being considered. Ecological Zone IV (mountainous, with dense semi-deciduous forests) is the one region (of five) that has best conserved its forest area over this period. This study contributes to the mission of forestry and territorial administration in Togo by providing methods and historical data regarding land cover that would help to control the factors involved in forest area reductions, reinforcing the system of measurement, notification, and verification within the REDD+ framework, and ensuring better, long-lasting management of forest ecosystems.

Keywords: land-cover change; REDD+; Google Earth Engine; random forest; landsat; Togo

Citation: Kombate, A.; Folega, F.; Atakpama, W.; Dourma, M.; Wala, K.; Goïta, K. Characterization of Land-Cover Changes and Forest-Cover Dynamics in Togo between 1985 and 2020 from Landsat Images Using Google Earth Engine. *Land* **2022**, *11*, 1889. https://doi.org/10.3390/land11111889

Academic Editors: Carmine Serio, Guido Masiello and Sara Venafra

Received: 2 September 2022
Accepted: 21 October 2022
Published: 25 October 2022

1. Introduction

Forests contribute greatly to soil conservation and climate change mitigation and represent one of the simplest and most effective means of establishing or maintaining carbon sinks [1]. As one of the most important global carbon reservoirs, tropical forests are home to between half and two-thirds of the Earth's species [2]. Unfortunately, these forest carbon stocks are not stable, given that conversion to other land cover is occurring at an alarming rate despite the increased awareness of climate change [3,4]. Between 2000 and 2005, land-use and land-cover (LULC) changes resulted in forest cover reductions of 0.6% per annum worldwide [5]. Between 2015 and 2020, annual deforestation rates were estimated at 10 million hectares globally [6]. Such land-cover changes occur mainly as a result of anthropogenic disturbances, including deforestation, together with the expansion of croplands and urban areas [7]. LULC changes, mostly caused by agriculture and deforestation, contribute to about one-third of global greenhouse gas (GHG) and worsen the adverse effects of climate change [8,9]. Faced with these increasingly significant effects of climate change, ongoing demands for action are becoming more urgent to curb the extent of deforestation and forest degradation, while enhancing carbon storage through better accounting of carbon sources and sinks. To this end, the United Nations Framework Convention on Climate Change (UNFCCC) has established the REDD+ (Reducing Emissions from Deforestation and forest Degradation Plus) mechanism, which is seen as a global system of centralized forest governance. Aimed primarily at developing nations, REDD+ provides financial compensation for these countries to preserve their forests to reduce carbon emissions and, thus, mitigate the risks of climate change [10,11].

In order to qualify for financial offsets by implementing REDD+, these countries are required to establish National Measurement, Reporting, and Verification (MRV) systems within a national forest monitoring system (NFMS) that must provide national estimates of changes in forest carbon stocks and emissions every two years. The Intergovernmental Panel on Climate Change (IPCC) recommends a combination of Earth observation data and field inventories to estimate forest area, carbon stocks, and changes that follow disturbance [12]. Regular analysis of forest dynamics and LULC changes using satellite data could effectively establish the baseline for the MRV reporting requirement in this context. However, many concerned developing countries are generally faced with a lack of quantitative data on forest degradation-induced changes and limited technical capabilities and material capacity to produce such data for GHG emissions monitoring [12].

The aforementioned challenges beset the West African nation of Togo (République Togolaise), which is the subject of our study, in its quest to meet reporting requirement needs within the framework of the REDD+ strategy, and to guide strategies for monitoring the evolution of forest ecosystems and land cover. A few studies based on observational data have made it possible to monitor changes in land cover in certain parts of the country, but they generally have a starting and an ending year for a period that occasionally spans several decades. The coarse temporal frequency of sampling does not make it possible to detect changes that have been incurred within these periods or to discern which main factors drive their behavior. Furthermore, the spatial extent of these studies is often very limited (i.e., river basins, protected zones, and administrative jurisdictions, among others), whereby changes are not perceived across an entire ecological region or on a national scale. Land and vegetation cover have been studied, but these changes are mainly in protected areas [13–16]. Other studies have focused on watersheds [17,18], while some have been carried out at regional or prefectural scales [19,20]. To a much lesser extent, few comprehensive studies have spanned several ecological zones [21]. These studies have generally covered about 1 to 10% of the national territory, and there are regrettably very few studies quantifying the LULC changes observed over time or analyzing the drivers of these changes.

The spatial and temporal limitations of these previous studies in detecting land-cover changes are related to the difficulties in finding sufficient cloud-free satellite images over large areas. This problem could be overcome by using Synthetic Aperture Radar (SAR) images which, even when acquired in all atmospheric and solar conditions, allow change detection analyses [22], but SAR long historical data does not exist in our study area. These limitations are also related to computational resource problems (large storage capacity and access to high computing power), together with the labor-intensive nature of processing these mega-data [23,24]. Furthermore, global-scale mapping projects often use satellite data with a variable spatial resolution (1 km to 30 m), and generally do not involve local experts; therefore, these approaches do not meet the standards of accuracy that are sought at the national level [25]. With the availability of the new geospatial technology of the Google Earth Engine (GEE), it is now possible to apply very advanced machine-learning algorithms in an efficient manner [26]. The GEE is a cloud-computing platform with a JavaScript code editor that integrates a long-time series of satellite imagery, thereby allowing the classification of large volumes of data and the production of multi-date land-cover changes. It should be further noted that relatively few studies in the scientific literature have focused on the use of these methods to advance operational forest monitoring in MRV systems [27].

The major challenges to implementing Togo's national REDD+ strategy are reversing the process of forest degradation and savannization, while spatially containing agricultural pressure and constraining urban expansion. These measures should eventually increase carbon stocks and reduce greenhouse gas emissions [28]. Unfortunately, most studies that have been conducted in Togo on progressive LULC changes are incomplete, and forest inventories over the last three decades are very limited. The availability of historical LULC data at a national scale is necessary to meet the challenge of better understanding the LULC dynamics and forest developmental trends over time. This study aims to answer the question of whether the use of multi-temporal images in the GEE would provide a picture of land-cover changes, particularly forest cover, at the national scale. Its main objective is to characterize vegetation dynamics over the entire national territory using a long-time series of Landsat images from 1985 to 2020. More specifically, the study aims to quantify the evolution of spatiotemporal changes and to analyze their effects on forest cover during this period.

2. Study Area and Data Used

2.1. Study Area

The study area was Togo (Figure 1A). It is a coastal country in West Africa that is bordered by Burkina Faso to the north, the Atlantic Ocean to the south, Benin to the east, and Ghana to the west. It belongs to the Sudano-Guinean zone, which is a climatic zone that is located south of the Sahara Desert in the continental and coastal areas, which extend from West Africa to Central Africa. With an area of 56,600 km^2, Togo has a population of 7,264,637 inhabitants unequally distributed in the administrative regions with proportions of 42.16% in Maritime, 22.16% in Plateaux, 9.99% in Centrale, 12.44% in Kara and 13.26% in Savanes [29]. It experiences a tropical Sudano-Guinean climate with rainfall ranging from 900 to 1100 mm year^{-1} in the northern regions (distinct wet and dry seasons), and from 1000 to 1600 mm year^{-1} in the southern regions (with four seasons), and an average temperature of 27 °C [30].

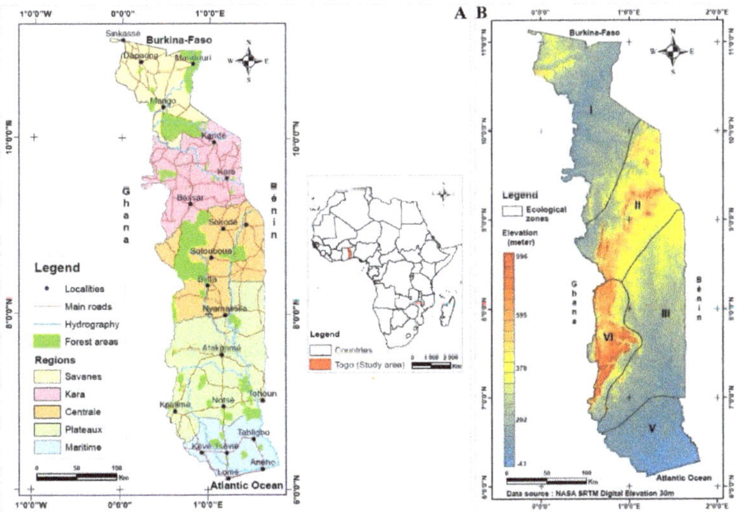

Figure 1. (**A**) Geographical location of the study area; (**B**) ecological zones and elevations.

Due to its position in the Dahomey Gap (a remarkable interruption in the extent of continuous tropical rainforest covering Central to West Africa), Togo has a low forest cover with a deforestation rate of 0.73% per year for the period from 1990 to 2000 [31]. To ensure the protection of the country's forest resources, 14.2% of its territory was classified between 1939 and 1957 as 83 protected areas (classified forests, national parks, and reserves). Yet, human populations seeking arable land and wood for energy have encroached upon nearly one-third of these areas [32]. Vegetation formations are composed of the Sudano-Guinean forest that is located in the mountainous areas of the country, gallery forest along main rivers, dry forest or dense tree savannah in the northern half, and tree savannah in the south and center. The landscape variability of these ecosystems led [33] to subdivision of the country into five ecological zones (Figure 1B).

Ecological zones correspond to distinct ecosystems that are characterized by various plant formations and topographies. Following an update of their descriptions, these ecological zones have been summarized in [34] as follows:

- Zone I (or Northern Plains Zone): This zone extends from the Dapaong peneplain to the southern limit of the Volta Basin, approximately following the Bendjeli-Kpessidè axis. This area is essentially dominated by agro-ecosystems; however, there are relic mosaics of savannahs, dry forests, degraded riparian forests, and swamp vegetation adjacent to the hydrographic network. The main spontaneous ligneous species found in this zone are *Vitellaria paradoxa, Anogeisus leiocarpus, Borassus aethiopum, Parkia biglobosa, Balanites aegyptiaca, Lannea microcarpa,* and *Detarium microcarpa*. The natural ecosystems of this area are highly degraded (80%), given the strong propensity of the inhabitants to practice unsustainable cultivation (68%) and fuel wood exploitation (28%). The zone is heavily disturbed by vegetation fires (40%), which have then been followed by extensive grazing (28%) [34].
- Zone II (or Northern Mountains Zone): This zone encompasses the Northern Mountain Range and extends between 8° and 10° N northeast under the influence of a Sudanian mountain climate. This zone is dominated by agrosystems, yet dry forests, open forests, and savannah mosaics can be found. Its main spontaneous ligneous species are *Parkia biglobosa, Vitellaria paradoxa, Nauclea latifolia, Daniellia oliveri, Elaeis guineensis, Piliostigma thonningii, Terminalia laxiflora,* and *Isoberlinia doka*. In this zone, natural ecosystems are also degraded (58%) and heavily disturbed by extensive grazing (31%), followed by vegetation fires (25%), floods (19%), and transhumance (seasonal livestock relocation, 17%). Activities such as working crop fields (41%), logging (22%), and grazing (20%) strongly contribute to ecosystem degradation [34].
- Zone III (or Central Plains Zone): This zone occupies the Benin-Togolese plain east of the Atakora Mountain Chain; it is characterized by a Guinean Lowland climate and is dominated by a diversity of agrosystems. This matrix of agroforestry parks combines patches of mosaic savannah, semi-deciduous forest, and degraded riparian formations. This zone is characterized by the following main spontaneous ligneous species: *Daniellia oliveri, Parkia biglobosa, Vitellaria paradoxa, Pterocarpus erinaceus, Anogeissus leiocarpus,* and *Adansonia digitata*. The natural ecosystems of this agro-ecological zone are 96% degraded. This degradation of ecosystems is the consequence of the exploitation of wood energy (46%) and cultivation practices (41%) and is not very sustainable. Ecosystems in this zone are strongly disturbed by vegetation fires (31%), transhumance (31%), and erosion (24%) [34].
- Zone IV (or Southern Zone of the Togo Mountains): This zone corresponds to the southern portion of the Togo Mountains. It has a sub-equatorial climate with a rainy season. Its main spontaneous ligneous species are *Cola gigantea, Millettia thoningii, Morinda lucida, Sterculia tragacantha, Antiaris fricana, Holarrhena floribunda,* and *Margaritaria dioscoidea*. Today, it is the domain par excellence of agroforestry that is interspersed with semi-deciduous forests and mosaics of Guinean savannah. The natural ecosystems of the southern zone of the Togo Mountains are highly degraded (70%), given that they are heavily disturbed by vegetation fires (55%), often followed by extensive grazing (15%), and logging (10%). Activities such as working the crop fields (59%) and logging (18%) contribute to the substantial degradation of ecosystems [34].

- Zone V (or Southern Coastal Zone): This zone corresponds to the country's coastline with a sub-equatorial climate with two rainy seasons. The very degraded natural environment is strongly dominated by agrosystems, with relic mosaics of savannahs, halophytic or swampy grasslands, and mangroves. The main spontaneous ligneous species found there are *Lonchocarpus sericeus, Parkia biglobosa, Piliostigma thonningii, Dialium guineense, Holarrhena floribunda, Bridelia ferruginea Millettia thonningii,* and *Vitellaria paradoxa*. These natural ecosystems are highly degraded (85%) due to cultivation practices (59%) and the unsustainable exploitation of wood energy (18%) and urbanization (10%). Lands in the Coastal Zone have been heavily disturbed by vegetation fires (55%), which are often followed by extensive grazing (15%), and transhumance, woodcutting, and flooding (5%) [34].

The aforementioned descriptions indicate the continuation of high-intensity land degradation that has been observed across most of these zones since the 1990s [35]. Even in Zone IV, which is known as being the most extensively forested of the ecological zones, deforestation and forest degradation have been occurring in recent years due to the combined effect of the advancing agricultural front with slash-and-burn agriculture, wildfires, and logging [36].

2.2. Data Used

Data used in this study included Landsat TM, ETM+ (Enhanced Thematic Mapper Plus), OLI (Operational Land Imager) satellite imagery, land-cover reference data, and vector data. The satellite images are from Landsat 5, 7, and 8 sensors with a spatial resolution of 30 m, which have been archived in the GEE (Table 1). Image selections were made for the level-1 scenes, which are the best quality images in terms of radiometric consistency and atmospheric correction [37]. These are surface reflectance data that were accompanied by meta-data and per-pixel quality information, which was intercalibrated between different Landsat sensors, and are considered suitable for time-series processing analysis [38].

Table 1. Information on Landsat images that were entered into composites from 1985 to 2020.

Sensors	Composite Target Years	Composite Image Acquisition Period	Admissible Cloud Threshold	Number of Images that Were Concerned
Landsat 5	1985	1983-01-01 to 1986-12-31	10%	57
Landsat 5	1990	1987-10-01 to 1988-03-31 1988-10-01 to 1989-03-31 1989-10-01 to 1990-03-31 1990-10-01 to 1991-03-31 1991-10-01 to 1992-03-31 1992-10-01 to 1992-12-31	10%	49
Landsat 7	2000	1999-04-16 to 2002-12-31	10%	95
Landsat 7	2005	2003-01-01 to 2007-12-31	20%	322
Landsat 8	2015	2013-01-01 to 2017-12-31	10%	265
Landsat 8	2020	2018-01-01 to 2020-12-31	10%	171

Land cover reference data consisted of data that were collected in the field, points that were sampled on image composites, and high-resolution Google Earth images. During the field campaign that was conducted from October 2020 to February 2021, we sampled 101 land occupancy points on the ICESat (Ice, Cloud, and land Elevation Satellite) data footprints, 303 points on the ICESat-2 data footprints, and 114 points elsewhere. These ICESat and ICESat-2 footprint data are dendrometric data that are intended for further studies on estimating aboveground biomass. Given that the land occupancies of these sites were known, they were used with other data as references for training and validation of classifications that were made during this study. Vector data mainly concerned forest areas, administrative regions, ecological zones, and jurisdictional boundaries in Togo. Large-scale international boundary data for Togo (i.e., the study area) that were also available in the GEE were used for delineation during the selection of these images and the final mapping.

3. Methodology

The methodological approach of this study involved the acquisition and pre-processing of satellite data, selection of training and validation data, supervised classification of the images with the Random Forest (RF) algorithm, evaluation of classification accuracies, and mapping and analysis of the results. The following flowchart (Figure 2) illustrates the methodological approach which is summarized in three main points in the description.

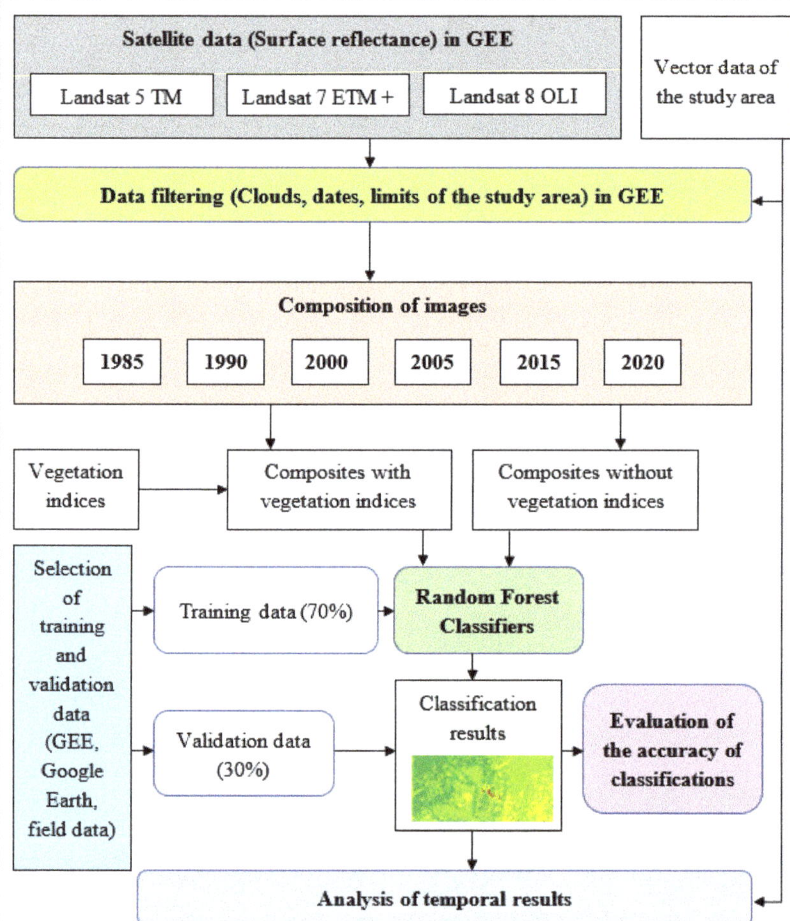

Figure 2. Methodological flowchart of the study.

3.1. Selection and Pre-Processing of Satellite Images

Since cloud-free images providing complete coverage of the study area for the target year were difficult to find, image composition was performed. This consisted of filtering all images with admissible cloud cover set to a certain threshold (Table 1) to create a mosaic of images around each target year. Referring to methods that are frequently used in the literature, several authors had performed image composition based on the temporal aggregation of data, applying the calculation of statistical parameters (mean, median, and maximum or minimum values) on the pixels of a pre-defined image time series [39,40]. Others simply used all available Landsat images in their study area to compose image time series [41,42]. According to [24], the most popular strategy for selecting input images for an annual cloud-free composite is using images that have been acquired over three years. In this study, the annual data composite was targeted from the year 1985, with a five-year step to better perceive disturbance lapse times. Unfortunately, there were problems of poor quality and insufficiently filtered images below the set cloud thresholds, together with gaps in the data covering the study area. Faced with these difficulties, only six image composites were created, from all available data in the target years, or occasionally, one or two years on either side of the target years (i.e., 1985, 1990, 2000, 2005, 2015, and 2020). Image composites were formed by applying a cloud mask *QA_PIXEL Bitmask* (provided with the data) to the image collections. Cloudy pixels were maintained (by removing the mask) when no other non-cloudy pixels were available to replace them from the entire time period around the target years. These were placed into the cloud class so that the entire extent of the study area could be considered when facilitating later surface analyses. We initially composited these images only from the best-available pixels derived from Landsat data [43]. Nevertheless, given that some parts of the area remained without data under the constraints of the filters, we calculated the median of all pixels that met these imposed filters.

Several vegetation indices were also calculated and added as bands to the image composites to see what improvements they could bring to the classification process. These were NDVI, NDBI, NDWI, and BSI (Table 2).

Table 2. Formulas of the used vegetation indices.

Acronym	Designation	Equation	References
NDVI	Normalized Difference Vegetation Index	$\frac{\rho_{NIR}-\rho_R}{\rho_{NIR}+\rho_R}$	[44,45]
NDBI	Normalized Difference Built-up Index	$\frac{\rho_{SWIR1}-\rho_{NIR}}{\rho_{SWIR1}+\rho_{NIR}}$	[46,47]
NDWI	Normalized Difference Water Index	$\frac{\rho_G-\rho_{NIR}}{\rho_G+\rho_{NIR}}$	[48,49]
BSI	Bare Soil Index	$\left[\frac{(\rho_{SWIR1}+\rho_R)-(\rho_{NIR}+\rho_B)}{(\rho_{SWIR1}+\rho_R)+(\rho_{NIR}+\rho_B)}\right]$	[50,51]

Note: ρ_R, ρ_G, ρ_B, ρ_{NIR}, and ρ_{SWIR1} represent the reflectance of red, green, blue, near-infrared, and short-wave infrared bands, respectively.

Since the study area was characterized by major land-cover classes, including vegetation (dense dry forest, open forest, and savannah), crops and fallow land, buildings and bare soil, and water bodies, we selected these vegetation indices to better characterize them. NDVI has been widely used over many decades to monitor vegetation dynamics in terrestrial ecosystems and remains the most popular index that is used for vegetation assessment [52,53]. Using NIR and SWIR bands, NDWI is commonly and successfully used in the detection and mapping of surface water bodies [54] and the improvement of terrain illumination differences and atmospheric effects. Furthermore, the BSI has been proposed as a more reliable estimator of vegetation status where vegetation covers less than half of an area [51,55]. Ref. [56] has shown that combining NDVI, NDWI, and NDBI data could refine several aspects of urban features and appearance while removing cloud-related noise in image classifications. Based upon these findings, these indices were combined with the classic bands of Landsat data, given that the former are expected to contribute to the development of a more nuanced classification scheme [57]. Using the vector data, the resulting image composites were then clipped with the study area to limit processing within this area.

3.2. Selection of Training and Validation Data

For each target year, sample points were selected based on the land cover that was detected through visual interpretation or by relying upon archival high-resolution Google

Earth data from periods as close as possible to the target years. Reference data was collected from various sources for the different target years. For the year 2020, we used data collected in the field as explained in Section 2.2. Reference data based on high-resolution archive images mainly concerned the years 2000, 2005, and 2015. For the years 1985 and 1990, when high-resolution images were not available on Google Earth, we relied on samples of land cover directly collected by visual interpretation on filtered Landsat images of these years. In addition to these data, the pixel values of the added vegetation indices bands were used to guide the selection of samples. Therefore, both visual interpretation and consultation of the pixels that were provided by these additional vegetation indices bands were used to make these selections.

In applying these sample selection methods to image composites of the target years 1985, 1990, 2000, 2005, 2015, and 2020, a total of 1007, 1102, 1219, 1278, 1372, and 1521 points were sampled per composite, respectively, to serve as training points. Each group of points represented the different land cover types. For the six target years, 7499 sample points were thus collected, some to serve as training samples (70%) during the classification of the composite images, and the remainder to validate the classification results (30%).

3.3. Image Classification and Evaluation of Accuracy

Following the identification and pre-processing of images, we proceeded to classify the image composites with the classic Landsat bands, followed by a second classification with these same bands to which were added the vegetation indices to determine their effect on the quality of these classifications. As for the pre-processing, image classifications were performed using JavaScript codes in the GEE. For the selection of the appropriate classification method, several classification algorithms related to supervised machine learning have been used in the literature. These include Support Vector Machines (SVM), Classification and Regression Trees (CART), Stepwise Multiple Linear Regression (SML), and Random Forests (RF). We determined that supervised machine learning classifiers, such as Classification and Regression Trees (CART) and Random Forests (RF), were the most frequently used for this purpose. Furthermore, the use of RF classifiers leads to greater classification accuracy, even when applied to the analysis of data with higher noise levels [58–60]. This is confirmed in studies by [61], who evaluated 179 relevant classifiers from 17 families using 121 datasets. The authors concluded that RF provided the best classifiers. Therefore, we selected the RF algorithm because it yields results with excellent accuracies and can work efficiently on large datasets [62].

The different image composites that resulted from filtering according to the previously mentioned parameters were then classified in the GEE using the RF algorithm. The number of decision trees that were selected for this algorithm was made with reference to the literature, which generally indicates that the greater the number of trees, the better the results. According to [63], it is unclear whether the number of trees should simply be set to the largest computationally manageable value or whether a smaller number of trees might be sufficient or provide better results. [64] compared the performance of the RF model with different numbers of trees on 29 datasets and noted that a forest with 512 trees performs better than one with 1024 trees. They concluded that forest performance does not always improve substantially as the number of trees increases beyond a certain level. While it is commonly thought that tuning hyper-parameters can improve RF performance, [65] acknowledged that improvements achieved by adding trees decreases as more and more trees are added. Generally, RF works quite well with default values of hyper-parameters, and, according to these authors, typical default values for the number of trees for RF are 500 and 1000. Therefore, we chose to use 500 trees in the RF classification algorithm that was applied to the image composite classifications in this study as this number of trees has been widely used in the literature in various fields and mainly in land cover classification with very good results [60,66–70].

The image classifications for this study were based on seven main land cover classes (Table 3). The definition of these classes was based on the Yangambi land classification system [34] appropriate to the West African context which was used during the 2016 National Forest Inventory (IFN) [71]. However, to take into account the limited capacity of available images to discriminate between different land cover, some classes have been aggregated into other larger classes.

Table 3. Description of LULC categories used in the classification.

LULC Categories	Description
Dense dry forest	Dense semi-deciduous forests, plantations, gallery forests, and agroforests
Open forest	Forests with open canopies and wooded savannahs
Savannahs	Tree savannahs, shrubby savannahs, and grassy savannahs
Crop and fallow	Areas with crops and abandoned agricultural land
Buildings and bare land	Infrastructure related to human settlements and commercial centers, roads, burnt or turned soil, and mining quarry
Water bodies	Continental water surfaces (lake, lagoon, water, dam, and river)
Clouds	Surface covered by clouds and their shadows

The original spectral bands B1, B2, B3, B4, B5, and B7 from Landsat 5 and 7, together with B2, B3, B4, B5, B6, and B8 from Landsat 8, were used as inputs to the RF model for the first classification. For the second classification, an ensemble combining these same bands with the four aforementioned vegetation indices was used as input, but with the same training and validation samples.

Based upon random selection in the model, 70% of the collected data were used as training samples when classifying the composite images, while 30% were used as validation data for the classification results. The accuracy of the classifications that were performed on each image composite was then evaluated. For each image composite, we calculated traditional metrics for evaluating the accuracy of image classification, which are the producer accuracy (PA), the user accuracy (UA), the overall accuracy (OA), and Cohen's kappa coefficient (***K***) [72].

The different metrics are defined by the following equations [73]:

$$OA = (1/N) \sum_{i=1}^{r} n_{ii} \tag{1}$$

$$PA = n_{ii} / n_{icol} \tag{2}$$

$$UA = n_{ii} / n_{irow} \tag{3}$$

$$K = N \sum_{i=1}^{r} n_{ii} - \sum_{i=1}^{r} (n_{icol}\ n_{irow} / N^2) - \sum_{i=1}^{r} n_{icol}\ n_{irow} \tag{4}$$

where n_{ii} is the number of correctly classified pixels in a category; N is the total number of pixels in the confusion matrix; r is the number of classes; n_{icol} is the column total (reference data); and n_{irow} is the row total (predicted classes).

Ref. [74] defines the main parameters of classification accuracies, such as OA, as the ratio of the number of correctly classified pixels to the total number of pixels in the class, and Kappa, which refers to the proportion of error reduction between the considered classification and a completely random classification. According to [73], OA represents the ground truth classes that are correctly classified by the analyst (error of omission), while UA is the percentage of pixels that do not really belong to the reference class but are engaged in other ground truth classes (error of commission).

Following these evaluations of the classification accuracies of the image composites, the results were exported from the GEE for formatting in mapping software. The land-cover maps were finalized in ArcMap 10.6.1, while land-cover conversion maps were produced using the semi-automatic classification extension that was recently developed with python code by [75], and which is usable in QGIS 3.6. The annual rate of forest cover change (r) and annual deforestation (R), which have been defined by [76], were also calculated for the periods between the selected target years of this study and between 1985 and 2020 by applying Equations (5) and (6), as follows:

$$r = \left(\frac{1}{(t_2 - t_1)}\right) * ln\left(\frac{A_2}{A_1}\right) * 100 \quad (5)$$

$$R = \frac{A_2 - A_1}{t_2 - t_1} \quad (6)$$

where t_1 is year 1, t_2 is year 2, A_1 is forest area in year 1, and A_2 is forest area in year 2.

4. Results

4.1. Assessing the Accuracy of Image Classifications

Seven land cover classes were generated in a supervised manner. Using the RF algorithm in the GEE, the accuracy of the results was evaluated when vegetation indices were not used (Table 4) and when indices were used (Table 5). Overall accuracies for image composites with and without vegetation indices range from 0.91 to 0.98, while Kappa ranges from 0.86 to 0.96.

Table 4. Accuracies obtained when classifications were made without vegetation indices (PA = Producer accuracy, UA = user accuracy, OA = overall accuracy, and K = Kappa coefficient).

Image Composite	Accuracy	Clouds	Water	Dense Dry Forest	Open Forest	Crops + Fallow	Savannah	Bldg. + Soil	OA	K
1985	UA	0.94	1.00	0.95	0.85	0.94	0.98	0.93	0.95	0.93
	PA	0.99	1.00	0.95	0.75	0.97	0.97	0.89		
1990	UA	1.00	0.99	0.94	0.95	0.92	0.93	0.99	0.96	0.95
	PA	1.00	0.99	0.95	0.95	0.95	0.97	0.92		
2000	UA	0.94	0.96	0.95	0.97	0.95	0.94	0.97	0.96	0.95
	PA	0.97	1.00	0.95	0.94	0.96	0.96	0.92		
2005	UA	0.78	0.99	0.50	0.81	0.83	0.97	0.84	0.91	0.86
	PA	0.77	1.00	0.95	0.83	0.87	0.92	0.73		
2015	UA		1.00	0.95	0.65	0.90	0.95	0.97	0.98	0.96
	PA		1.00	0.98	0.83	0.93	0.90	0.95		
2020	UA		1.00	0.96	0.29	0.89	0.90	0.98	0.93	0.91
	PA		1.00	0.99	0.58	0.87	0.89	0.91		

Table 5. Accuracies obtained when classifications were made with vegetation indices (PA = Producer accuracy, UA = user accuracy, OA = overall accuracy, and K = Kappa coefficient).

Image Composite	Accuracy	Clouds	Water	Dense Dry Forest	Open Forest	Crops + Fallow	Savannah	Bldg. + Soil	OA	K
1985	UA	0.92	1.00	1.00	0.84	0.95	0.97	0.91	0.94	0.93
	PA	0.99	1.00	0.97	0.76	0.96	0.96	0.89		
1990	UA	0.99	0.99	0.95	0.95	0.92	0.93	0.98	0.96	0.95
	PA	1.00	0.99	0.95	0.85	0.95	0.96	0.93		
2000	UA	0.94	0.97	0.95	0.97	0.95	0.94	0.96	0.96	0.95
	PA	0.96	1.00	0.97	0.95	0.95	0.96	0.92		
2005	UA	0.78	0.99	0.52	0.82	0.83	0.97	0.84	0.91	0.86
	PA	0.75	1.00	0.98	0.82	0.87	0.93	0,73		
2015	UA		1.00	0.97	0.62	0.90	0.97	0.96	0.98	0.96
	PA		1.00	0.98	0.79	0.92	0.90	0.96		
2020	UA		1.00	0.95	0.25	0.90	0.91	0.98	0.93	0.91
	PA		1.00	1.00	0.50	0.88	0,89	0.93		

After extracting these precision parameters from the confusion matrices of the composite classification of each target year, one of the target years (1985) without vegetation indices was presented as an example (Table A1) in Appendix A. Overall accuracies and Kappa coefficients for the classification of composite images with the original bands was very similar to those of composites made with the original bands and vegetation indices. Nevertheless, under the null hypothesis that their slopes do not differ from a 1:1 relationship, linear regressions between the values of these two types of data yield *p*-values much less than 0.001 for the OAs and *K*s. This indicates that these values for the original band classifications of the image composites are significantly different from those including the vegetation indices. In Appendix B, this same finding of a significant difference was verified between the UA and PA accuracies for all land cover classes in all image composites (Table A2).

4.2. Distribution of Land Cover

Classifications made on the basis of the different land-cover classes that were identified made it possible to produce a land-cover map of the entire study area for each of the composite images, i.e., 1985, 1990, 2000, 2005, 2015, and 2020. The results of classifications without vegetation indices for the six targeted years were mapped (Figure 3). Regarding the results of the classifications with vegetation indices, predictions of the water body class and those of the built-up and bare land (building + soil) class were overestimated. With regard to the visual interpretation of the image composites before classification and the land cover contained in the field data, it was noted that these results of classifications with vegetation indices were not improved compared to the others and reflected the field realities less. Therefore, we decided to continue the other analyses with only those classifications without vegetation indices, considering that further, more specific studies involving the combination of other data could better elucidate the real impacts of these indices on the image classifications.

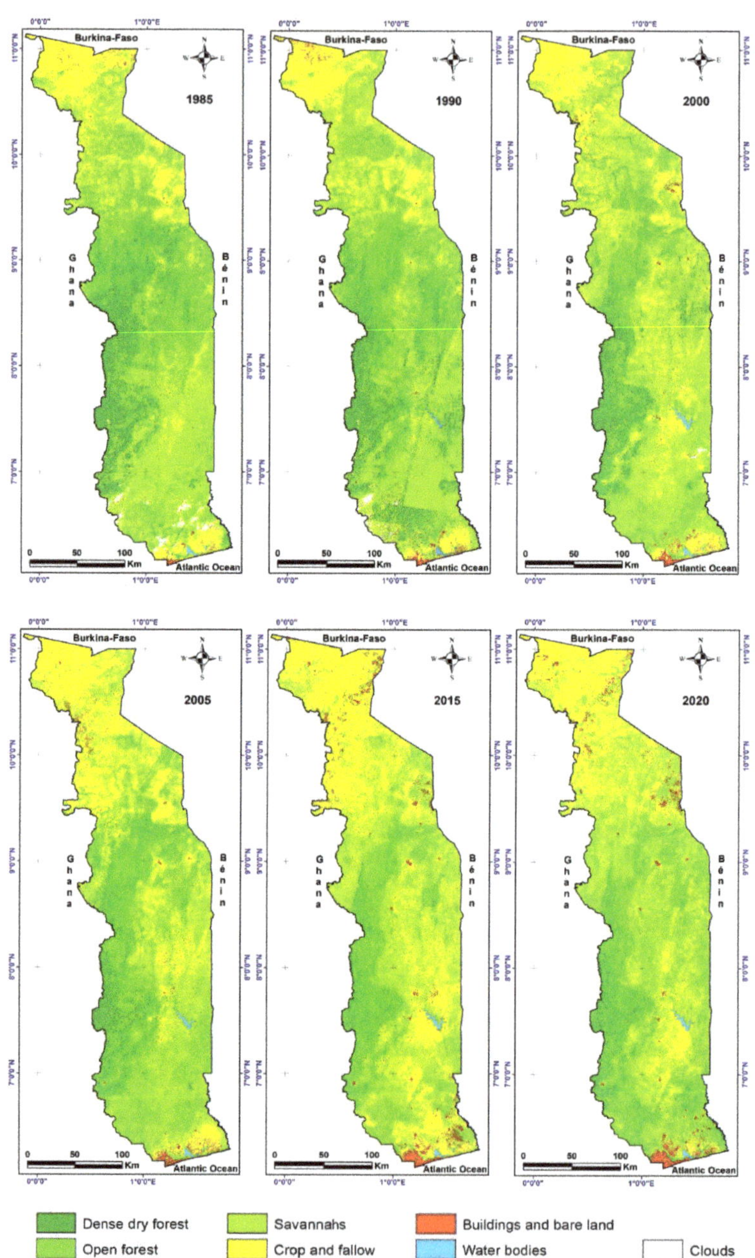

Figure 3. Land cover maps of the classifications of the six composites without vegetation indices.

Since the study area is located in tropical regions where the availability of optical data is very often limited by cloud cover [77], we included this latter as a land-cover class (but which is not presented in the following analyses). Apart from clouds, results of the classifications indicate that in 1985, there were four main land-cover classes, viz., dense dry forest (10,722.53 km^2), open forest (17,547.75 km^2), crops and fallow land (11,940.55 km^2), and savannah (14,533.13 km^2), which represented 18.92%, 30.97%, 21.07%, and 25.65%, respectively, of the nation's land surface. The lowest land-cover percentages were water bodies (0.09%) and built-up and bare soil (0.50%) classes. A quantitative evaluation of these land-cover changes and conversions between target years, as well as those between starting and ending years, was provided (Table 6).

Table 6. Land-cover change and conversions between the target years 1985-2020.

Year	LULC	Clouds	Water	Dense Dry Forest	Open Forest	Crops + Fallows	Savannah	Bldg. + Soil	Total
1985	Sup. (km^2)	1592.42	50.73	10,722.53	17,547.75	11,940.55	14,533.13	281.79	56,668.90
	Sup. (%)	2.81	0.09	18.92	30.97	21.07	25.65	0.50	100.00
1990	Sup. (km^2)	1029.65	163.44	9095.25	14,378.62	11,641.92	20,012.41	347.62	56,668.90
	Sup. (%)	1.82	0.29	16.05	25.37	20.54	35.31	0,61	100.00
	Conv. (km^2)	−562.77	112.70	−1627.27	−3169.13	−298.63	5479.28	65.83	
	Conv. (%)	−35.30	222.10	-15.20	−18.10	−2.50	37.70	23.40	
2000	Sup. (km^2)	211.48	256.37	7704.97	10,515.96	14,179.42	23,379.66	421.06	56,668.90
	Sup. (%)	0.37	0.45	13.60	18.56	25.02	41.26	0.74	100.00
	Conv. (km^2)	−818.17	92.94	−1390.29	−3862.67	2537.50	3367.25	73.44	
	Conv. (%)	−79.50	56.90	−15.30	−26.90	21.80	16.80	21.10	
2005	Sup. (km^2)	176.23	332.72	8505.64	10,439.68	19,577.14	16,956.20	681.29	56,668.90
	Sup. (%)	0.31	0.59	15.01	18.42	34.55	29.92	1.20	100.00
	Conv. (km^2)	−35.25	76.35	800.67	−76.27	5397.73	−6423.46	260.23	
	Conv. (%)	−16.70	29.8	10.40	−0.70	38.10	−27.50	61.80	
2015	Sup. (km^2)	0.00	196.67	4186.70	8549.65	20,522.50	22,045.29	1168.10	56,668.90
	Sup. (%)	0.00	0.35	7.39	15.09	36.21	38.90	2.06	100.00
	Conv. (km^2)	−176.23	−136.05	−4318.94	−1890.04	945.36	5089.09	486.81	
	Conv. (%)	−100.00	−40.90	−50.80	−18.10	4.80	30.00	71.50	
2020	Sup. (km^2)	0.00	192.02	3785.27	9709.70	21,677.56	20,146.17	1158.19	56,668.90
	Sup. (%)	0.00	0.34	6.68	17.13	38.25	35.55	2.04	100.00
	Conv. (km^2)	0.00	−4.66	−401.43	1160.05	1155.06	−1899.12	9.91	
	Conv. (%)	0.00	−2.37	−9.59	13.57	5.63	−8.61	−0.85	
1985–2020	Sup. (km^2)	0.00	192.02	3785.27	9709.70	21,677.56	20,146.17	1158.19	56,668.90
	Sup. (%)	0.00	0.34	6.68	17.13	38.25	35.55	2.04	100.00
	Conv. (km^2)	−1592.42	141.28	−6937.26	−7838.06	9737.01	5613.04	876.40	
	Conv. (%)	−100.00	278.47	−64.70	−44.67	81.55	38.62	311.01	

Note: LULC = Land Use and Land Cover; Conv. = Conversions; Sup. = Area (Superficie); Bldg. + soil = buildings and bare land.

The area of each land cover has changed slightly for some and greatly for others in different directions in all the target years during the period considered by this study (Figure 4).

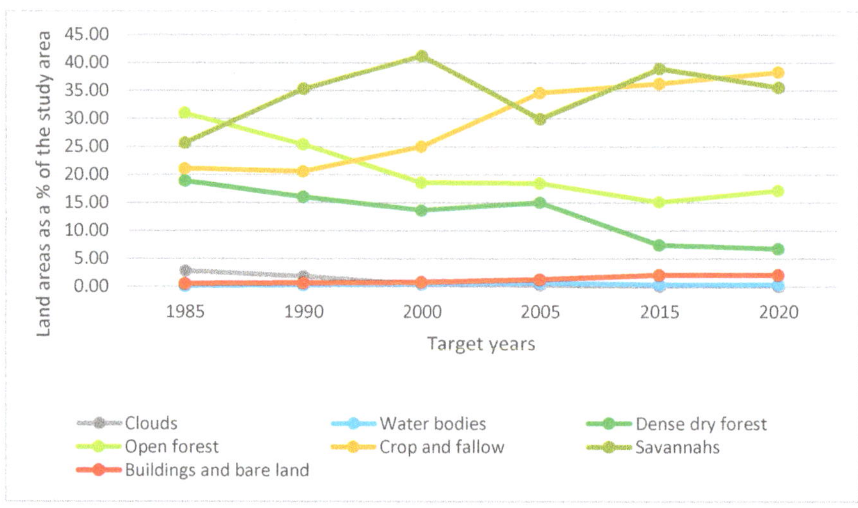

Figure 4. LULC changes as percentages of the study area.

4.3. Land-Cover Conversions

From the outset, areas of dense dry forest and open forest have decreased from 10,722.53 km^2 and 17,547.75 km^2 to 7704.97 km^2 and 10,515.96 km^2, respectively, between 1985 and 2000. During the same period, areas of crops/fallow and savannah have increased from 11,940.55 km^2 and 14,533.13 km^2 to 14,179.42 km^2 and 23,379.66 km^2, respectively. Thus, we note a 3017.56 km^2 contraction for dry dense forests and 7031.80 km^2 of open forests, while crops/fallow lands expanded by 2238.86 km^2 and savannahs by 8846.53 km^2. In 2020, these main classes occupied only 3785.27 km^2 for dense dry forests and 9709.70 km^2 for open forests, but 21,677.56 km^2 for crops/fallows and 20,146.17 km^2 for savannahs. These classes represent 6.68%, 17.13%, 38.25%, and 35.55%, respectively, of Togo's land surface area.

These changes correspond to a reduction of 64.70% of dense dry forests and 44.67% of open forests, versus an 81.55% increase in crops/fallows and 38.62% in savannahs compared to their respective starting areas. The water body area increased considerably between 1985 and 1990, through the construction of a large hydroelectric dam at Nangbeto in the southeastern part of the country (1987), together with the creation of other small water reservoirs. Built-up (buildings) and bare land (bare soil) class areas increased by +300%, from 281.79 km^2 in 1985 to 876.40 km^2 in 2020. In short, all land cover has changed during the period covered by the study, with a decrease in areas of dense dry forest and open forest, accompanied by a sharp increase in the areas of crops/fallow lands and savannahs. For illustrative purposes, the conversions from one land cover to another, as well as areas that were retained and not changed during the 2015 to 2020 period, are shown in Figure 5. The same types of charts for other time periods (1985 to 1990, 1990 to 2000, and 2000 to 2005) are provided (Figures A1–A3) in Appendix C.

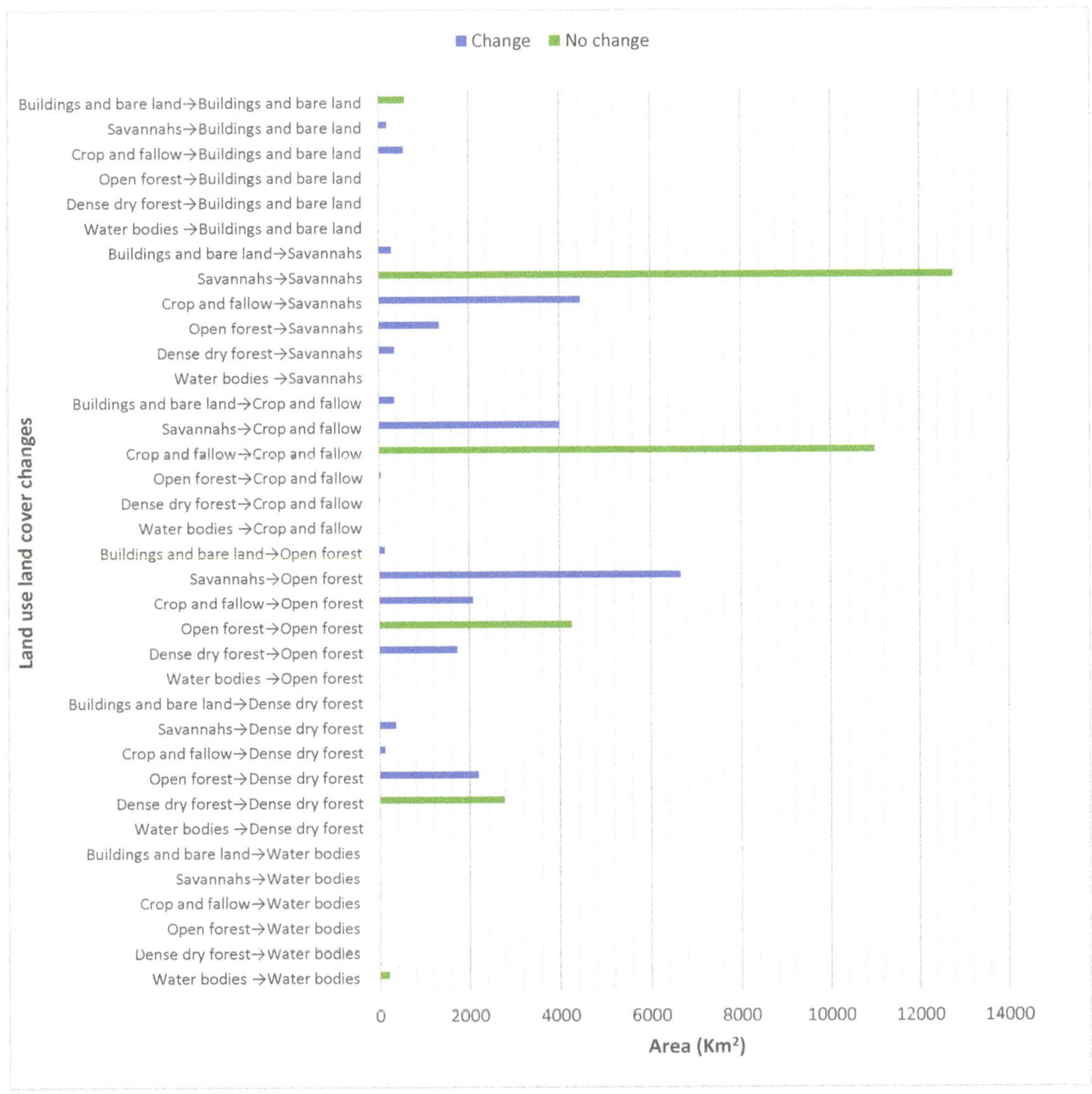

Figure 5. Land-cover conversions that occurred between 2015 and 2020.

Maps of the changes were then produced (Figure 6) by combining all classes that had undergone conversions on one hand, and all those that had not undergone conversion during the periods that were considered on the other. The change map between the 2005 and 2015 classifications was not produced because images of the first four target years have one more land-cover class (i.e., clouds) than the last two. Therefore, the application of the change detection algorithm between these two years (with a different number of land-cover classes) generates several hybrid classes that do not reflect the situation on the ground.

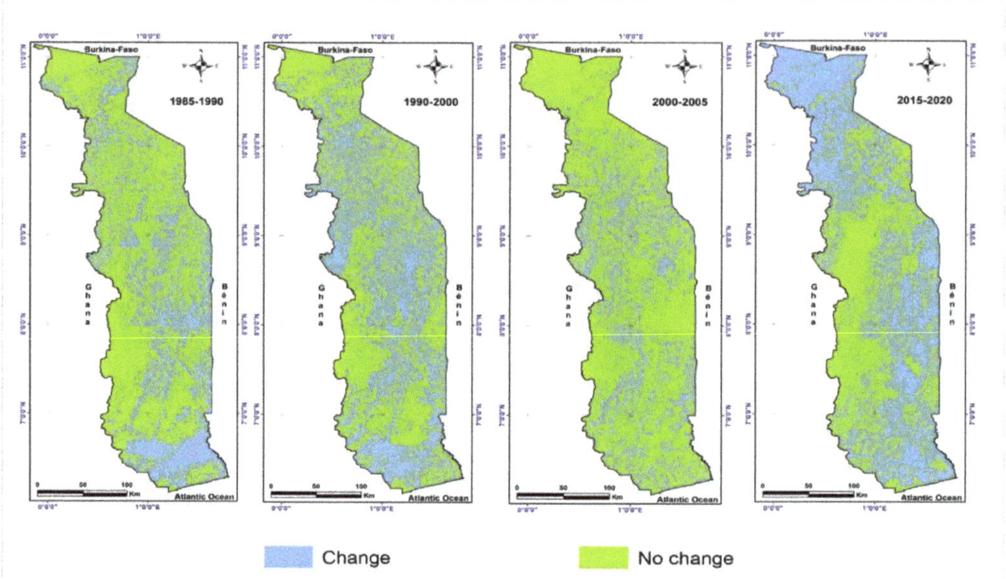

Figure 6. LULC change maps.

4.4. Evolution of Forest Cover

When considering only dense dry forest and open forest classes, their respective starting areas were 10,722.53 km^2 and 17,547.75 km^2 in 1985, i.e., 18.92% and 30.97% of the nation's total territory. Under the effects of land-cover change, they have decreased to 16.05% and 25.37% in 1990, 13.60% and 18.56% in 2000, 15.01% and 18.42% in 2005, 7.39% and 15.09% in 2015, and to 6.68% and 17.13% in 2020. With an area of 3785.27 km^2 for dense dry forests and 9709.70 km^2 for open forests in 2020, forest areas have thus declined by 12.24% for the first category and 13.83% for the second, i.e., a total of 26.07% at the national level during the 35 years covered by this study. Details on the quantification of these two land covers in the different ecological zones and their changes over time are indicated in Appendix D (Table A3).

To facilitate the subsequent quantitative analysis of forest cover change, we have cumulated the two aforementioned occupancy classes to form the forest class. The trend line (Figure 7) that summarizes the percentage change in forest area relative to that of the country illustrates the degree of deforestation and forest degradation over the period that was considered. Forest area distributions as a land-cover percentage by ecological zone and by target year were estimated (Figure 8).

When we explored the data at the level of ecological zones to determine how these forest areas have changed through time, we noted that the deforestation or degradation of these forests has not proceeded at the same rate in these ecosystems. The evolutions of forest areas in the different ecological zones were illustrated by the distribution maps of forest cover of the target years from the period from 1985 to 2020 (Figure 9).

In ecological zones I, II, and III, these forested areas declined almost continuously from 1985 with a cumulative loss until 2020 of 16.73%, 48.62%, and 28.66%, respectively, compared to their starting size in these areas. We can, nevertheless, note a forest area recovery in the 2015 to 2020 period in zone I and between 2000 and 2005 in zone II. Zone IV (the smallest ecological zone) experienced a sharp decline in forest area (18.35%) between 1985 and 1990, followed by a smaller loss (7.41%) between 1990 and 2000, prior to its recovery and then contraction (to 1.58%) from 2015 to 2020. Zone V is characterized by a 30.49% loss of forest area between 1985 and 1990, then a rapid increase in area (21.73%)

for a decade (1990–2000). These areas continued to increase until 2005 and then declined slightly from 2005 to 2015 before increasing again to 29.2% of the total area in 2020.

The finer-scale examination (zooming) of the maps produced from the results (Figure A4) in Appendix E shows the development of two towns (Sokodé and Tchamba), as well as the Abdoulaye Forest Reserve between 1985 and 2020. We noted the expansion over time of both these towns and agriculture, as well as the appearance of small new settlements at the expense of wooded areas. As a result of these two main factors, the average annual rate of change of forest cover to other land cover is about −2.11% between 1985 and 2020, leading to the disappearance of more than half of the forest areas during this period.

The results of calculating the annual rate of change in forest cover and annual deforestation between individual target years, and from the beginning to the end of the study period are shown (Table 7).

4.5. Land-Cover Changes at the Administrative Regions Scale

Following the analysis of land-cover conversions at the national level and the evolution of forest cover in the ecological zones, the quantification of all changes that have occurred at the level of the administrative regions was mapped (Figure 10).

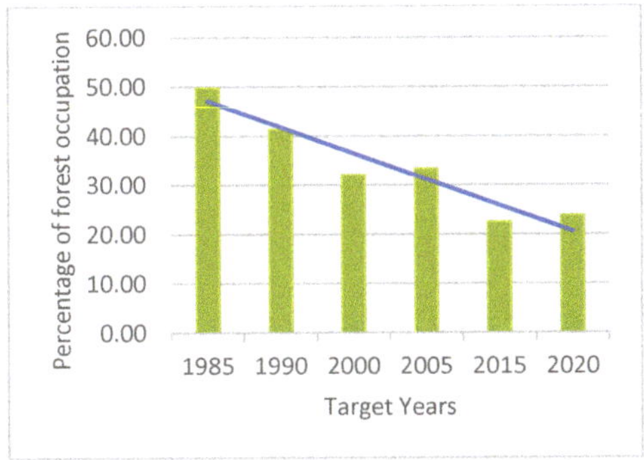

Figure 7. Countrywide forest percentage changes.

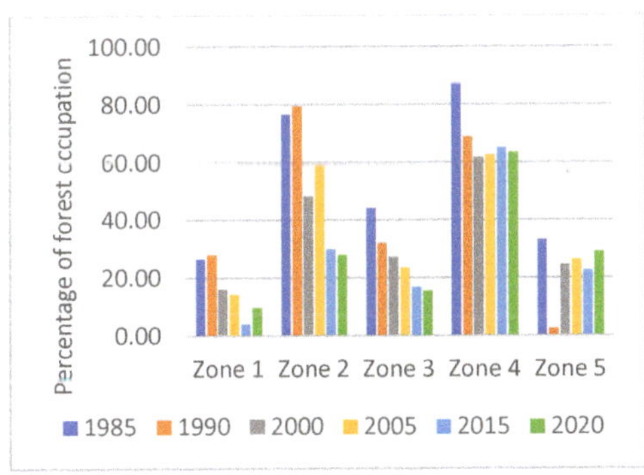

Figure 8. Forest change by ecological zone.

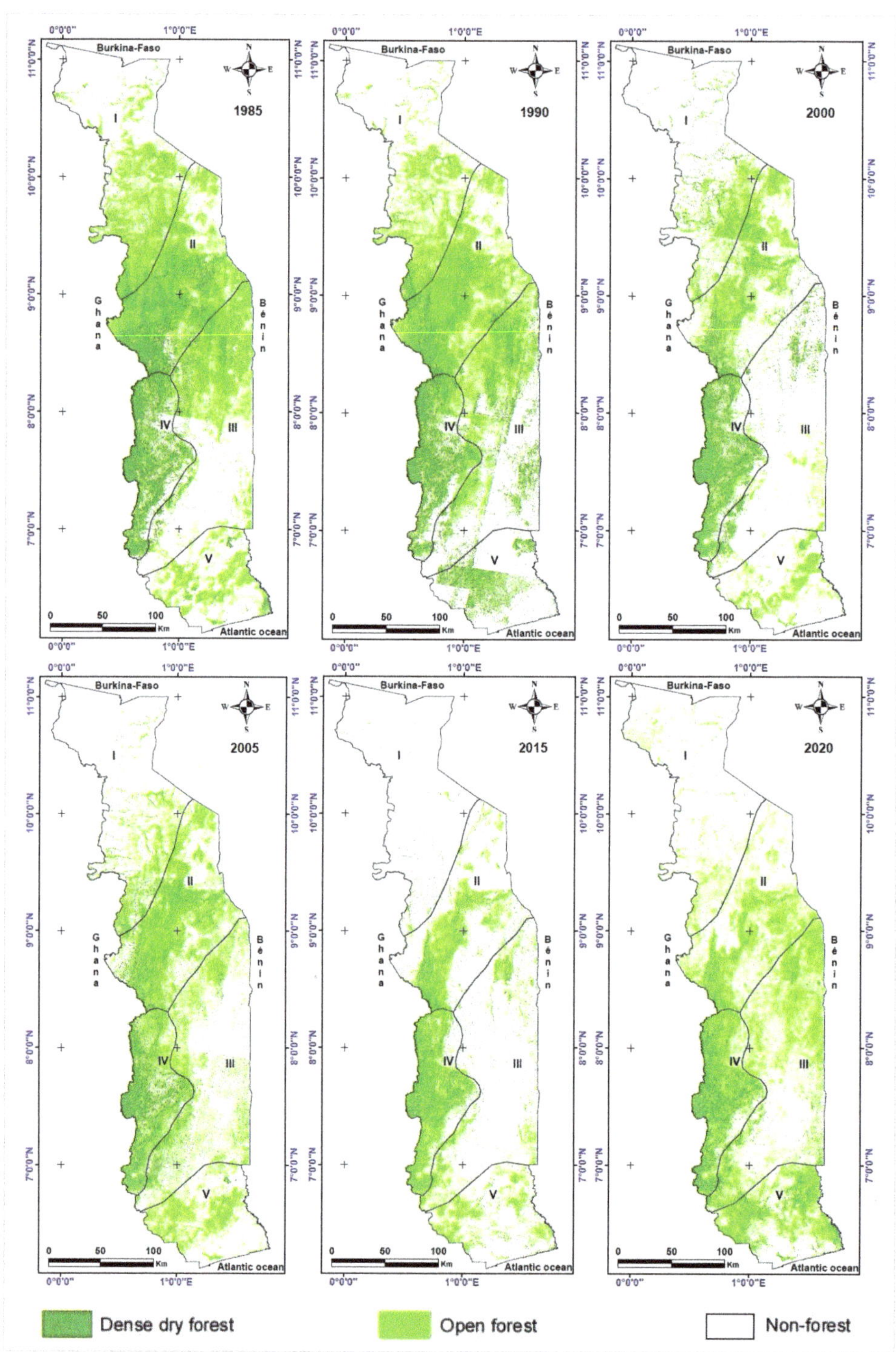

Figure 9. Forest evolution across ecological zones by target year.

Table 7. Evolution of forested areas between 1985 and 2020.

Year	Area (km^2)	Forest Area (% of Togo)	r (% y^{-1})	R (km^2 y^{-1})
1985	28,270.28	49.89		
1990	23,473.88	41.42	−3.72	959.28
2000	18,220.92	32.15	−2.53	525.30
2005	18,945.32	33.43	0.78	−144.88
2015	12,736.35	22.48	−3.97	620.90
2020	13,494.97	23.81	1.16	−151.72
1985–2020			−2.11	422.15

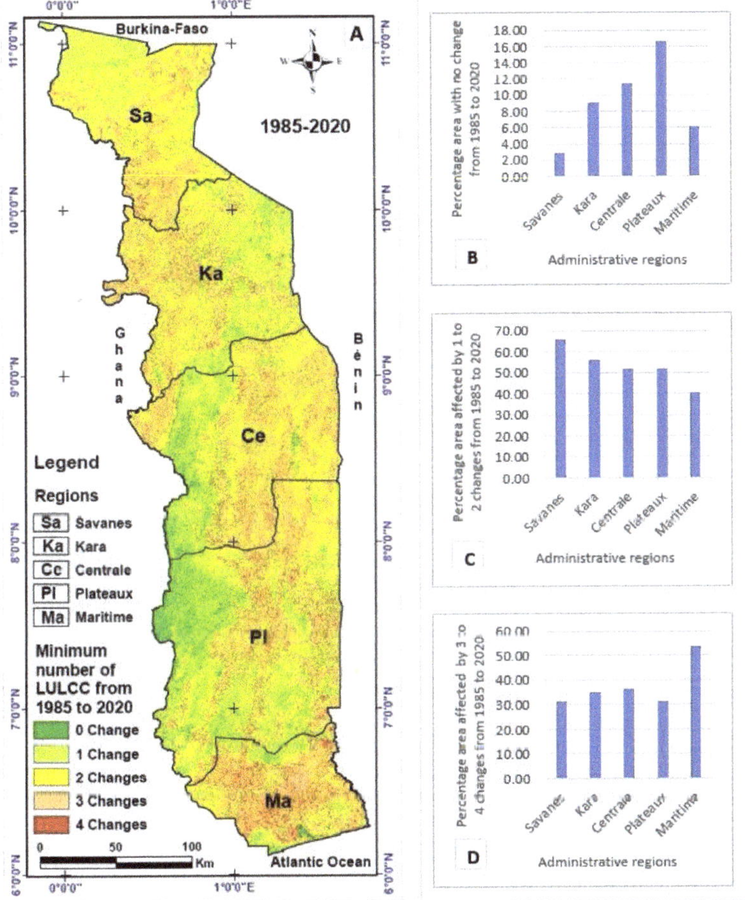

Figure 10. (A) Land-cover change gradient by region from 1985 to 2020; (B) area unchanged; (C) area with one to two changes; and (D) area with three to four changes.

In all of these administrative regions, original land covers were retained in part during the period covered by the study (Figure 10A). For those remaining parts where the land cover was altered, they had undergone at least one, two, three, or four changes between

1985 and 2020. Visual inspection reveals that parts where little or no change had been experienced were mostly forested areas (e.g., from the northeast to southwest), urban areas such as the national capital Lomé, and large bodies of water such as Lake Togo in the extreme south of the country. The Plateaux administrative region retained the most area (16.69%) of this land cover that had never changed (Figure 10B). This region is followed by the Centrale (11.48%), Kara (9.09%), and Maritime (6.12%) regions, while the Savanes region has the smallest proportion (2.84%) of its area not being affected by change over the 35-year period.

It can be observed that 65.75% of the Savanes region has undergone at least one to two changes in land cover (Figure 10C). In the Kara, Centrale, and Plateaux regions, slightly more than half of their respective areas have been similarly affected. In contrast to the areas by region that have never undergone change, the Maritime region has the largest percentage of the regional area (53.50%) that has undergone at least three to four land-cover changes (Figure 10D). For the same locations, land-cover changes have occurred more rapidly in the Maritime, followed by the Centrale (36.59%), Kara (34.98%), (31.44%), and Savanes (31.41%) regions.

5. Discussion

5.1. Quality of Results from Composite Image Classifications

During this study, data from Landsat 5, 7, and 8 archives were used to form different image composites, the supervised classifications of which (under the GEE platform) led to the production of land cover maps of Togo. Despite difficulties that were encountered in finding the best quality images, the results that were obtained indicate relatively high overall accuracies of 91% to 98% for composites with the original bands and 86% to 96% for those including the vegetation indices. However, the classification results including vegetation indices tended to overestimate the built-up and bare land (buildings + soil) class and the water body class. We believe that this is likely due to the simultaneous presence of NDBI, which captures residential areas and bare soil, the BSI, which is a bare soil-specific index, and NDWI, which would have difficulty distinguishing water bodies from shadows. These results are consistent with those of [78] and [24], who found that the NDBI and modified NDWI yielded image classification results with very low accuracies, despite being two popular indices in the literature.

The results have shown that OA and *Ks* for the original composite band classifications are significantly different from those with vegetation indices, but the latter did not improve the image classification results as one would have expected. Nevertheless, the spontaneous decrease in overall accuracy and *Ks* for the 2005 composite classification (Tables 4 and 5) could be primarily related to deficiencies in the Landsat 7 data that are observed as fine stripes on the 2005 map (Figure 9). It should be noted that this sensor suffered hardware failure in its Scan Line Corrector (SLC) in 2003, resulting in the loss of about 22–25% of the data in each scene [79]. Additional research could be done on the impact of these indices on the quality of image classification results and also test new indices such as the Emissivity Contrast Index (ECI), which have overcome the NDVI limitation concerning its capability to distinguish bare soil from senescent vegetation [80]. Another thing that could be tested in future research using RF in order to improve image classification accuracy is to tune the hyper-parameters of this model to improve its performance [65], instead of using the default number of trees.

5.2. Land-Cover Changes

The classifications indicate fairly rapid changes in land cover over the 35 years that are covered by this study and the rapid deforestation or degradation of forest cover, the area of which fell countrywide from 49.89% in 1985 to 23.81% in 2020. These changes have favored crops and fallow lands, savannahs, urban areas, and bare soil. In considering the evolution of forest areas in the different ecological zones, we found that zones II and IV, which cover 32.55% of the national area, contained 55.10% of the national forest cover in

2020. This could be explained by the fact that these two zones are mountainous with very steep relief (Figure 1B), making it very difficult to access forest resources and land in these zones. Zone IV, in particular, has retained most of its original forest area (72.77%), even though it is the smallest of the five ecological zones. Furthermore, ecological zones I and III are areas par excellence in terms of agriculture and housing, as can be seen in our mapped results. Zone V is home to more than one-third of the country's population; the relatively broad extent of forest that was found in this zone would have more to do with poor image quality than with the actual area.

In Table 7, we note that r is negative and R is positive when there is a contraction of forested areas, while the opposite occurs when there is an expansion of forested areas. From these two indicators of forest cover change, we further note that the study area experienced a substantial loss of forest area between 1985 and 2000 and, again, between 2005 and 2015. In contrast, only small increases in the area occurred until 2005 and, again, between 2015 and 2020. Current forest area declines are most likely related to agricultural expansion and rapid human population growth in Togo (2.84% y^{-1}), which exert strong pressures on natural resources and land. The national REDD+ Togo study of 2018 on the causes and consequences of deforestation and forest degradation across the nation has confirmed that agricultural development, including associated management practices (notably, the use of fire), is the main cause of forest disturbance, ahead of timber exploitation (timber and energy) and urban expansion. Furthermore, the dynamics of urbanization, which underlies the country's population growth, are driving rapid changes in LULC and are contributing to forest loss, both directly and indirectly [32,33].

Nevertheless, the increase in forest area in 2005 could be attributed simply to the aforementioned poor quality of Landsat 7 data, which would influence the classifiers during processing. The 2020 increase could be due to an overestimation by classifiers of the open forest class at the expense of savannah, but this could also be due to the results of conservation policies and programs that have been recently implemented by the government (forest inventory and REDD + strategy). In order to achieve the state's objective of increasing forest cover to 30% of the territory by 2030, these factors of forest degradation would have to be reconsidered in terms of governmental actions at the social, environmental, and political levels. In addition, the rate of land-cover conservation and the speed of change that has been quantified at the level of administrative regions indicate that the Plateaux and Centrale regions are better conserved, while the Maritime region records the highest frequency of change. The Savanes region is intermediate between these two extremes; most land cover has only changed once or twice. Yet, it should be noted that most of the plant formations of the Savanes region were very early transformed into crops and remained in this class. This explains why this region has a relatively low rate of land cover change for a given location despite its higher rate of degraded area. The Maritime region has experienced the most land-cover changes over the period, i.e., three to four times. These conditions would thus need to be monitored when making land-cover planning or development decisions. Given that forest management across the study area is based more on administrative subdivisions, our results should enable centralized administrative and forestry authorities to prioritize actions for a much more balanced environmental governance.

5.3. Advantages and Limitations of the Method Used

For the selection, pre-processing, and classifications of satellite images during this study, we used the RF algorithm, which can take into account even disparate data to make a fairly accurate classification of heterogeneous land cover such as in forest-savannah mosaics [60]. This algorithm has been used on the GEE platform containing a vast catalog of Earth observational data. It is based upon millions of servers around the world that allow for the rapid processing and analysis of satellite data over large areas, without the need to download them [81]. The GEE has a user-friendly programming environment with high computational efficiency, which allows less time to be spent on usual satellite data processing steps that are frequently quite time-consuming when using dedicated software.

A further advantage of this method is the possibility of making enormous savings in both time and money when conducting regional or national forest inventories. For example, when considering the results that were obtained for several land-cover classes through methods requiring very few means that were applied in this study, we note that they are more or less comparable to those that were obtained from the national forest inventory (NFI), which had mobilized many more human and financial resources. For the 2015 results (the year closest to the NFI), we obtained 22.48% for the forest class, 38.90% for the savannah class, and 38.27% for the grouping of agriculture and infrastructure classes versus 24.24%, 34.86%, and 40.90, respectively, for the 2016 NFI [71]. With this method of processing satellite data in the GEE, once the processing code is completed, it can be easily optimized and applied for the long-term monitoring of LULC changes when incorporating newly acquired images [62].

However, it must be noted that this processing power is not available on demand for all types of operations, given that a quota is allocated to each user and, thus, the GEE system sometimes limits or aborts certain code executions that are computationally demanding [26]. Furthermore, despite having millions of images, some areas have long periods when cloud-free data are absent, especially in tropical environments. This is a particularly lamentable state of affairs, given that research in this region has calculated the probability of acquiring Landsat MSS or Landsat TM images with <70% cloud cover in a year to be only 26% [79]. In these cases, the GEE permits the selection of pixels from multiple images exhibiting large temporal differences in acquisition dates to form the composite, as was the case in our study. Unfortunately, such selections do not allow for estimates of seasonal differences or phenologies, thereby introducing potential classification errors. A further limitation is that during satellite data processing, code execution errors that are encountered can be difficult to debug, given that scripts in the GEE run in the Google Cloud. As confirmed by [62], errors also can occur in the JavaScript code, either on the client side, which is manageable with some effort, or during server-side execution, a situation that can be very difficult to manage.

6. Conclusions

The LULC changes that are attributable to anthropogenic disturbance are leading to reductions in forest cover, contributing significantly to global carbon emissions. In this study, we employed the median satellite image composition method with historical Landsat sensor data in the GEE to quantify changes across the nation of Togo between 1985 and 2020 using the Random Forest algorithm. Our results indicate that all land-cover classes identified from the 1985 composite image were affected to varying degrees by these land-cover changes. Furthermore, forests lost about 52.28% of their original area from 1985 to 2020 through the expansion of crop and fallow lands, savannah, and urbanization. Ecological zones I, III, and V cover more than two-thirds of the total area of the country and contain less than half of the forest cover. The changes are mainly reflected by a strong increase in agricultural activity, deforestation through timber exploitation, and the urban expansion of a burgeoning human population. Easier accessibility of the areas and a greater human presence favor all of these activities. In contrast, ecological zones II and IV, which cover less than one-third of the total area of the country, contain more than 55% of the national forest cover in 2020. These are very mountainous areas, the steep slopes of which limit the adverse effects of human activities and, consequently, their effects on natural resources.

The methods that were applied in this study and the results that were obtained could help forestry and territorial administrators to better understand the factors that are involved in land-cover change and forest area reduction. They could also help the national coordination of REDD+ in Togo to better operate or to boost the measurement, reporting, and verification system, as part of the nation's forest monitoring system. For similar future studies in Togo, more reliable satellite data (Landsat 8 and 9) with lower cloud cover or higher spatial resolution (Sentinel 2 and greater) could be used when sufficient time-series images become available on the GEE platform over the study area, as well as other countries in Sub-Saharan Africa.

Author Contributions: Conceptualization: A.K. and K.G.; Methodology: A.K., K.G. and F.F.; Data collection: A.K., F.F., W.A., M.D. and K.W.; Data analysis: A.K. and K.G., Preparation of the manuscript draft: A.K.; Supervision: K.G.; Writing and editing: all authors. All authors have read and agreed to the published version of the manuscript.

Funding: This research has been funded by the *Programme Canadien de bourses de la Francophonie* (Government of Canada, Department of Foreign Affairs, Trade and Development, Canadian Partnership Branch), as well as the *Natural Sciences and Engineering Research Council of Canada* (NSERC Discovery grant RGPIN-2018-06101 and NSERC CREATE Grant 543360-2020).

Institutional Review Board Statement: Not applicable.

Informed Consent Statement: Not applicable.

Data Availability Statement: Data are available upon request and can be obtained by contacting the lead author.

Acknowledgments: We thank all of the reviewers for their guidance and contributions to writing and commenting on this manuscript. W.F.J. Parsons translated the French text.

Conflicts of Interest: The authors declare no conflict of interest. The organization providing the funding has played no role in the design of the study, collection, analysis, interpretation of the data, writing of the manuscript, or in the decision to publish the results.

Appendix A

Table A1. Confusion matrix for the target year 1985 without vegetation indices. Perfectly predicted values for each category are highlighted in bold along the diagonal.

	Clouds	Water	Dense Dry Forest	Open Forest	Crops + Fallows	Savannah	Bldg. + Soil	Producer Accuracy
Clouds	25,986	0	0	0	0	147	0	0.99
Water	0	8805	1	0	0	14	2	1.00
Dense dry forest	0	0	129,893	3690	249	138	0	0.97
Open forest	0	0	7144	26,071	1390	138	0	0.75
Crops + fallows	171	0	167	570	123,841	893	2649	0.97
Savannah	1053	0	30	174	1715	117,812	818	0.97
Bldg. + Soil	297	0	72	111	3974	1523	46,131	0.89
User Accuracy	0.94	1.00	0.95	0.85	0.94	0.98	0.93	
Overall Accuracy				0.95				
Kappa				0.93				

Appendix B

Table A2. Comparison of accuracies by land-cover class; with vs. without vegetation indices.

Classes	Accuracy	*p*-Value
Water	UA	0.000
	PA	0.001
Dense dry forest	UA	0.000
	PA	0.001

Table A2. *Cont.*

Classes	Accuracy	p-Value
Open forest	UA	0.000
	PA	0.001
Crops + fallows	UA	0.000
	PA	0.000
Savannah	UA	0.001
	PA	0.000
Bldg. + soil	UA	0.000
	PA	0.000

Appendix C

Conversion of Land-Cover Classes

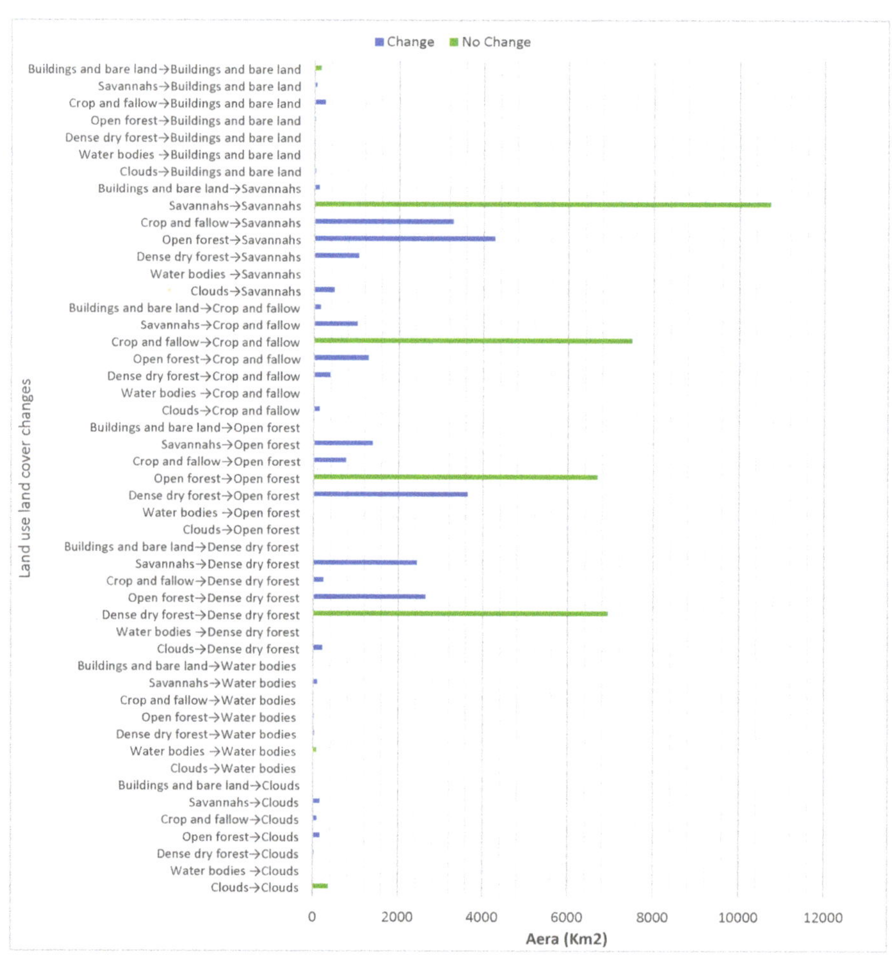

Figure A1. Land-cover conversions between 1985 and 1990.

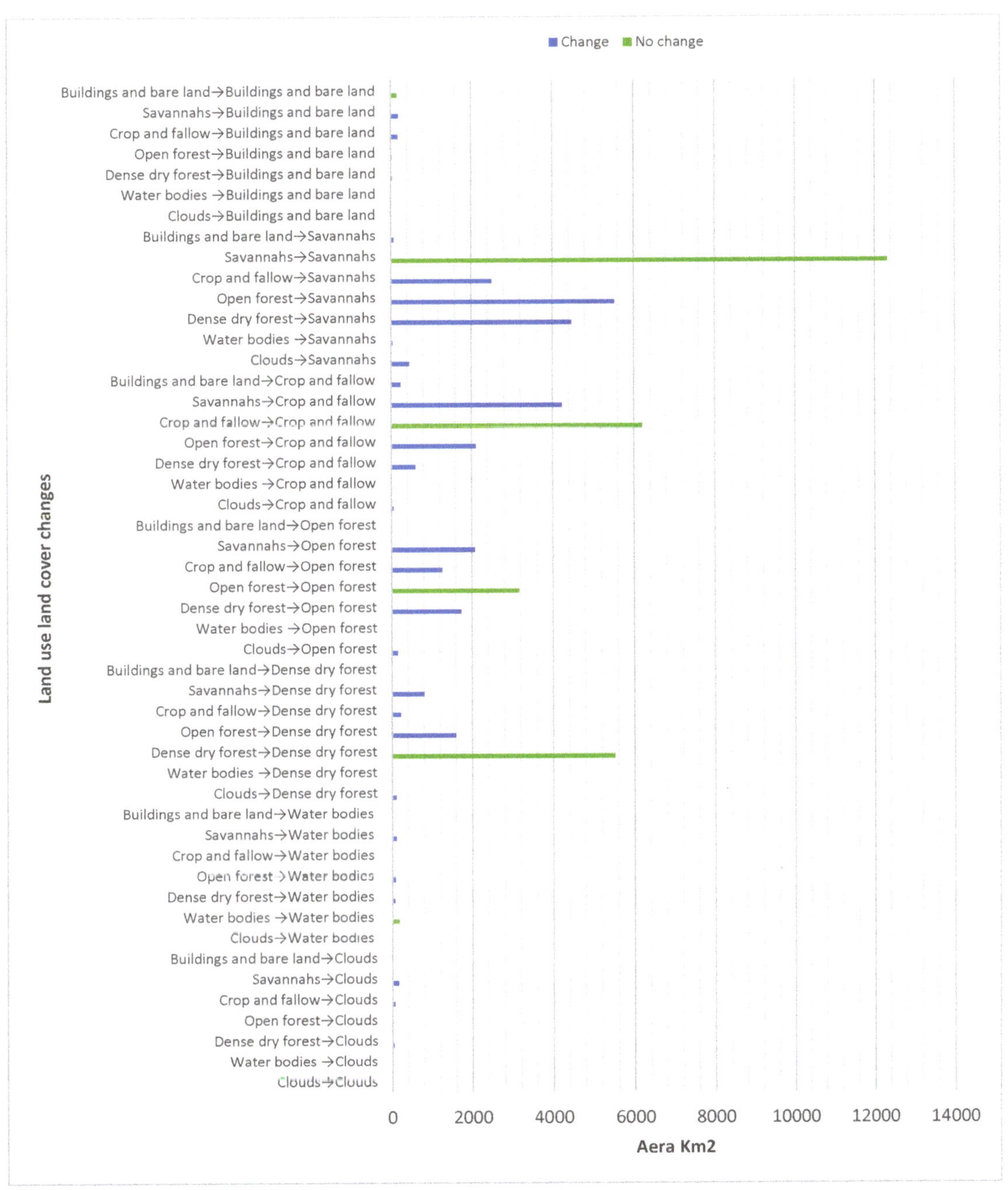

Figure A2. Land-cover conversions between 1990 and 2000.

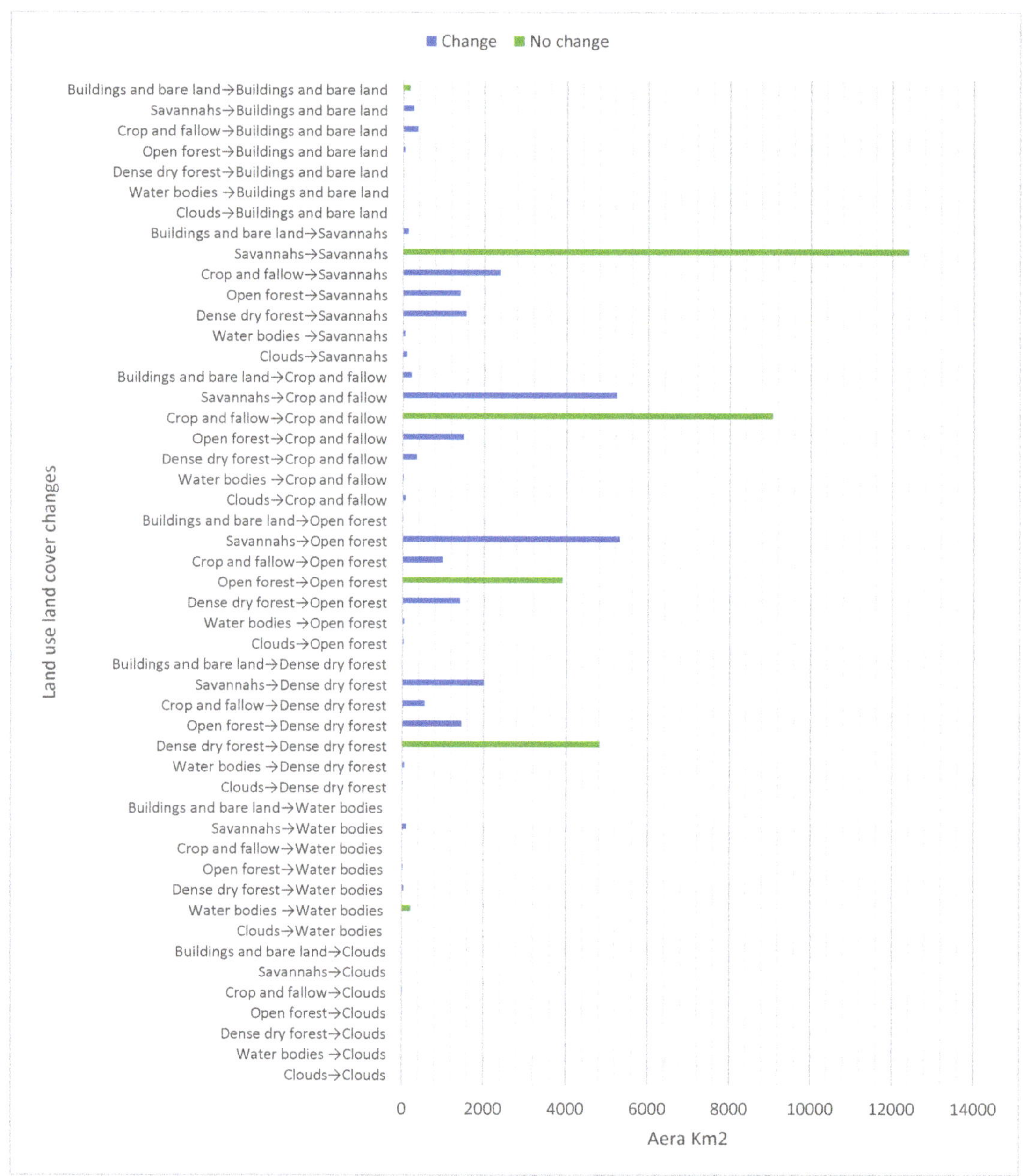

Figure A3. Land-cover conversions between 2000 and 2005.

Appendix D

Table A3. Changes in forest-covered areas.

Years	Classes	Zone I	Zone II	Zone III	Zone IV	Zone V	Total
1985	Dense dry forest	878.81	3977.56	1491.35	4192.75	182,06	10,722.53
	Open forest	3039.83	5253.77	5634.57	1392.29	2227,29	17,547.75
	Forest areas	3918.64	9231.32	7125.92	5585.04	2409.35	28,270.28
	%/Country	6.91	16.29	12.57	9.86	4.25	49.89
	%/Zone	26.47	76.60	44.17	87.37	33.07	
1990	Dense dry forest	542.65	3054.40	1562.54	3752.49	183.18	9095.25
	Open forest	3599.47	6522.88	3592.65	659.38	4.25	14,378.62
	Forest areas	4142.12	9577.28	5155.19	4411.86	187.43	23,473.88
	%/Country	7.31	16.90	9.10	7.79	0.33	41.42
	%/Zone	27.98	79.47	31.95	69.02	2.57	
2000	Dense dry forest	370.09	2463.65	1407.70	3316.65	146.87	7704.97
	Open forest	1989.39	3313.71	2967.11	621.64	1624.10	10515.96
	Forest areas	2359.48	5777.36	4374.81	3938.29	1770.98	18,220.92
	%/Country	4.16	10.19	7.72	6.95	3.13	32.15
	%/Zone	15.94	47.94	27.12	61.61	24.30	
2005	Dense dry forest	392.21	3148.15	1577.43	3228.93	158.92	8505.64
	Open forest	1714.80	3996.35	2206.50	770.48	1751.55	10,439.68
	Forest areas	2107.01	7144.50	3783.93	3999.41	1910.47	18,945.32
	%/Country	3.72	12.61	6.68	7.06	3.37	33.43
	%/Zone	14.23	59.28	23.45	62.57	26.22	
2015	Dense dry forest	145.76	1025.65	752.59	2136.58	126.13	4186.70
	Open forest	463.68	2578.21	1957.75	2028.74	1521.27	8549.65
	Forest areas	609.44	3603.86	2710.34	4165.31	1647.40	12,736.35
	%/Country	1.08	6.36	4.78	7.35	2.91	22.48
	%/Zone	4.12	29.90	16.80	65.16	22.61	
2020	Dense dry forest	128.64	975.70	365.86	2076.96	238.12	3785.27
	Open forest	1313.22	2395.81	2136.68	1987.47	1876.52	9709.70
	Forest areas	1441.85	3371.51	2502.53	4064.43	2114.63	13,494.97
	%/Country	2.54	5.95	4.42	7.17	3.73	23.81
	%/Zone	9.74	27.98	15.51	63.59	29.02	
Ecological Zone areas		14,805.30	12,051.40	16,133.60	6392.10	7286.50	56,668.90

Appendix E

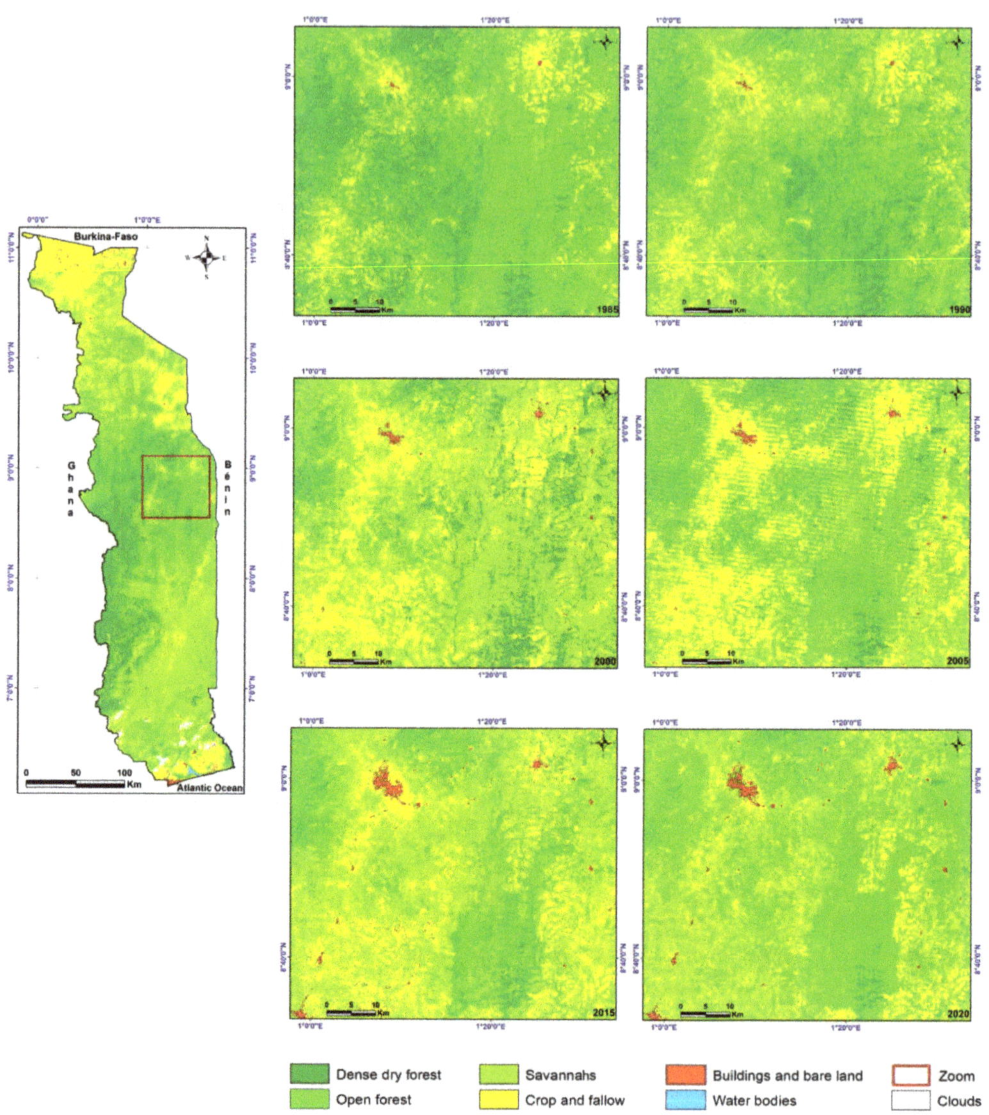

Figure A4. Enlarged ("zoomed") insets in East-Central Togo for all target years.

References

1. Nunes, L.J.R.; Meireles, C.I.R.; Pinto Gomes, C.J.; Almeida Ribeiro, N.M.C. Forest Contribution to Climate Change Mitigation: Management Oriented to Carbon Capture and Storage. *Climate* **2020**, *8*, 21. [CrossRef]
2. Cardil, A.; de-Miguel, S.; Silva, C.A.; Reich, P.B.; Calkin, D.; Brancalion, P.H.S.; Vibrans, A.C.; Gamarra, J.G.P.; Zhou, M.; Pijanowski, B.C.; et al. Recent Deforestation Drove the Spike in Amazonian Fires. *Environ. Res. Lett.* **2020**, *15*, 121003. [CrossRef]
3. Laumonier, Y.; Edin, A.; Kanninen, M.; Munandar, A.W. Landscape-Scale Variation in the Structure and Biomass of the Hill Dipterocarp Forest of Sumatra: Implications for Carbon Stock Assessments. *For. Ecol. Manag.* **2010**, *259*, 505–513. [CrossRef]

4. Gogoi, A.; Ahirwal, J.; Sahoo, U.K. Plant Biodiversity and Carbon Sequestration Potential of the Planted Forest in Brahmaputra Flood Plains. *J. Environ. Manag.* **2021**, *280*, 111671. [CrossRef]
5. Kundu, K.; Halder, P.; Mandal, J.K. Forest Cover Change Analysis in Sundarban Delta Using Remote Sensing Data and GIS. In *Intelligent Computing Paradigm: Recent Trends*; Mandal, J.K., Sinha, D., Eds.; Studies in Computational Intelligence; Springer Singapore: Singapore, 2020; pp. 85–101, ISBN 9789811373343.
6. FAO and UNEP. *The State of the World's Forests 2020: Forests, Biodiversity and People*; The State of the World's Forests (SOFO); FAO and UNEP: Rome, Italy, 2020; ISBN 978-92-5-132419-6.
7. Chen, Y.-Y.; Huang, W.; Wang, W.-H.; Juang, J.-Y.; Hong, J.-S.; Kato, T.; Luyssaert, S. Reconstructing Taiwan's Land Cover Changes between 1904 and 2015 from Historical Maps and Satellite Images. *Sci. Rep.* **2019**, *9*, 12. [CrossRef]
8. Deo, R.K.; Russell, M.B.; Domke, G.M.; Andersen, H.-E.; Cohen, W.B.; Woodall, C.W. Evaluating Site-Specific and Generic Spatial Models of Aboveground Forest Biomass Based on Landsat Time-Series and LiDAR Strip Samples in the Eastern USA. *Remote Sens.* **2017**, *9*, 598. [CrossRef]
9. Olorunfemi, I.E.; Olufayo, A.A.; Fasinmirin, J.T.; Komolafe, A.A. Dynamics of Land Use Land Cover and Its Impact on Carbon Stocks in Sub-Saharan Africa: An Overview. *Environ. Dev. Sustain.* **2021**, *24*, 40–76. [CrossRef]
10. Angelsen, A.; Brockhaus, M.; Sunderlin, W.D.; Verchot, L.V. *Analyse de la REDD+ Les Enjeux et les Choix*; CIFOR: India Nishiya, Momono, 2013; ISBN 978-602-1504-00-0.
11. Minh, D.H.T.; Ndikumana, E.; Vieilledent, G.; McKey, D.; Baghdadi, N. Potential Value of Combining ALOS PALSAR and Landsat-Derived Tree Cover Data for Forest Biomass Retrieval in Madagascar. *Remote Sens. Environ.* **2018**, *213*, 206–214. [CrossRef]
12. Mitchell, A.L.; Rosenqvist, A.; Mora, B. Current Remote Sensing Approaches to Monitoring Forest Degradation in Support of Countries Measurement, Reporting and Verification (MRV) Systems for REDD+. *Carbon Balance Manag.* **2017**, *12*, 9. [CrossRef]
13. Folega, F.; Zhang, C.; Zhao, X.; Wala, K.; Batawila, K.; Huang, H.; Dourma, M.; Akpagana, K. Satellite Monitoring of Land-Use and Land-Cover Changes in Northern Togo Protected Areas. *J. For. Res.* **2014**, *25*, 385–392. [CrossRef]
14. Akakpo, K.M.; Quensière, J.; Gadal, S.; Kossi, A.; Kokou, K. Caractérisation et Dynamique Spatiale de La Couverture Végétale Dans Les Aires Protégées Du Togo: Étude Par Télédétection Satellitaire de La Forêt Classée de Missahoé Dans La Région Des Plateaux. *Rev. Int. De Géomatique Aménagement Et Gest. Des Ressour.* **2017**, *1*, 181–194.
15. Atsri, H.K.; Konko, Y.; Cuni-Sanchez, A.; Abotsi, K.E.; Kokou, K. Changes in the West African Forest-Savanna Mosaic, Insights from Central Togo. *PLoS ONE* **2018**, *13*, e020399. [CrossRef]
16. Polo-Akpisso, A.; Wala, K.; Soulemane, O.; Folega, F.; Akpagana, K.; Tano, Y. Assessment of Habitat Change Processes within the Oti-Keran-Mandouri Network of Protected Areas in Togo (West Africa) from 1987 to 2013 Using Decision Tree Analysis. *Science* **2020**, *2*, 1. [CrossRef]
17. Badjana, H.M.; Helmschrot, J.; Selsam, P.; Wala, K.; Flügel, W.-A.; Afouda, A.; Akpagana, K. Land Cover Changes Assessment Using Object-Based Image Analysis in the Binah River Watershed (Togo and Benin). *Earth Space Sci.* **2015**, *2*, 403–416. [CrossRef]
18. Diwediga, B.; Agodzo, S.; Wala, K.; Le, Q.B. Assessment of Multifunctional Landscapes Dynamics in the Mountainous Basin of the Mo River (Togo, West Africa). *J. Geogr. Sci.* **2017**, *27*, 579–605. [CrossRef]
19. Koumoi, Z.; Boukpessi, T.; Kpedenou, K.D. Principaux Facteurs Explicatifs de La Dynamique de l'occupation Du Sol Dans Le Centre-Togo: Apport Des SIG et Des Statistiques Spatiales. *Rev. Ivoir. Géographie Savanes* **2017**, *3*, 252–273.
20. Koglo, Y.S.; Gaiser, T.; Agyare, W.A.; Sogbedji, J.M.; Kouami, K. Implications of Some Major Human-Induced Activities on Forest Cover Using Extended Change Matrix Quantity and Intensity Analysis Based on Historical Landsat Data from the Kloto District, Togo. *Ecol. Indic.* **2019**, *96*, 628–634. [CrossRef]
21. Folega, F.; Woegan, Y.A.; Marra, D.; Wala, K.; Batawila, K.; Seburanga, J.L.; Zhang, C.; Peng, D.; Zhao, X.; Akpagana, K. Long Term Evaluation of Green Vegetation Cover Dynamic in the Atacora Mountain Chain (Togo) and Its Relation to Carbon Sequestration in West Africa. *J. Mt. Sci.* **2015**, *12*, 921–934. [CrossRef]
22. Mastro, P.; Masiello, G.; Serio, C.; Pepe, A. Change Detection Techniques with Synthetic Aperture Radar Images: Experiments with Random Forests and Sentinel-1 Observations. *Remote Sens.* **2022**, *14*, 3323. [CrossRef]
23. Mahdianpari, M.; Jafarzadeh, H.; Granger, J.E.; Mohammadimanesh, F.; Brisco, B.; Salehi, B.; Homayouni, S.; Weng, Q. A Large-Scale Change Monitoring of Wetlands Using Time Series Landsat Imagery on Google Earth Engine: A Case Study in Newfoundland. *GIScience Remote Sens.* **2020**, *57*, 1102–1124. [CrossRef]
24. Phan, T.N.; Kuch, V.; Lehnert, L.W. Land Cover Classification Using Google Earth Engine and Random Forest Classifier—The Role of Image Composition. *Remote Sens.* **2020**, *12*, 2411. [CrossRef]
25. Souza, C.M.; Shimbo, J.Z.; Rosa, M.R.; Parente, L.L.; Alencar, A.A.; Rudorff, B.F.T.; Hasenack, H.; Matsumoto, M.; Ferreira, L.G.; Souza-Filho, P.W.M.; et al. Reconstructing Three Decades of Land Use and Land Cover Changes in Brazilian Biomes with Landsat Archive and Earth Engine. *Remote Sens.* **2020**, *12*, 2735. [CrossRef]
26. Gorelick, N.; Hancher, M.; Dixon, M.; Ilyushchenko, S.; Thau, D.; Moore, R. Google Earth Engine: Planetary-Scale Geospatial Analysis for Everyone. *Remote Sens. Environ.* **2017**, *202*, 18–27. [CrossRef]
27. Arévalo, P.; Olofsson, P.; Woodcock, C.E. Continuous Monitoring of Land Change Activities and Post-Disturbance Dynamics from Landsat Time Series: A Test Methodology for REDD+ Reporting. *Remote Sens. Environ.* **2020**, *238*, 111051. [CrossRef]
28. REDD+ Togo. *Plan D'actions de Mise En Oeuvre de La Stratégie Nationale de Réduction Des Émissions Dues à La Déforestation et à La Dégradation Des Forêts (REDD+) 2020–2029*; Coordination Nationale REDD+ du Togo: Lomé, Togo, 2020.

29. INSEED et AFRISTAT. *Enquête Régionale Intégrée Sur l'Emploi et Le Secteur Informel*; 2017; Institut National de la Statistique et des Etudes Economiques et Démographiques et AFRISTAT: Lomé, Togo; Bamako, Mali, 2019; p. 73.
30. PANA. *Plan d'Action National d'Adaptation Au Changement Climatique*; Ministère de l'Environnement et des Ressources Forestières (MERF): Lomé, Togo, 2009; p. 113.
31. REDD+ Togo. *Définition et Calcul Du Taux National de Défloration Annuel Du Togo Entre 1990 et 2015*; Coordination Nationale REDD+ du Togo: Lomé, Togo, 2018.
32. REDD+ Togo. *Étude Sur Les Causes et Conséquences de La Déforestation et La Dégradation Des Forets Au Togo et Identification Des Axes d'intervention Appropries*; Coordination Nationale REDD+ du Togo: Lomé, Togo, 2018.
33. Ern, H. Die Vegetation Togos. Gliederung, Gefährdung, Erhaltung. *Willdenowia* **1979**, *9*, 295–312.
34. MEDDPN. *Analyse Cartographique de l'occupation Des Zones Agroécologiques et Bassins de Concentration Des Populations Au Togo, Folega F., Consultant Sous Ordre de La Coordination Nationale Sur Les Changements Climatiques*; MEDDPN: Lomé, Togo, 2019; p. 66.
35. Brabant, P.; Darracq, S.; Egué, K.; Simonneaux, V.; Aing, A.; Auberton-Habert, E. Togo: État de Dégradation Des Terres Résultant Des Activités Humaines (Note Explicative de La Carte Au 1: 500 000 Des Indices de Dégradation). In *Notice Explicative*; Éditions de I'ORSTOM: Paris, France, 1996; p. 66. ISBN 2-7099-1 348-8.
36. Atakpama, W.; Amegnaglo, K.; Afelu, B.; Folega, F.; Batawila, K.; Akpagana, K. Biodiversité et biomasse pyrophyte au Togo. *Vertigo* **2019**, *19*, 22. [CrossRef]
37. Alboabidallah, A.; Martin, J.; Lavender, S.; Abbott, V. Using Landsat-8 and Sentinel-1 Data for Above Ground Biomass Assessment in the Tamar Valley and Dartmoor. In Proceedings of the 2017 9th International Workshop on the Analysis of Multitemporal Remote Sensing Images (MultiTemp), Bruges, Belgium, 27–29 June 2017; IEEE: Piscataway Township, NJ, USA, 2017; pp. 1–7.
38. Wulder, M.A.; Loveland, T.R.; Roy, D.P.; Crawford, C.J.; Masek, J.G.; Woodcock, C.E.; Allen, R.G.; Anderson, M.C.; Belward, A.S.; Cohen, W.B.; et al. Current Status of Landsat Program, Science, and Applications. *Remote Sens. Environ.* **2019**, *225*, 127–147. [CrossRef]
39. Hu, Y.; Hu, Y. Land Cover Changes and Their Driving Mechanisms in Central Asia from 2001 to 2017 Supported by Google Earth Engine. *Remote Sens.* **2019**, *11*, 554. [CrossRef]
40. Xie, S.; Liu, L.; Zhang, X.; Yang, J.; Chen, X.; Gao, Y. Automatic Land-Cover Mapping Using Landsat Time-Series Data Based on Google Earth Engine. *Remote Sens.* **2019**, *11*, 3023. [CrossRef]
41. Griffiths, P.; van der Linden, S.; Kuemmerle, T.; Hostert, P. A Pixel-Based Landsat Compositing Algorithm for Large Area Land Cover Mapping. *IEEE J. Sel. Top. Appl. Earth Obs. Remote Sens.* **2013**, *6*, 2088–2101. [CrossRef]
42. Zhu, Z.; Woodcock, C.E. Continuous Change Detection and Classification of Land Cover Using All Available Landsat Data. *Remote Sens. Environ.* **2014**, *144*, 152–171. [CrossRef]
43. Hermosilla, T.; Wulder, M.A.; White, J.C.; Coops, N.C.; Hobart, G.W. Disturbance-Informed Annual Land Cover Classification Maps of Canada's Forested Ecosystems for a 29-Year Landsat Time Series. *Can. J. Remote Sens.* **2018**, *44*, 67–87. [CrossRef]
44. Giovos, R.; Tassopoulos, D.; Kalivas, D.; Lougkos, N.; Priovolou, A. Remote Sensing Vegetation Indices in Viticulture: A Critical Review. *Agriculture* **2021**, *11*, 457. [CrossRef]
45. Sozzi, M.; Kayad, A.; Marinello, F.; Taylor, J.; Tisseyre, B. Comparing Vineyard Imagery Acquired from Sentinel-2 and Unmanned Aerial Vehicle (UAV) Platform. *Oeno One* **2020**, *54*, 189–197. [CrossRef]
46. Li, Y.; Zhao, Z.; Xin, Y.; Xu, A.; Xie, S.; Yan, Y.; Wang, L. How Are Land-Use/Land-Cover Indices and Daytime and Nighttime Land Surface Temperatures Related in Eleven Urban Centres in Different Global Climatic Zones? *Land* **2022**, *11*, 1312. [CrossRef]
47. Khan, M.S.; Ullah, S.; Chen, L. Comparison on Land-Use/Land-Cover Indices in Explaining Land Surface Temperature Variations in the City of Beijing, China. *Land* **2021**, *10*, 1018. [CrossRef]
48. Xue, J.; Su, B. Significant Remote Sensing Vegetation Indices: A Review of Developments and Applications. *J. Sens.* **2017**, *2017*, 1353691. [CrossRef]
49. McFeeters, S.K. The Use of the Normalized Difference Water Index (NDWI) in the Delineation of Open Water Features. *Int. J. Remote Sens.* **1996**, *17*, 1425–1432. [CrossRef]
50. Vaudour, E.; Gomez, C.; Lagacherie, P.; Loiseau, T.; Baghdadi, N.; Urbina-Salazar, D.; Loubet, B.; Arrouays, D. Temporal Mosaicking Approaches of Sentinel-2 Images for Extending Topsoil Organic Carbon Content Mapping in Croplands. *Int. J. Appl. Earth Obs. Geoinf.* **2021**, *96*, 102277. [CrossRef]
51. Xu, N.; Tian, J.; Tian, Q.; Xu, K.; Tang, S. Analysis of Vegetation Red Edge with Different Illuminated/Shaded Canopy Proportions and to Construct Normalized Difference Canopy Shadow Index. *Remote Sens.* **2019**, *11*, 1192. [CrossRef]
52. Chen, Y.; Cao, R.; Chen, J.; Liu, L.; Matsushita, B. A Practical Approach to Reconstruct High-Quality Landsat NDVI Time-Series Data by Gap Filling and the Savitzky–Golay Filter. *ISPRS J. Photogramm. Remote Sens.* **2021**, *180*, 174–190. [CrossRef]
53. Huang, S.; Tang, L.; Hupy, J.P.; Wang, Y.; Shao, G. A Commentary Review on the Use of Normalized Difference Vegetation Index (NDVI) in the Era of Popular Remote Sensing. *J. For. Res.* **2021**, *32*, 1–6. [CrossRef]
54. Özelkan, E. Water Body Detection Analysis Using NDWI Indices Derived from Landsat-8 OLI. *Pol. J. Environ. Stud.* **2020**, *29*, 1759–1769. [CrossRef]
55. Rikimaru, A.; Roy, P.S.; Miyatake, S. Tropical Forest Cover Density Mapping. *Trop. Ecol.* **2002**, *43*, 39–47.
56. Zheng, Y.; Tang, L.; Wang, H. An Improved Approach for Monitoring Urban Built-up Areas by Combining NPP-VIIRS Nighttime Light, NDVI, NDWI, and NDBI. *J. Clean. Prod.* **2021**, *328*, 129488. [CrossRef]

57. Nyland, K.E.; Gunn, G.I.; Shiklomanov, N.N.; Engstrom, R.A.; Streletskiy, D. Land Cover Change in the Lower Yenisei River Using Dense Stacking of Landsat Imagery in Google Earth Engine. *Remote Sens.* **2018**, *10*, 1226. [CrossRef]
58. Tian, S.; Zhang, X.; Tian, J.; Sun, Q. Random Forest Classification of Wetland Landcovers from Multi-Sensor Data in the Arid Region of Xinjiang, China. *Remote Sens.* **2016**, *8*, 954. [CrossRef]
59. Wingate, V.R.; Phinn, S.R.; Kuhn, N.; Bloemertz, L.; Dhanjal-Adams, K.L. Mapping Decadal Land Cover Changes in the Woodlands of North Eastern Namibia from 1975 to 2014 Using the Landsat Satellite Archived Data. *Remote Sens.* **2016**, *8*, 681. [CrossRef]
60. Zurqani, H.A.; Post, C.J.; Mikhailova, E.A.; Schlautman, M.A.; Sharp, J.L. Geospatial Analysis of Land Use Change in the Savannah River Basin Using Google Earth Engine. *Int. J. Appl. Earth Obs. Geoinf.* **2018**, *69*, 175–185. [CrossRef]
61. Fernández-Delgado, M.; Cernadas, E.; Barro, S.; Amorim, D. Do We Need Hundreds of Classifiers to Solve Real World Classification Problems? *J. Mach. Learn. Res.* **2014**, *15*, 3133–3181.
62. Zhang, D.-D.; Zhang, L. Land Cover Change in the Central Region of the Lower Yangtze River Based on Landsat Imagery and the Google Earth Engine: A Case Study in Nanjing, China. *Sensors* **2020**, *20*, 2091. [CrossRef] [PubMed]
63. Probst, P.; Boulesteix, A.-L. To Tune or Not to Tune the Number of Trees in Random Forest. *J. Mach. Learn. Res.* **2017**, *18*, 6673–6690.
64. Oshiro, T.M.; Perez, P.S.; Baranauskas, J.A. How Many Trees in a Random Forest? In Proceedings of the Machine Learning and Data Mining in Pattern Recognition, New York, NY, USA, 30 August–3 September 2011; Perner, P., Ed.; Springer: Berlin/Heidelberg, Germany, 2012; pp. 154–168.
65. Probst, P.; Wright, M.N.; Boulesteix, A.-L. Hyperparameters and Tuning Strategies for Random Forest. *WIREs Data Min. Knowl. Discov.* **2019**, *9*, e1301. [CrossRef]
66. Guo, L.; Ma, Y.; Cukic, B.; Singh, H. Robust Prediction of Fault-Proneness by Random Forests. In Proceedings of the 15th International Symposium on Software Reliability Engineering, Saint-Malo, France, 2–5 November 2004; pp. 417–428.
67. Gislason, P.O.; Benediktsson, J.A.; Sveinsson, J.R. Random Forests for Land Cover Classification. *Pattern Recognit. Lett.* **2006**, *27*, 294–300. [CrossRef]
68. Bernard, S.; Adam, S.; Heutte, L. Dynamic Random Forests. *Pattern Recognit. Lett.* **2012**, *33*, 1580–1586. [CrossRef]
69. Kulkarni, A.; Lowe, B. Random Forest Algorithm for Land Cover Classification. *Comput. Sci. Fac. Publ. Present.* **2016**, *4*, 58–63.
70. Lind, A.P.; Anderson, P.C. Predicting Drug Activity against Cancer Cells by Random Forest Models Based on Minimal Genomic Information and Chemical Properties. *PLoS ONE* **2019**, *14*, e0219774. [CrossRef]
71. MERF. *Rapport Inventaire Forestier National Du Togo 2015–2016*; Ministère de l'Environnement et des Ressources Forestières (MERF): Lomé, Togo, 2016; p. 79.
72. Verma, P.; Raghubanshi, A.; Srivastava, P.K.; Raghubanshi, A.S. Appraisal of Kappa-Based Metrics and Disagreement Indices of Accuracy Assessment for Parametric and Nonparametric Techniques Used in LULC Classification and Change Detection. *Model. Earth Syst. Environ.* **2020**, *6*, 1045–1059. [CrossRef]
73. Petropoulos, G.P.; Kalivas, D.P.; Georgopoulou, I.A.; Srivastava, P.K. Urban Vegetation Cover Extraction from Hyperspectral Imagery and Geographic Information System Spatial Analysis Techniques: Case of Athens, Greece. *J. Appl. Remote Sens.* **2015**, *9*, 096088. [CrossRef]
74. Tang, H.; Hu, Z. Research on Medical Image Classification Based on Machine Learning. *IEEE Access* **2020**, *8*, 93145–93154. [CrossRef]
75. Congedo, L. Semi-Automatic Classification Plugin: A Python Tool for the Download and Processing of Remote Sensing Images in QGIS. *J. Open Source Softw.* **2021**, *6*, 3172. [CrossRef]
76. Puyravaud, J.-P. Standardizing the Calculation of the Annual Rate of Deforestation. *For. Ecol. Manag.* **2003**, *177*, 593–596. [CrossRef]
77. Lopes, M.; Frison, P.-L.; Crowson, M.; Warren-Thomas, E.; Hariyadi, B.; Kartika, W.D.; Agus, F.; Hamer, K.C.; Stringer, L.; Hill, J.K.; et al. Improving the Accuracy of Land Cover Classification in Cloud Persistent Areas Using Optical and Radar Satellite Image Time Series. *Methods Ecol. Evol.* **2020**, *11*, 532–541. [CrossRef]
78. Feyisa, G.L.; Meilby, H.; Fensholt, R.; Proud, S.R. Automated Water Extraction Index: A New Technique for Surface Water Mapping Using Landsat Imagery. *Remote Sens. Environ.* **2014**, *140*, 23–35. [CrossRef]
79. Wijedasa, L.S.; Sloan, S.; Michelakis, D.G.; Clements, G.R. Overcoming Limitations with Landsat Imagery for Mapping of Peat Swamp Forests in Sundaland. *Remote Sens.* **2012**, *4*, 2595–2618. [CrossRef]
80. Masiello, G.; Cersosimo, A.; Mastro, P.; Serio, C.; Venafra, S.; Pasquariello, P. Emissivity-Based Vegetation Indices to Monitor Deforestation and Forest Degradation in the Congo Basin Rainforest. In Proceedings of the Remote Sensing for Agriculture, Ecosystems, and Hydrology XXII, Online, 20 September 2020; SPIE: Bellingham, WA, USA, 2020; Volume 11528, pp. 125–138.
81. Dong, J.; Xiao, X.; Menarguez, M.A.; Zhang, G.; Qin, Y.; Thau, D.; Biradar, C.; Moore, B. Mapping Paddy Rice Planting Area in Northeastern Asia with Landsat 8 Images, Phenology-Based Algorithm and Google Earth Engine. *Remote Sens. Environ.* **2016**, *185*, 142–154. [CrossRef] [PubMed]

 *land*

Article

Innovative Fusion-Based Strategy for Crop Residue Modeling

Solmaz Fathololoumi [1], Mohammad Karimi Firozjaei [2] and Asim Biswas [1,*]

[1] School of Environmental Sciences, University of Guelph, Guelph, ON N1G 2W1, Canada
[2] Department of Remote Sensing and GIS, Faculty of Geography, University of Tehran, Tehran 1417853933, Iran
* Correspondence: biswas@uoguelph.ca; Tel.: +1-519-731-6252

Abstract: The purpose of this study was to present a new strategy based on fusion at the decision level for modeling the crop residue. To this end, a set of satellite imagery and field data, including the Residue Cover Fraction (RCF) of corn, wheat and soybean was used. Firstly, the efficiency of Random Forest Regression (RFR), Support Vector Regression (SVR), Artificial Neural Networks (ANN) and Partial-Least-Squares Regression (PLSR) in RCF modeling was evaluated. Furthermore, to increase the accuracy of RCF modeling, different algorithms results were combined based on their modeling error, which is called the decision-based fusion strategy. The R2 (RMSE) between the actual and modeled RCF based on ANN, RFR, SVR and PLSR algorithms for corn were 0.83 (3.89), 0.86 (3.25), 0.76 (4.56) and 0.75 (4.81%), respectively. These values were 0.81 (4.86), 0.85 (4.22), 0.78 (5.45) and 0.74 (6.20%) for wheat and 0.81 (3.96), 0.83 (3.38), 0.76 (5.01) and 0.72 (5.65%) for soybean, respectively. The error of corn, wheat and soybean RCF estimating decision-based fusion strategy was reduced by 0.90, 0.96 and 0.99%, respectively. The results showed that by implementing the decision-based fusion strategy, the accuracy of the RCF modeling was significantly improved.

Keywords: crop residue; fusion; machine learning algorithm; reflective and radar bands

Citation: Fathololoumi, S.; Karimi Firozjaei, M.; Biswas, A. Innovative Fusion-Based Strategy for Crop Residue Modeling. *Land* **2022**, *11*, 1638. https://doi.org/10.3390/land11101638

Academic Editors: Carmine Serio, Guido Masiello and Sara Venafra

Received: 22 August 2022
Accepted: 20 September 2022
Published: 23 September 2022

1. Introduction

Modern agricultural activities, such as plowing and using heavy machinery known as tillage, can damage soil health [1,2]. In this case, the soil is more easily leaching by rain and loses its top layer, which is crucial for crop growth. The leached soil will flow downstream into the rivers and pollute the water due to elements such as phosphorus [3]. On the other hand, with decreasing soil quality, precipitated carbon is released [4]. The release of carbon from the soil plays an important role in increasing the carbon dioxide in the Earth's atmosphere [5,6]. Gases are one of the parameters affecting climate change. Therefore, maintaining soil quality in the agricultural process is very important [7,8].

One of the possible, cheap and feasible ways to reduce the damage caused by wind and water erosion and increase water storage to soil productivity is to maintain the remaining vegetation on the soil surface of agricultural lands at harvest time [9–11]. Crop residues consist of various components of the crop, including leaves, seeds, stems, etc., after harvest on agricultural land [12,13]. The presence of these residues on the soil surface can strengthen soil organic matter, better the absorption of nutrients by the plant and increase the efficiency of chemical fertilizers. Crop residues also have a large effect on soil, crop and environmental factors, such as water permeability, evaporation, crop yield and erosion [7,14–16]. They can improve the physical, chemical and biological condition of the soil and ultimately lead to a healthier crop due to its desirable and nutritious composition. Preserving residues at the soil surface by preventing the emission of gases, such as NH_3, CO_2 and SO_2, can reduce air pollution, while burning plant residues emits these gases [17,18].

Due to the importance of preserving crop residues on the soil surface, modeling and mapping residues as an indicator of tillage intensity are of great importance in agricultural management and achieving sustainable agricultural goals, including maintaining environmental health [13,19]. Mapping crop residues for agricultural areas can be a criterion for evaluating the efficiency and quality of methods and tools used in harvesting. It is practically impossible to use traditional methods, such as field visits and sampling, to determine the amount of residues on a large scale and in a short time [12,20]. Utilizing the capabilities of remote sensing techniques and data can be useful in quantifying crop residues on a large spatial and temporal scales and higher accuracy [18,21]. Previous studies have used various satellite imagery to model the amount of residue cover fraction (RCF), including reflective multispectral imagery, such as Landsat; Sentinel 2 [16,22,23]; radar imagery, such as RADARSAT [13,24]; and reflective hyper Spectral imagery, such as Probe-1 [25]. Each of these types of images has advantages and disadvantages [13].

In previous studies, several remote sensing-based indices, such as Normalized Difference Senescent Vegetation Index (NDSVI) [26], Normalized Difference Residue Index (NDRI) [27], Normalized Difference Tillage Index (NDTI) [28], Shortwave Green Normalized Difference Index (SGNDI) [29], Shortwave Infrared Normalized Difference Residue Index (SINDRI) [30], Broadband spectral Angle Index (BAI) [8], Dead Fuel Index (DFI) [31], Normalized Difference Index (NDI) [32], Normalized Difference Vegetation Index (NDVI) [8], Simulated crop residue cover (MCRC) [29], Simple tillage index (STI) [28], Simulated cellulose absorption index (3BI1) [33], Simulated lignin Cellulose Absorption Index (3BI2) [33], Simulated NDRI (3BI3) [33] and Short-wave near-infrared Normalized Difference residue Index (SRNDI) [34], etc., have been proposed to identify and quantify the RCF [14,34]. In some studies, the efficiency of different spectral indices was compared [29,33]. The results showed that each of these indicators can have different performances, some of them are suitable for dry areas and some for wet areas. A number of indicators do not perform well in areas with high vegetation. Each of the developed indicators has advantages and disadvantages. Yue, Tian, Dong, Xu and Zhou [29] showed that single indices are not highly capable of modeling RCF in the complex surface conditions of agricultural areas. Hence, in some studies, multivariate regression based methods, such as Random Forest Regression (RFR), Support Vector Regression (SVR), Artificial Neural Networks (ANN), etc., were proposed to model RCF [33]. The researchers used experimental regression methods to examine the linear or nonlinear relationship between actual RCF and remote sensing-based indices related to RCF [7,14,33], the spectral angle [8] or spectral unmixing [35] used to estimate the amount of RCF. Raoufat, Dehghani, Abdolabbas, Kazemeini and Nazemossadat [9] utilized Landsat 8 and drone data for RCF mapping and found that Landsat 8 data was more accurate than drone data although the drone data had its own advantages. Yue and Tian [36] used the spectral and laboratory data for RCF mapping. They evaluated RS data and triangle technique using RFR method in their study. They concluded that their proposed method was very effective in the accurate modeling of RCF and decreased the negative effect of soil moisture on it. Wang et al. [37] used MODIS and ground data to quantify some crop-related indices in a large-scale area using ground data and building a linear regression relationship between them. They showed that their used method was successful in monitoring soil conditions, including soil erosion.

Although the research in RCF modeling is limited, summarizing previous studies show that several models have been developed over the years to estimate the RCF, each with its own advantages and disadvantages. Selecting the appropriate model to estimate the amount of RCF has a high impact on the modeling accuracy of this parameter. Therefore, providing an integrated model based on using the capabilities of different models and indicators in estimating RCF can be useful in improving the modeling accuracy of this parameter. Sentinel 1 satellite imagery is one of the well-known and widely used radar data in various agricultural applications. However, our knowledge shows that the capability of this radar image's bands in estimating the amount of RCF has not been evaluated. Therefore, evaluating the performance of the satellites and combining the capabilities of these bands with the indicators presented in previous studies to improve the accuracy of modeling RCF can be useful and crucial.

The purpose of this study was to present a new strategy based on fusion at the decision level for modeling the RCF. In this study, (1) the efficiency of spectral indices based on reflective multispectral images presented in previous studies in modeling RCF were compared. (2) The importance of using Sentinel 1 radar satellite imagery bands in improving the accuracy of RCF modeling was assessed. (3) The efficiency of RFR, SVR, ANN and Partial-Least-Squares Regression (PLSR) algorithms in modeling RCF was evaluated and compared. A new strategy was proposed to integrate the results obtained from different algorithms at the decision level to improve the accuracy of modeling RCF.

2. Study Area

Ontario is one of Canada's most important agricultural centers, with a wide variety of agricultural products and orchards. Major agricultural products in Ontario include corn, wheat and soybeans, barley, forage, oat, canola, etc. A region in southern of Ontario was selected as the study area. This region is located at 82.5 degrees West and 42.5 degree North. The agricultural products of this area include the three main crops of corn, wheat and soybeans (Figure 1). The area of the study region is about 4324 km^2 and the area under corn, wheat and soybean cultivation in this area is 1218, 284 and 1087 km^2, respectively. The study area in terms of climatic division, has a warm summer humid continental based on the Köppen climate classification. The average annual temperature and rainfall are 17.0 °C and 424 mm, respectively. Therefore, due to the weather conditions and long winters in this area, most agricultural products are harvested by September. Criteria for selecting the study area were (1) high diversity of agricultural products, especially important crops in Canada; (2) proximity of the harvest time of the three types of agricultural products available in the region and the different farms available for each product; (3) access to the suitable cloud-free satellite image corresponding to the harvest date of all three crops for that area; and (4) easy access of the area in order to conduct field studies.

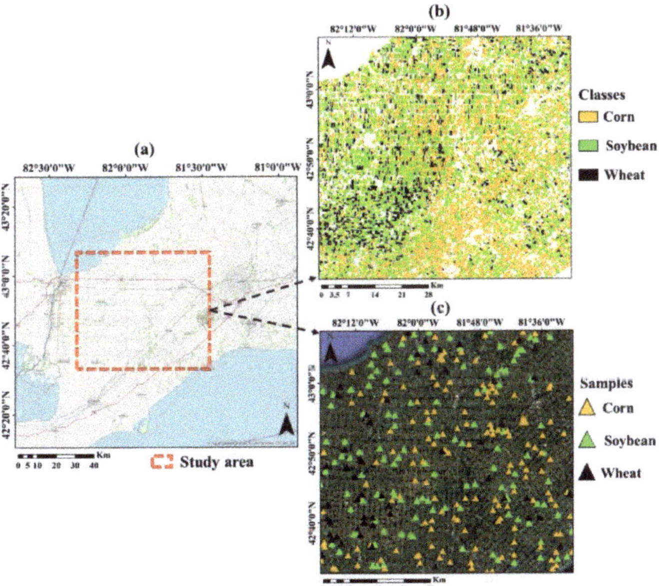

Figure 1. Maps of (**a**) the geographical location of the study area, (**b**) the land crop classes of the study area and (**c**) the geographical location and spatial distribution of RCF samples of different land crops.

3. Data and Methods

3.1. Data

A set of satellite images and ground data were used to evaluate different models and data in RCF modeling. Satellite imagery included the Landsat 8 Collection 2 Surface Reflectance image for 13 October 2020 and the Sentinel 1 Ground Range Detected (GRD) image for 7 October 2020. The spatial resolution of the Landsat 8 and Sentinel 1 image bands is 30 and 10 m, respectively. The Landsat 8 bands, including Blue, Green, Red, NIR, SWIR 1 and SWIR 2, and Sentinel 1 bands, including VV and VH, were used in this study. Some criteria were considered when selecting the date of the images: (1) a lack of cloud cover in the area during the Landsat 8 overpass time, (2) the absence of precipitation in the study area a few days before the satellite overpass time and (3) proximity to the date of harvest of agricultural products. Landsat 8 image downloaded from https://earthexplorer.usgs.gov/ (accessed on 13 February 2021) and sentinel 1 downloaded from https://scihub.copernicus.eu/ (accessed on 16 February 2021). The land crop map prepared by Agriculture and Agri-Food Canada (AAFC) in 2020 with the spatial resolution of 30 m was used to mask various agricultural products. This data is produced annually and can be downloaded from https://open.canada.ca/data/en/dataset?q (accessed on 2 March 2021) website.

Ground data collection includes determining the RCF in autumn from a number of corn, soybean and wheat fields that were performed after harvest on the 9 October 2020, 10 October 2020 and 9 November 2020 dates. For this purpose, RCF values were determined at 57 land points for wheat, 149 points for corn and 128 points for soybeans. The selection of land areas was carried out in such a way that there was a suitable distribution in the fields of these crops in the whole study area (Figure 1b). The minimum distance between sampling points was 500 m. Additionally, based on the initial and complete field visit of the study area, the suitable distribution of absolute values between the highest and lowest actual values of RCF in the study area was also considered when selecting the sampling points. Little to no rain (less than 0.25 cm) had fallen the week before the sampling, and the soil moisture levels were constantly dry. When field sampling took place, the weather was identical to the weather when taking images the day before. Ordinary camera images were used for sampling. A Phantom 3 SE drone was used to produce images from each selected sampling point. The flight height of the drone for imaging was 20 m, and a digital orthophoto with a spatial resolution of 20 cm was prepared for each sampling point. Then, for each image, the position of product residues was manually digitized. After processing the images taken by the camera, the remaining coverage fraction was calculated for an area of 900 m^2 around each sampling point. Agisoft Metashape 1.8.3, developed by Agisoft LLC (St. Petersburg, Russia) and ArcMap 10.6.1, developed by Esri (Environmental Systems Research Institute), (Redlands, California, United States) software were used to determine the RCF based on the images prepared with the drone (digital orthophoto preparation and digitization).

3.2. *Methods*

Firstly, a map of common spectral indices in RCF modeling was prepared based on the Landsat 8 and Sentinel 1 VV and VH spectral bands. Then, these maps were masked to areas with NDVI < 0.3 to limit analysis to areas without significant green vegetative ground cover [16] and was also masked to agricultural crop fields using the AAFC map. Then, the efficiency of each of these indicators in modeling the RCF was evaluated for each agricultural product. Secondly, the efficiency of different algorithms, including RFR, SVM, ANN and PLSR in modeling RCF, was evaluated and they were compared with each other. Furthermore, different algorithms' results were combined based on the modeling error in order to increase the accuracy of receipt modeling. This strategy used the fusion capability at the decision level to improve the accuracy of RCF modeling.

3.2.1. Effective Variables

In previous studies, various spectral indices, including MCRC, SGNDI, DFI, STI, NDSVI, NDTI, NDI5, NDI7, NDVI, 3BI1, 3BI2, 3BI3, BAI and SRNDI, have been developed for RCF modeling [16,25,30,32–34]. Reflective band information was used to calculate these indices. In this study, in addition to spectral indices based on reflective band information, backscatter data obtained from VV and VH bands of Sentinel 1 were also used in RCF modeling. The details of spectral indices and radar bands used as independent and primary variables in RCF modeling are shown in Table 1. To compare the efficiency of each of these independent variables in RCF modeling, their correlation coefficient with the values of RCF at the validation sampling points was calculated.

Table 1. Spectral indices and radar bands used as independent and primary variables in RCF modeling.

Type	Spectral Indices	Equation	Reference
Reflective-based index	MCRC	(OLI6 – OLI3)/(OLI6 + OLI3)	[29]
	SGNDI	(OLI3 – OLI7)/(OLI3 + OLI7)	[29]
	DFI	100 × (1 – OLI7/OLI6) × (OLI4/OLI5)	[31]
	STI	OLI6/OLI7	[28]
	NDSVI	(OLI6 – OLI4)/(OLI6 + OLI4)	[26]
	NDTI	(OLI6 – OLI7)/(OLI6 + OLI7)	[28]
	NDI5	(OLI5 – OLI6)/(OLI5 + OLI6)	[32]
	NDI7	(OLI5 – OLI7)/(OLI5 + OLI7)	[32]
	SRNDI	(OLI7 – OLI4)/(OLI7 + OLI4)	[34]
	NDVI	(OLI5 – OLI4)/(OLI5 + OLI4)	[33]
	3BI1	100 × (0.5 × (OLI2 + OLI7) – OLI4)	[33]
	3BI2	(OLI2 – OLI4)/(OLI2 – OLI7)	[33]
	3BI3	(OLI7 – OLI4)/(OLI7 + OLI6)	[33]
	BAI	-	[8]
Backscatter bands	VV	-	-
	VH	-	-

3.2.2. Machine Learning Methods

Multivariate modeling based on four common algorithms in agricultural and environmental modeling, including RFR, SVM, ANN and PLSR, was used to model the RCF. In the first scenario, all reflective band-based spectral indices were used as independent variables. In the second scenario, in addition to spectral indices based on reflective bands, VV and VH band information was also used in the modeling process. Each RFR, SVM, ANN and PLSR algorithm was calibrated based on training data (96 samples for corn, 38 samples for wheat and 84 samples for soybean). Then the efficiency of each of these algorithms in estimating the RCF fraction using test data (53 samples for corn, 20 samples for wheat and 43 samples for soybean) was evaluated.

PLSR breaks down both dependent and independent variables into a number of major components. PLSR is a two-line calibration algorithm that converts a large number of correlated linear variables into several non-correlated variables based on data compression. Hence, this algorithm can solve the challenges of high correlation between variables and overfitting in the modeling process [38].

In recent years, artificial neural networks have been widely used to estimate various environmental variables based on satellite data [39,40]. In this study, a back-propagation ANN was used to model the residue. This algorithm consists of input, hidden and output layers. Sigmoid and linear functions were used for activation in hidden and output nodes, respectively. To calibrate synaptic coefficients, the Levenberg–Marquardt minimization algorithm was used [41]. The number of nodes in the hidden layers varied between 4 to 8. To optimize the structural parameters of the ANN algorithm for the network, we changed the momentum coefficient and learning rate from 0.1 to 1.0 with a step of 0.05. The number of nodes in the hidden layer varied from 3 to 7. Mean squared error was used as a measure of the performance threshold and the determination of a network with optimal structure in receipt fraction modeling. The optimal network was selected in terms of mean absolute error between validation and predictions data.

The SVR model is a widely used algorithm for solving nonlinear problems [42]. In the SVR method, n-dimensional input variables are transferred to the new feature space with higher dimensions using the core functions and, as a result, optimal separator super planes are developed [43]. In this study, different Gaussian, linear, nonlinear quadratic, cubic, etc., kernels were evaluated, and finally the Gaussian kernel was selected and used as a function in the receipt fraction estimation process. The optimal values of box constraint, kernel scale and epsilon were set to 909, 857 and 0.04, respectively.

RFR is an ensemble-learning algorithm that combines a large set of decision trees to improve the accuracy of estimating a variable [44]. RFR has several advantages in modeling environmental variables, including (1) low sensitivity to noise and over-fitting, and (2) the use of a large number of quantitative and qualitative variables in the modeling process [45,46]. To implement this algorithm, two parameters, the number of trees and the number of attributes, must be set. The number of trees varied from 30 to 300 with step size 30 and the number of trees 150 was selected as the optimal value.

The optimal model for each of these four algorithms was selected to estimate the RCF and the mean absolute error between the validation data and the predictions.

3.2.3. Decision-Based Fusion Approach

To reduce the error of modeled RCF based on remote sensing data due to the weaknesses of different algorithms, in the proposed strategy, the results of four RFR, SVM, ANN and PLSR algorithms were combined based on Equation (1).

$$\mathrm{RCF_f} = \sum_{i=1}^{n} \mathrm{W_i RCF_{model(i)}} \quad (1)$$

In Equation (1), $\mathrm{RCF_f}$ is the modeled RCF based on the remote sensing data by combining the results of different algorithms, $\mathrm{RCF_{model(i)}}$ is the fraction of the modeled RCF based on the remote sensing data obtained from the ith algorithm, Wi is the degree of importance of the ith algorithm and n is the number of used algorithms. Equation (2) is used to calculate the significance of the ith algorithm.

$$\mathrm{W_i} = \frac{\mathrm{RMSE_{model(i)}}}{\sum_{i=1}^{n} \mathrm{RMSE_{model(i)}}} \quad (2)$$

In Equation (2), $\mathrm{RMSE_{model(i)}}$ is the mean squares root of the of the estimated fraction and is based on the ith algorithm. The lower the RMSE of an algorithm in estimating the RCF, the greater its impact and importance in the result of the RCF estimation. MATLAB 2019a software was used to implement various indices and algorithms for RCF modeling.

4. Results

The mean (sd) of the RCF for the calibration data of corn, wheat and soybean were 17.8 (9.6), 20.1 (11.2) and 19.2 (7.7)%, respectively (Figure 2). These values were 17.4 (9.4), 18.2 (10.3) and 18.6 (8.5)% for validation data, respectively. The RCF means and sd values based on the calibration and validation datasets were close to each other. The highest frequency of calibration data for corn, wheat and soybean crops was in the 10.9–18.3, 10.5–17.3 and 17.4–23.4% categories, respectively. For validation data, the highest frequency for these products was in the 12.2–18.4, 11.1–16.6 and 18.8–25.0% categories, respectively.

The efficiency of spectral indices based on reflective bands in RCF modeling was different. The efficiency of NDI5, NDI7, NDTI, NDVI, STI, DFI and BAI was higher than other indices, such as 3BI1, 3BI2, MCRC, NDSVI, SGNDI, and SRNDI (Table 2). The R2 between BAI and corn, wheat and soybean residues were 0.63, 0.66 and 0.61, respectively, which was higher than other spectral indices. The R^2 between the VV (VH) bands and the corn, wheat and soybean residues were 0.25 (0.29), 0.28 (0.36) and 0.20 (0.25), respectively. The efficiency of radar bands in RCF modeling was less than the spectral indices.

In the first scenario (dataset including reflective band-based spectral indices), based on calibration data, the R^2 (RMSE) between the actual and modeled RCF using ANN, RFR, SVR and PLSR algorithms for corn crop were 0.86 (3.13), 0.91 (2.63), 0.82 (3.92) and 0.79 (4.22%), respectively (Figure 3). These values were 0.84 (4.22), 0.88 (3.72), 0.81 (5.25) and 0.77 (5.85%) for wheat and 0.85 (3.14), 0.89 (2.73), 0.79 (3.81) and 0.75 (4.04%), respectively, for soybean. The efficiency of RCF modeling using different machine learning algorithms based on spectral indices was different. The RFR and PLSR algorithms had the highest and lowest accuracy in forming an optimal network for RCF modeling, respectively.

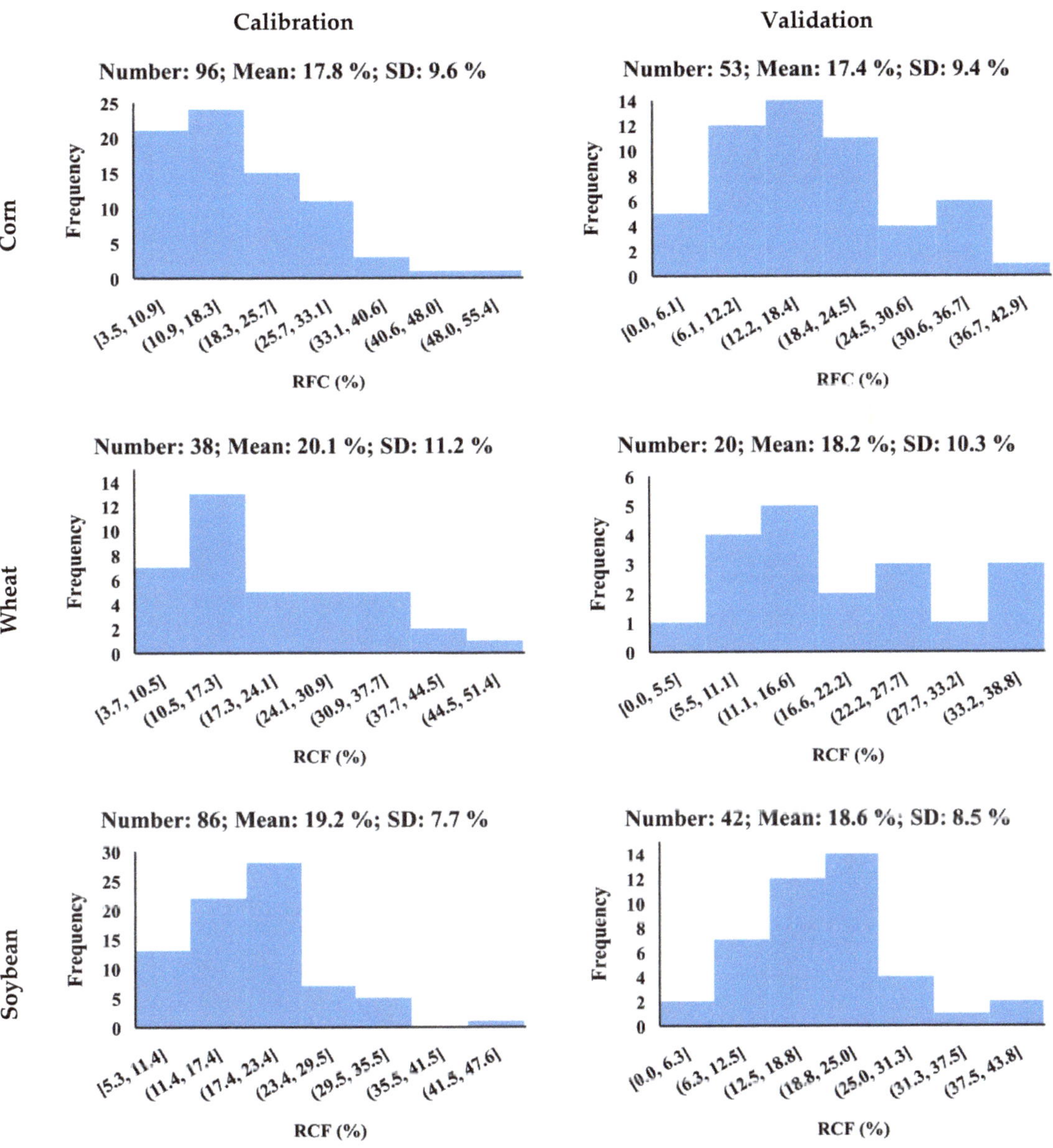

Figure 2. Frequency distribution of RCF values of corn, wheat and soybean crops for calibration and validation data in different classes.

Table 2. The R^2 between the effective variables and the residues for different crops.

	Corn	Wheat	Soybean
3BI1	0.09	0.08	0.11
3BI2	0.15	0.10	0.33
3BI3	0.46	0.35	0.56
MCRC	0.13	0.38	0.13
NDI5	0.54	0.48	0.41
NDI7	0.60	0.51	0.45
NDSVI	0.07	0.11	0.11
NDTI	0.42	0.48	0.61
NDVI	0.43	0.51	0.22
SGNDI	0.20	0.08	0.22
SRNDI	0.24	0.15	0.21
STI	0.43	0.50	0.60
DFI	0.55	0.59	0.52
BAI	0.63	0.66	0.61
VV	0.25	0.28	0.20
VH	0.29	0.36	0.25

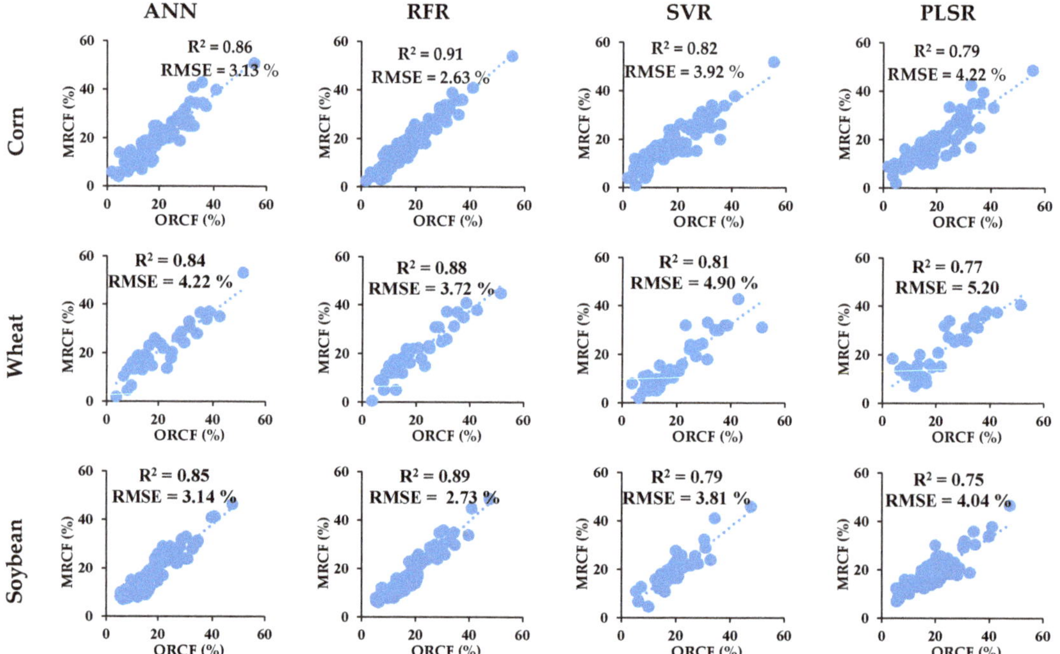

Figure 3. R^2 and RMSE between real and modeled RCF based on calibration data.

For the validation data, the R^2 (RMSE) between the actual and modeled RCF based on ANN, RFR, SVR and PLSR algorithms for corn were 0.83 (3.89), 0.86 (3.25), 0.76 (4.56) and 0.75 (4.81%) (Figure 4). These values were 0.81 (4.86), 0.85 (4.22), 0.78 (5.45) and 0.74 (6.20%) for wheat and 0.81 (3.96), 0.83 (3.38), 0.76 (5.01) and 0.72 (5.65%), respectively, for soybean. The results showed that the RFR algorithm had the highest accuracy in RCF modeling. The efficiency of this algorithm in corn RCF modeling was higher than soybean and wheat.

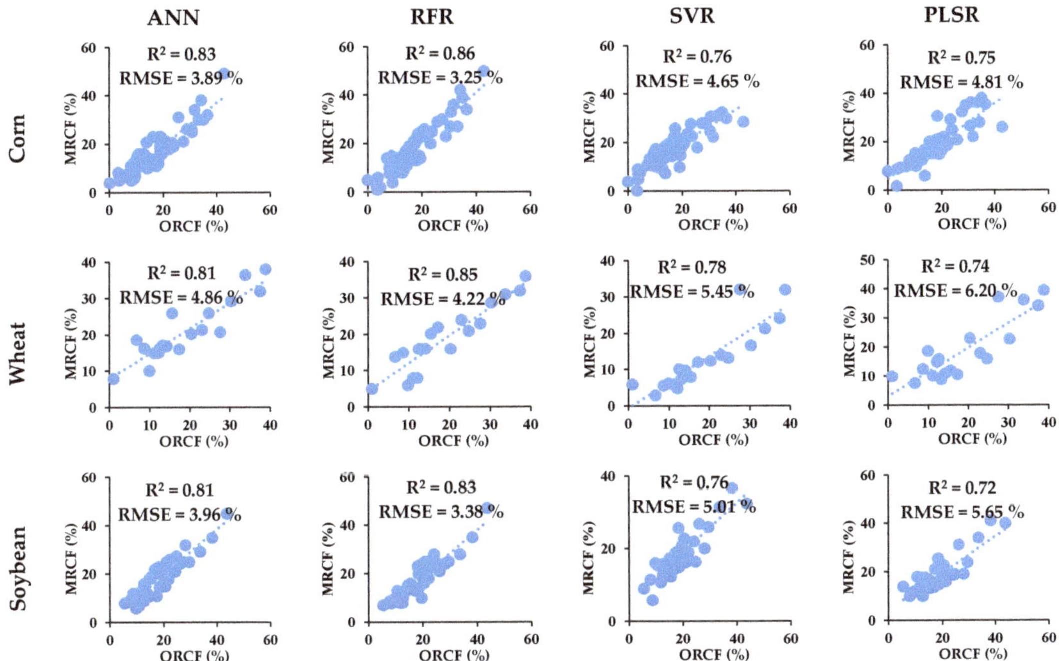

Figure 4. R^2 and RMSE between actual and model RCF based on validation data.

The addition of radar bands to the spectral indices dataset in the corn RCF modeling, caused an increase in the accuracy of RCF estimation using machine learning algorithms (Table 3). Considering the radar bands, the RMSE of corn RCF modeling using ANN, RFR, SVR and PLSR decreased by 0.44, 0.57, 0.54 and 0.30%, respectively. The reduction rates of RMSE for wheat (soybean) were 0.71 (0.37), 0.61 (0.49), 0.55 (0.51) and 0.38 (0.64), respectively.

Table 3. R^2 (RMSE) between the actual and modeled values of the RCF based on different machine learning algorithms, considering spectral indices and radar bands as dependent variables.

	Corn	Wheat	Soybean
ANN	0.85 (3.45)	0.84 (4.15)	0.85 (3.59)
RFR	0.89 (2.68)	0.87 (3.61)	0.89 (2.89)
SVR	0.80 (4.02)	0.80 (4.90)	0.80 (4.50)
PLSR	0.77 (4.51)	0.77 (5.58)	0.75 (5.01)

The R2 between the actual and modeled RCF based on the fusion strategy at the decision level for corn, wheat and soybean crops was 0.92, 0.89 and 0.88, respectively (Figure 5). RMSE values were 1.78, 2.65 and 1.90%, respectively. The error of estimating the RCF of corn, wheat and soybean products based on the proposed strategy was reduced by 0.90, 0.96 and 0.99%, respectively, compared to the results of the best machine learning algorithm.

The RCF map of corn, wheat and soybean crops prepared based on the fusion strategy at the decision level showed that the spatial distribution of the residue varied across the study area (Figure 6). The RCF of three crops varied between 0 and 62%. The RCF of corn on farms located in the eastern parts of the study area was less than the western part. Corn fields located in the northwestern parts of the study area had the highest values of residue. The lowest values of soybean RCF were in farms located in the central parts of the study area. The number of wheat fields in the study area was less than corn and soybean fields. The number of wheat fields with low RCF was lower than wheat fields with high RCF.

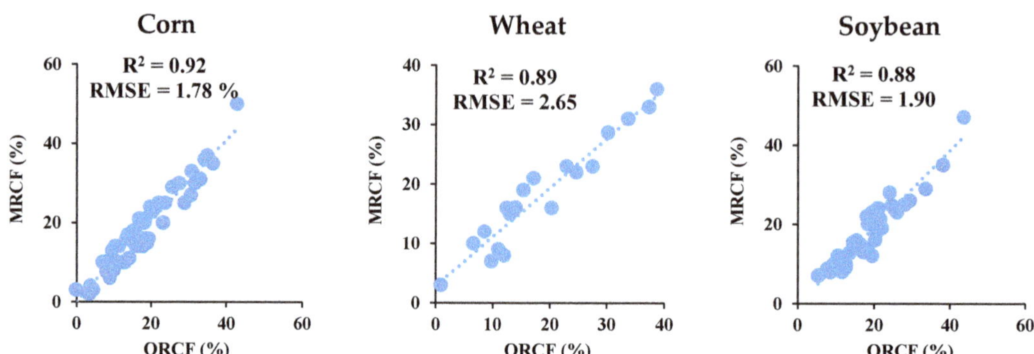

Figure 5. R^2 and RMSE between the actual and modeled RCF based on the fusion strategy at the decision level for corn, wheat and soybean crops. ORCF: observed residual cover fraction; MRCF: modeled residual cover fraction.

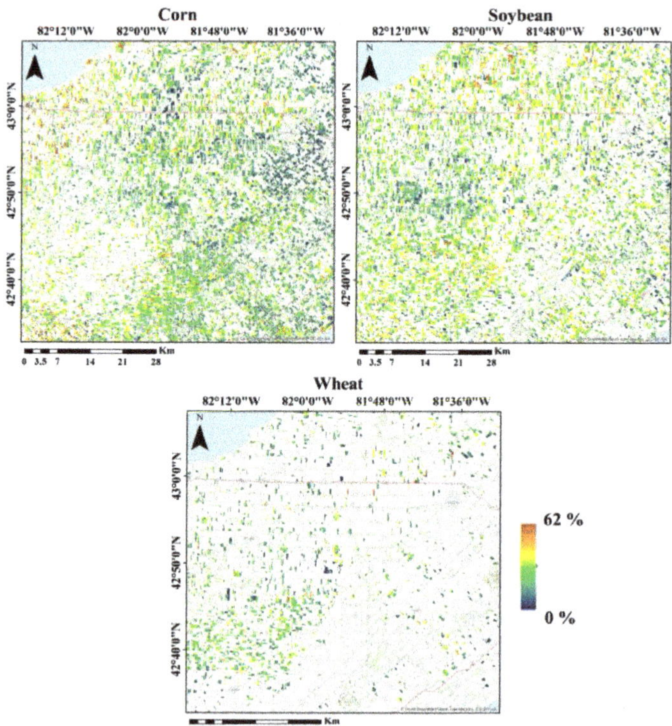

Figure 6. RCF maps for different land crop in the study area.

The mean RCF for corn, wheat and soybean crops in the study area were 18.2%, 19.39% and 17.7%, respectively (Figure 7). RCF was higher in wheat fields than in corn and soybean fields. The values of the standard deviation (Sd) of the RCF for corn, wheat and soybean fields in the study area were 8.3%, 10.23% and 7.4%, respectively. The highest and lowest variation of RCF in this study area was related to wheat and soybean crops. The range of variation in the RCF amount for corn fields as greater than wheat and corn crops.

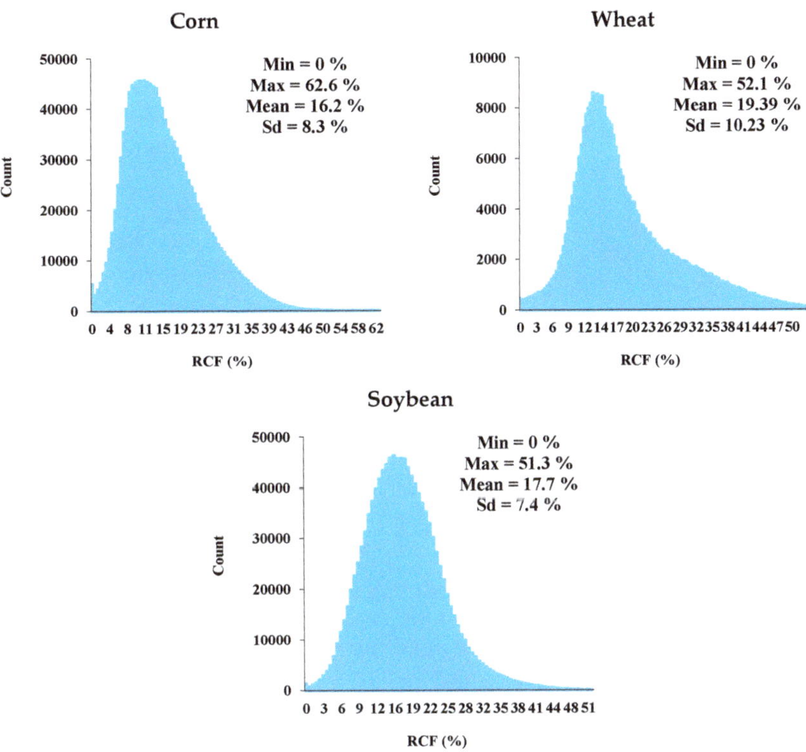

Figure 7. Frequency histogram and statistical parameters of the modeled RCF values of corn, wheat and soybean crops in the study area based on the proposed strategy.

5. Discussion

An accurate RCF map is crucial in agricultural planning and management [19,22,33]. Unlike terrestrial methods, satellite images have high application and high efficiency in preparing these maps due to their extensive spatiotemporal coverage and their low cost [8,29,30,34]. However, the accuracy of different crops' RCF mapping using satellite data is dependent on various factors, including (1) dependent variables used in the modeling process, (2) the quality of calibration and validation data and (3) the algorithms used to construct the appropriate model between the effective variables and the RCF [8,29,33]. In this study, the effect of dependent variables and algorithms used in the RCF modeling process was investigated.

In previous studies, various spectral indices based on satellite images have been provided to prepare this map [29,33]. In a number of these studies, the efficiencies of these spectral indices in RCF modeling were evaluated and compared [8,33]. The results showed a different performance of each of these indicators in different conditions. In this study, it was also shown that the efficiency of a number of indices, such as NDI5, NDI7, NDTI, NDVI, STI, DFI and BAI, was higher than other spectral indices when preparing the RCF map. Even the efficiency of each spectral index in preparing a RCF map for different crops is different. The results showed that for modeling the RCF of corn, wheat and soybean crops, BAI efficiency was higher than other indicators. The BAI is designed to minimize the effect of background soil moisture in the process of RCF estimating. Yue, Tian, Dong and Xu [8] showed that BAI had a high efficiency in estimating the RCF of wheat and maize products by reducing the effect of field soil moisture. In general, the spectral behavior of different crops at different wavelengths and their differences with the background soil and vegetation in the study area can affect the accuracy of the developed spectral indices. NDVI preparing the receipt map and agricultural products are highly sensitive to vegetation in the study area. As a result, for areas with high vegetation cover, such as forests and pastures, the RCF values are overestimated.

The focus of previous studies has been on the development of optimal indicators and methods in RCF modeling and on the use of reflective bands of satellite images, including Landsat, Sentinel 2, ASTER, etc. [16,30,33,34]. In a limited number of studies, the efficiency of the information obtained from radar images, including RADARSAT, in RCF modeling was evaluated [24]. However, in this study, for the first time, the efficiency of Sentinel 1 VV and VH bands in modeling this parameter was evaluated and compared with the efficiency of spectral indices. The results showed that considering spectral indices and VV and VH radar bands simultaneously as effective variables increases the accuracy of RCF modeling based on machine learning algorithms. Due to the simple and free access to Sentinel 1 images with a high frequency for different agricultural regions around the world, the use of these images in RCF modeling can be of great practical importance. In general, the accuracy of multivariate RCF modeling is higher than univariate modeling using machine learning algorithms. The results of this study showed that the efficiencies of machine learning algorithms in RCF modeling were different to one another. The efficiency of RFR algorithm in modeling this parameter was higher than ANN, SVR and PLSR. Ding, Zhang, Wang, Xie, Wang, Liu and Hall [33] also showed that RFR was highly efficient in RCF mapping.

Each of these algorithms may have high or low performance under different conditions. Therefore, providing an integrated model based on the results of these algorithms can be useful. In various fields, such as land cover classification, improving the spatial resolution of land surface temperature, etc., the strategy of combining the results of different algorithms called fusion at the decision level has been used to improve the modeling accuracy of target variable. In this study, the results of the RCF estimation obtained from different machine learning algorithms were combined based on the degree of importance of each algorithm to improve the modeling accuracy of this variable. The results showed that by implementing the fusion strategy at the decision level, the accuracy of the RCF map was significantly improved.

6. Conclusions

In this study, to improve the accuracy of RCF modeling and mapping, a new strategy based on the fusion of different machine learning algorithms' results at the decision level was developed. The results showed that by considering both spectral indices based on reflective bands and radar bands as dependent variables in machine learning algorithms, the RCF modeling error is reduced by an average of 15%. Among the various machine learning algorithms in RCF modeling, RF accuracy is higher than other algorithms, including ANN, SVR and PLSR. The results of the proposed strategy showed that the integration of the capabilities of different machine learning algorithms increases the accuracy of RCF modeling. With the fusion of the results of different machine learning algorithms at the decision level, the accuracy of RCF modeling for corn, wheat and soybean crops compared to the most optimal algorithm has increased by more than 33, 25 and 34%, respectively. It is suggested that in future studies, the efficiency of deep learning algorithms in RCF modeling be evaluated. It could also be very useful to use the proposed algorithm to prepare a more accurate RCF map in agricultural areas around the world and implement optimal programs to improve the agricultural situation and conserve soil and environmental quality.

Author Contributions: Conceptualization, A.B. and S.F.; methodology, S.F., M.K.F. and A.B.; software, S.F. and M.K.F.; validation, S.F. and M.K.F.; formal analysis, S.F. and M.K.F.; data curation, S.F., M.K.F. and A.B.; writing—original draft preparation, S.F. and M.K.F.; writing—review and editing, A.B.; supervision, A.B.; project administration, A.B.; funding acquisition, A.B. All authors have read and agreed to the published version of the manuscript.

Funding: This project is supported by the Canada First Research Excellence Fund (CFREF)–Food from Thought project at the University of Guelph and Natural Sciences and Engineering Research Council of Canada (NSERC) (RGPIN-2014-4100). Funding for open access charge: is from Natural Sciences and Engineering Research Council of Canada (NSERC) (RGPIN-2014-4100).

Data Availability Statement: The data used to support the findings of this study are available from the corresponding author upon reasonable request.

Conflicts of Interest: The authors declare no conflict of interest.

References

1. Birkás, M.; Jolánkai, M.; Gyuricza, C.; Percze, A. Tillage effects on compaction, earthworms and other soil quality indicators in Hungary. *Soil Tillage Res.* **2004**, *78*, 185–196. [CrossRef]
2. Reicosky, D.; Allmaras, R. Advances in tillage research in North American cropping systems. *J. Crop Prod.* **2003**, *8*, 75–125. [CrossRef]
3. Yuce, G.; Pinarbasi, A.; Ozcelik, S.; Ugurluoglu, D. Soil and water pollution derived from anthropogenic activities in the Porsuk River Basin, Turkey. *Environ. Geol.* **2006**, *49*, 359–375. [CrossRef]
4. Jiang, Z.; Bian, H.; Xu, L.; Li, M.; He, N. Pulse effect of precipitation: Spatial patterns and mechanisms of soil carbon emissions. *Front. Ecol. Evol.* **2021**, *9*, 302. [CrossRef]
5. Isson, T.T.; Planavsky, N.J.; Coogan, L.; Stewart, E.; Ague, J.; Bolton, E.; Zhang, S.; McKenzie, N.; Kump, L. Evolution of the global carbon cycle and climate regulation on earth. *Glob. Biogeochem. Cycles* **2020**, *34*, e2018GB006061. [CrossRef]
6. Lal, R.; Kimble, J.; Follett, R. Pedospheric processes and the carbon cycle. In *Soil Processes and the Carbon Cycle*; CRC Press: Boca Raton, FL, USA, 2018; pp. 1–8.
7. Hively, W.D.; Lamb, B.T.; Daughtry, C.S.; Shermeyer, J.; McCarty, G.W.; Quemada, M. Mapping crop residue and tillage intensity using WorldView-3 satellite shortwave infrared residue indices. *Remote Sens.* **2018**, *10*, 1657. [CrossRef]
8. Yue, J.; Tian, Q.; Dong, X.; Xu, N. Using broadband crop residue angle index to estimate the fractional cover of vegetation, crop residue, and bare soil in cropland systems. *Remote Sens. Environ.* **2020**, *237*, 111538. [CrossRef]
9. Raoufat, M.H.; Dehghani, M.; Abdolabbas, J.; Kazemeini, S.A.; Nazemossadat, M.J. Feasibility of satellite and drone images for monitoring soil residue cover. *J. Saudi Soc. Agric. Sci.* **2020**, *19*, 56–64.
10. Page, K.L.; Dang, Y.P.; Dalal, R.C.; Reeves, S.; Thomas, G.; Wang, W.; Thompson, J.P. Changes in soil water storage with no-tillage and crop residue retention on a Vertisol: Impact on productivity and profitability over a 50 year period. *Soil Tillage Res.* **2019**, *194*, 104319. [CrossRef]
11. Alberts, E.E.; Neibling, W.H. Influence of crop residues on water erosion. *Manag. Agric. Residues* **1994**, *13*, 19–44.
12. Cai, W.; Zhao, S.; Wang, Y.; Peng, F.; Heo, J.; Duan, Z. Estimation of winter wheat residue coverage using optical and SAR remote sensing images. *Remote Sens.* **2019**, *11*, 1163. [CrossRef]
13. Zheng, B.; Campbell, J.B.; Serbin, G.; Galbraith, J.M. Remote sensing of crop residue and tillage practices: Present capabilities and future prospects. *Soil Tillage Res.* **2014**, *138*, 26–34. [CrossRef]
14. Najafi, P.; Navid, H.; Feizizadeh, B.; Eskandari, I. Remote sensing for crop residue cover recognition: A review. *Agric. Eng. Int. CIGR J.* **2018**, *20*, 63–69.
15. Dvorakova, K.; Shi, P.; Limbourg, Q.; van Wesemael, B. Soil organic carbon mapping from remote sensing: The effect of crop residues. *Remote Sens.* **2020**, *12*, 1913. [CrossRef]
16. Hively, W.D.; Shermeyer, J.; Lamb, B.T.; Daughtry, C.T.; Quemada, M.; Keppler, J. Mapping crop residue by combining Landsat and WorldView-3 satellite imagery. *Remote Sens.* **2019**, *11*, 1857. [CrossRef]
17. Bégué, A.; Arvor, D.; Bellon, B.; Betbeder, J.; De Abelleyra, D.; Ferraz, R.P.D.; Lebourgeois, V.; Lelong, C.; Simões, M.; Verón, S.R. Remote sensing and cropping practices: A review. *Remote Sens.* **2018**, *10*, 99. [CrossRef]
18. Sonmez, N.K.; Slater, B. Measuring intensity of tillage and plant residue cover using remote sensing. *Eur. J. Remote Sens.* **2016**, *49*, 121–135. [CrossRef]
19. Daughtry, C.S.; Hunt, E.; Doraiswamy, P.; McMurtrey, J. Remote sensing the spatial distribution of crop residues. *Agron. J.* **2005**, *97*, 864–871. [CrossRef]
20. Zhuang, Y.; Chen, D.; Li, R.; Chen, Z.; Cai, J.; He, B.; Gao, B.; Cheng, N.; Huang, Y. Understanding the influence of crop residue burning on PM2. 5 and PM10 concentrations in China from 2013 to 2017 using MODIS data. *Int. J. Environ. Res. Public Health* **2018**, *15*, 1504. [CrossRef]
21. Quemada, M.; Hively, W.; Daughtry, C.; Lamb, B.; Shermeyer, J. Improved crop residue cover estimates obtained by coupling spectral indices for residue and moisture. *Remote Sens. Environ.* **2018**, *206*, 33–44. [CrossRef]
22. Zheng, B.; Campbell, J.B.; de Beurs, K.M. Remote sensing of crop residue cover using multi-temporal Landsat imagery. *Remote Sens. Environ.* **2012**, *117*, 177–183. [CrossRef]
23. Yue, J.; Fu, Y.; Guo, W.; Feng, H.; Qiao, H. Estimating fractional coverage of crop, crop residue, and bare soil using shortwave infrared angle index and Sentinel-2 MSI. *Int. J. Remote Sens.* **2022**, *43*, 1253–1273. [CrossRef]
24. McNairn, H.; Wood, D.; Gwyn, Q.; Brown, R.; Charbonneau, F. Mapping tillage and crop residue management practices with RADARSAT. *Can. J. Remote Sens.* **1998**, *24*, 28–35. [CrossRef]

25. Bannari, A.; Pacheco, A.; Staenz, K.; McNairn, H.; Omari, K. Estimating and mapping crop residues cover on agricultural lands using hyperspectral and IKONOS data. *Remote Sens. Environ.* **2006**, *104*, 447–459. [CrossRef]
26. Qi, J.; Marsett, R.; Heilman, P.; Bieden-bender, S.; Moran, S.; Goodrich, D.; Weltz, M. RANGES improves satellite-based information and land cover assessments in southwest United States. *Eos Transact. Am. Geophys. Union* **2002**, *83*, 601–606. [CrossRef]
27. Gelder, B.; Kaleita, A.; Cruse, R. Estimating mean field residue cover on midwestern soils using satellite imagery. *Agron. J.* **2009**, *101*, 635–643. [CrossRef]
28. Van Deventer, A.; Ward, A.; Gowda, P.; Lyon, J. Using thematic mapper data to identify contrasting soil plains and tillage practices. *Photogramm. Eng. Remote Sens.* **1997**, *63*, 87–93.
29. Yue, J.; Tian, Q.; Dong, X.; Xu, K.; Zhou, C. Using hyperspectral crop residue angle index to estimate maize and winter-wheat residue cover: A laboratory study. *Remote Sens.* **2019**, *11*, 807. [CrossRef]
30. Serbin, G.; Hunt, E.R.; Daughtry, C.S.; McCarty, G.W.; Doraiswamy, P.C. An improved ASTER index for remote sensing of crop residue. *Remote Sens.* **2009**, *1*, 971–991. [CrossRef]
31. Bocco, M.; Sayago, S.; Willington, E. Neural network and crop residue index multiband models for estimating crop residue cover from Landsat TM and ETM+ images. *Int. J. Remote Sens.* **2014**, *35*, 3651–3663. [CrossRef]
32. McNairn, H.; Protz, R. Mapping corn residue cover on agricultural fields in Oxford County, Ontario, using Thematic Mapper. *Can. J. Remote Sens.* **1993**, *19*, 152–159. [CrossRef]
33. Ding, Y.; Zhang, H.; Wang, Z.; Xie, Q.; Wang, Y.; Liu, L.; Hall, C.C. A comparison of estimating crop residue cover from sentinel-2 data using empirical regressions and machine learning methods. *Remote Sens.* **2020**, *12*, 1470. [CrossRef]
34. Jin, X.; Ma, J.; Wen, Z.; Song, K. Estimation of maize residue cover using Landsat-8 OLI image spectral information and textural features. *Remote Sens.* **2015**, *7*, 14559–14575. [CrossRef]
35. Yue, J.; Tian, Q.; Tang, S.; Xu, K.; Zhou, C. A dynamic soil endmember spectrum selection approach for soil and crop residue linear spectral unmixing analysis. *Int. J. Appl. Earth Observ. Geoinf.* **2019**, *78*, 306–317. [CrossRef]
36. Yue, J.; Tian, Q. Estimating fractional cover of crop, crop residue, and soil in cropland using broadband remote sensing data and machine learning. *Int. J. Appl. Earth Observ. Geoinf.* **2020**, *89*, 102089. [CrossRef]
37. Wang, G.; Wang, J.; Zou, X.; Chai, G.; Wu, M.; Wang, Z. Estimating the fractional cover of photosynthetic vegetation, non-photosynthetic vegetation and bare soil from MODIS data: Assessing the applicability of the NDVI-DFI model in the typical Xilingol grasslands. *Int. J. Appl. Earth Observ. Geoinf.* **2019**, *76*, 154–166. [CrossRef]
38. Hansen, P.; Schjoerring, J. Reflectance measurement of canopy biomass and nitrogen status in wheat crops using normalized difference vegetation indices and partial least squares regression. *Remote Sens. Environ.* **2003**, *86*, 542–553. [CrossRef]
39. Verrelst, J.; Camps-Valls, G.; Muñoz-Marí, J.; Rivera, J.P.; Veroustraete, F.; Clevers, J.G.; Moreno, J. Optical remote sensing and the retrieval of terrestrial vegetation bio-geophysical properties—A review. *ISPRS J. Photogramm.* **2015**, *108*, 273–290. [CrossRef]
40. Baret, F.; Weiss, M.; Lacaze, R.; Camacho, F.; Makhmara, H.; Pacholcyzk, P.; Smets, B. GEOV1: LAI and FAPAR essential climate variables and FCOVER global time series capitalizing over existing products. Part1: Principles of development and production. *Remote Sens. Environ.* **2013**, *137*, 299–309. [CrossRef]
41. Hagan, M.T.; Menhaj, M.B. Training feedforward networks with the Marquardt algorithm. *IEEE Transact. Neural Netw.* **1994**, *5*, 989–993. [CrossRef] [PubMed]
42. Vapnik, V.; Golowich, S.; Smola, A. Support vector method for function approximation, regression estimation and signal processing. *Adv. Neural Inf. Process. Syst.* **1996**, *9*, 281–287.
43. Cristianini, N.; Shawe-Taylor, J. *An Introduction to Support Vector Machines and Other Kernel-Based Learning Methods*; Cambridge University Press: Cambridge, UK, 2000.
44. Breiman, L. Random forests. *Mach. Learn.* **2001**, *45*, 5–32. [CrossRef]
45. Shah, S.H.; Angel, Y.; Houborg, R.; Ali, S.; McCabe, M.F. A random forest machine learning approach for the retrieval of leaf chlorophyll content in wheat. *Remote Sens.* **2019**, *11*, 920. [CrossRef]
46. Belgiu, M.; Drăguţ, L. Random forest in remote sensing: A review of applications and future directions. *ISPRS J. Photogramm.* **2016**, *114*, 24–31. [CrossRef]

land

Article

Land Subsidence Detection in the Coastal Plain of Tabasco, Mexico Using Differential SAR Interferometry

Zenia Pérez-Falls [1], Guillermo Martínez-Flores [1,*] and Olga Sarychikhina [2]

[1] Instituto Politécnico Nacional, Centro Interdisciplinario de Ciencias Marinas, Av. Instituto Politécnico Nacional S/N, Col. Playa Palo de Santa Rita, La Paz 23096, BCS, Mexico

[2] CICESE, Carretera Ensenada-Tijuana No. 3918, Zona Playitas, Ensenada 22860, BC, Mexico

* Correspondence: gmflores@ipn.mx

Abstract: Land subsidence (LS) increases flood vulnerability in coastal areas, coastal plains, and river deltas. The coastal plain of Tabasco (TCP) has been the scene of recurring floods, which caused economic and social damage. Hydrocarbon extraction is the main economic activity in the TCP and could be one of the causes of LS in this region. This study aimed to investigate the potential of differential SAR interferometric techniques for LS detection in the TCP. For this purpose, Sentinel-1 SLC descending and ascending images from the 2018–2019 period were used. Conventional DInSAR, together with the differential interferograms stacking (DIS) approach, was applied. The causes of interferometric coherence degradation were analyzed. In addition, Sentinel-1 GRD images were used for delimitation of areas recurrently affected by floods. Based on the results of the interferometric processing, several subsiding zones were detected. The results indicate subsidence rates of up to −6 cm/yr in the urban centers of Villahermosa, Paraíso, Comalcalco, and other localities. The results indicate the possibility of an influence of LS on the flood vulnerability of the area south of Villahermosa city. They also suggest a possible relationship between hydrocarbon extraction and surface deformation.

Keywords: land subsidence; DInSAR; differential interferograms stacking; floods; coastal plain of Tabasco

Citation: Pérez-Falls, Z.; Martínez-Flores, G.; Sarychikhina, O. Land Subsidence Detection in the Coastal Plain of Tabasco, Mexico Using Differential SAR Interferometry. *Land* **2022**, *11*, 1473. https://doi.org/10.3390/land11091473

Academic Editors: Sara Venafra, Carmine Serio and Guido Masiello

Received: 3 July 2022
Accepted: 29 August 2022
Published: 3 September 2022

1. Introduction

Land subsidence (LS) is a major worldwide hazard, and it is defined as the downward, mainly vertical, displacement of the Earth's surface relative to a stable reference level [1,2]. LS is caused by a wide variety of processes of natural and anthropogenic origin. The natural-driven processes, such as glacial isostatic adjustment (GIA) [3,4], tectonic movements (except co-seismic displacement) [5,6], and sediment compaction [7,8], often cause a slow and steady motion (a few mm/yr). Human activities that cause subsidence include withdrawal of groundwater [9–12], hydrocarbons [13–15], geothermal water, and brine [16–18]; mining [19–21]; loading of engineered structures [22,23]; and wetland drainage [24]. Generally, the observed rates of human-induced subsidence greatly exceed the rates of natural subsidence, reaching centimeters per year, to even meters per year (e.g., mining activities [25]). LS damage to urban and civil infrastructure causes constant and considerable economic losses. However, the most notable impact of LS is produced in coastal areas, coastal plains, and river deltas, where LS increases flood vulnerability (flood frequency, inundation depth, and duration of floods) [26–28]. Identifying LS-prone areas and estimating their rate and spatial extension is essential in this phenomenon's assessment and management.

The use of satellite data and remote sensing (RS) techniques is a common practice in Earth surface observations. The advantages of satellite RS techniques are their comprehensive area coverage, non-invasiveness, and cost-effectiveness. In particular, the differential interferometric synthetic aperture radar (DInSAR) technique has become an effective RS tool for monitoring and assessing the Earth surface displacements induced by a variety of geophysical and geological processes, including earthquakes, volcanoes, landslides, LS, and sinkholes, among others [29]. The DInSAR technique is based on acquiring complex SAR images over the same area at different times using repeated passes. The standard DInSAR approach (or conventional DInSAR) exploits the phase difference of the SAR image pairs, providing a measurement of surface displacements occurring between the two acquisitions with a sub-centimetric accuracy and a decametric spatial resolution (e.g., [30–32]).

The uncertainties in the measurement of conventional DInSAR, due to the contribution of non-displacement signals, such as the digital elevation model (DEM) and orbitals errors, and atmospheric delay, are the handicaps of this approach [33]. In addition, the temporal and geometrical decorrelation limit its practical applications [34]. Advanced-DInSAR techniques, based on large stacks composed of many SAR images, partly overcome DInSAR limitations (e.g., [35–37]). Despite the considerable advances in DInSAR processing techniques, applying DInSAR for displacement measurements in areas where the conditions of the land surface change significantly, e.g., densely vegetated areas, remains challenging.

Tabasco is an oil-rich state located in the southeast of Mexico; its northern border runs along the Gulf of Mexico. Much of the state is a wide alluvial coastal plain, the so-called Tabasco Coastal Plain (TCP). Due to its climatic and hydro-geologic conditions, Tabasco is one of the most flood-prone Mexican states [38,39]. The state's high incidence of floods has been exacerbated by sea-level rises and possibly LS, through natural or anthropogenic effects. LS is not considered a high-risk phenomenon in the Tabasco state. The LS phenomenon has been poorly investigated, and its effects on the increase of TCP area's vulnerability to flooding and coastal erosion is unknown. Hydrocarbon production is the main economic activity in the region, with more than a thousand wells distributed in 106 oil fields, so the possibility of significant anthropogenic subsidence occurrence cannot be discarded and must be investigated in detail.

DInSAR techniques have proven practical LS detection and monitoring tools in coastal areas (e.g., [26,40–45]). However, to the authors' knowledge, there are not formal papers published or submitted to journals where DInSAR was applied to investigate LS in Tabasco. Early DInSAR results for the Tabasco region were published only as a conference paper [46]. Therefore, this study evaluates DInSAR's potential for land subsidence detection and monitoring in the TCP. Conventional DInSAR and the interferograms stacking procedure (A-DInSAR) were applied to identify the Earth's surface displacement in TCP. Sentinel-1 data from January 2018 to January 2020 were used. The achieved results allowed the identification of land sinking areas during the period covered by the study, which should be the target of more detailed investigations.

2. Materials and Methods

2.1. Study Area

This study's area of interest (AOI) belongs to the TCP, a tropical lowland on the Gulf of Mexico, in Tabasco State, southeastern Mexico (Figure 1). It is part of the Mexican physiographic province called the Southern Coastal Plain of the Gulf of Mexico [47]. It was formed by alluvial sediments brought by rivers from the mountains of Chiapas (Mexico) and Guatemala; the rivers cross the state to flow into the Gulf of Mexico. The land is largely covered with lakes, lagoons, and wetlands (floodable areas), one of the most important in Mexico. About 80% of the TCP surface is composed of marsh, alluvial, coastal, and lake deposits from the Quaternary period; corresponding with the development of the current environments, from the Pliocene to today, and about 20% is made up of sedimentary rock from the Tertiary period [48–50]. The soils of the TCP are predominantly of alluvial and organic origin, such as Gleysols, Histosols, and Fluvisols [50,51], and are characterized by a poor drainage capacity.

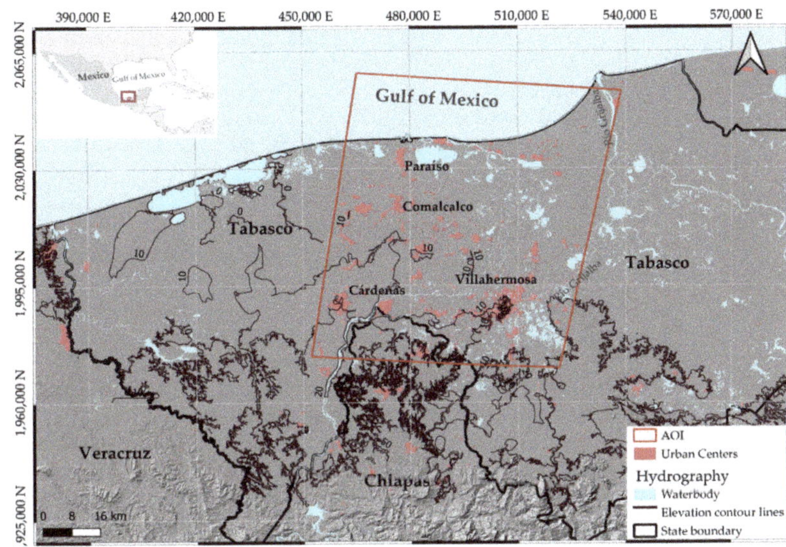

Figure 1. Central region of Tabasco (AOI) and its main urban areas, hydrography, and topography. The background image is a shaded relief based on the INEGI elevation model (www.inegi.org.mx, accessed on 11 July 2022).

The TCP region has a tropical rainforest climate, designated as Af under the Köppen-Geiger climate classification system. This region's average annual air temperature is 26 °C, with average monthly temperatures ranging between 22.7 °C (January) and 28.9 °C (May). The TCP receives 1500–2000 mm of annual precipitation, mainly in the rainy season between June and November [52]. Furthermore, the region is regularly subjected to tropical storms and hurricanes from the Gulf of Mexico and the Pacific Ocean. The monthly average precipitation in the analyzed period (2018–2019) is presented in Figure 2.

Due to TCP's climatic and hydro-geologic conditions, its territory is exposed to floods annually [39]. Some floods have been catastrophic, such as those of 1980, 1995, 2007 [53], and 2020 [54]. The extensive flooding that occurred in 2020, at the end of October and early November, affected over 62% of the Tabasco state and more than 1.2 million people [54].

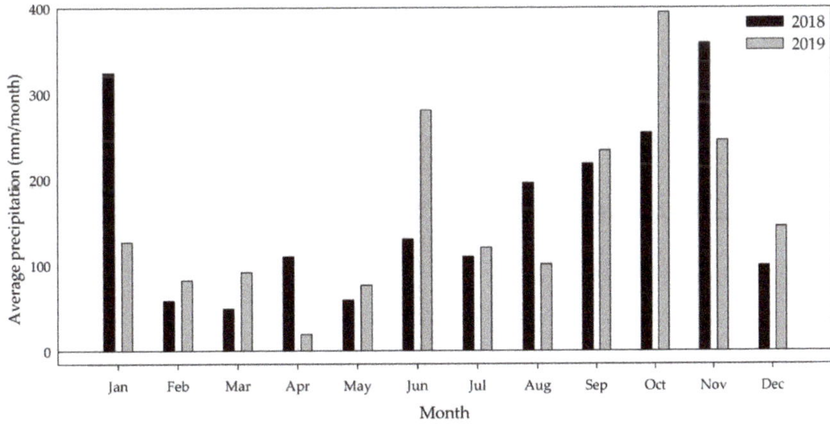

Figure 2. Monthly average precipitation (data available at [55]).

The AOI covers 596,573 ha, of which 69.76% are dedicated to economic activities, and 22.01% are covered by natural vegetation. The natural vegetation in TCP is represented by tropical rainforest and various wetland communities, including mangroves. The most important economic activities for the state of Tabasco are oil and gas production, agriculture, and livestock. Tabasco is a mainly rural state; agricultural fields and pastures cover approximately 69% of the area used for economic activities, and only 3.48% is urban (Figure 3a).

Since large-scale exploitation of hydrocarbon resources began at the end of the 1950s, oil and gas production has become Tabasco's economic mainstay. At present, Tabasco is a leader in hydrocarbon reserves and is one of Mexico's primary oil-producer states. Figure 3a shows the hydrocarbon extraction wells distribution over the AOI. Most wells have a depth ranging from 1500 to 3500 m. However, some wells reach up to 7615 m.

The study area is part of the Salina de Istmo, Pilar Reforma-Akal, and Macuspana basins (Figure 3b). The Salina de Istmo basin is Miocene-Pliocene and associated with a system of normal faults, including the Comalcalco sub-basin, associated with sediment loading and salt evacuation. The Macuspana basin is from the early Miocene-Pliocene. Sedimentary facies vary from fluviodeltaic to marine and are characterized by turbidite deposition. The Pilar Reforma-Akal Basin is the most representative of the study area where hydrocarbons are stored in limestone from the Upper Cretaceous and Upper Jurassic [56,57]. The hydrocarbon system's distribution corresponds to the Mesozoic oil fields and, to a lesser extent, to the Tertiary (Figure 3b) [57].

High volume extraction of hydrocarbons can cause LS around the producing wells. Land surface sinking due to oil and gas production depends on the geometrical shape and thickness of the reservoir, the compaction coefficient, the pressure drops in the reservoir, and the geomechanical behavior of the overburden [58]. The documented rates of LS caused by hydrocarbon extraction range from a few mm/yr [59] to up to 0.75 m/yr [60]. Even a small subsidence in plain areas could significantly increase flood vulnerability.

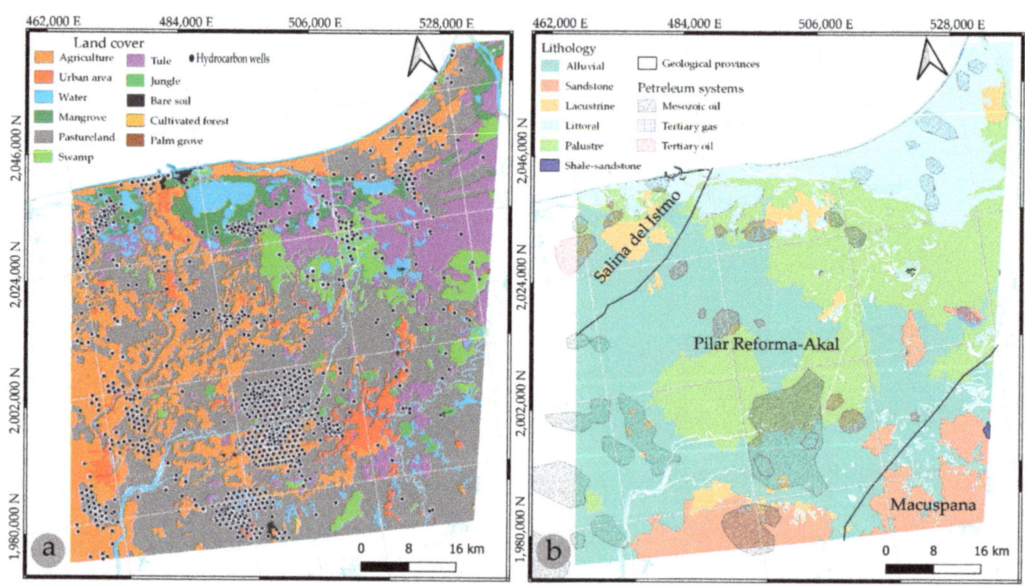

Figure 3. Physical-geographical characteristics of AOI; (**a**) land cover (INEGI [61] and distribution of hydrocarbon wells (IICNIH [62]). The area with the highest well density is located west of Villahermosa, with wells depth ranging between 1500 and 3500 m; (**b**) Lithology, geological provinces, distribution of the oil system, and its geological era [57,62].

2.2. Data

In this study, Sentinel-1 level-1 images provided in the Single Look Complex (SLC) format [63] by the Alaska Satellite Facility (https://asf.alaska.edu, accessed on 2 December 2019) were used in DInSAR processing. The Copernicus Sentinel-1 mission, developed by the European Space Agency (ESA), is based on a constellation of two satellites: Sentinel-1A, launched in April 2014, and Sentinel-1B, launched in April 2016. The Sentinel-1 constellation operates at the C-Band frequency (5.405 GHz, 5.5 cm wavelength), with a 12-day repeated acquisition for a single mission, and 6-days in the case of a two-satellite combination. Sentinel-1 imagery was selected in this study, thanks to its free accessibility with a regularly repeated acquisition at a 6-day interval.

A total of 115 (Sentinel-1A/1B) images were acquired on ascending Path 34, Frame 54/55, in Interferometric Wide (IW) swath mode, with dual (VV + VH) polarization between 2 January 2018 and 29 December 2019. Moreover, 108 images were acquired on descending Path 99, Frame 530/532, in IW swath mode [64–66], with dual polarization between 6 January 2018 and 27 December 2019. Only Sentinel-1 VV polarization bands were used, since co-polarized bands provide higher coherence than VH polarization. The main characteristics of the Sentinel-1 SLC data used in this study for DInSAR processing are listed in Table 1.

Table 1. Characteristics of the Sentinel-1 SLC data used in this study in DInSAR processing. The nominal spatial resolution is specified for single-look data. N is the number of used SAR images, and I is the number of interferograms for each dataset calculated and analyzed in this study.

Dataset	**1**	**2**
Orbit	Descending	Ascending
Mode	IW	IW
Sub-Swath	IW3	IW2 + IW3
Track	99	34
Frame	530/532	54/55
Wavelength (cm)	5.5 (C-band)	5.5 (C-band)
Polarization	VV	VV
Nominal ground resolution (Ground Range × Azimuth, m)	5 × 20	5 × 20
Time span	2 January 2018–29 December 2019	6 January 2018–27 December 2019
Number of images (N)	115	108
Number of Interferometric Pairs (I)	321	290

The external datasets used for the SAR SLC data interferometric processing include the Shuttle Radar Topographic Mission (SRTM) 1-arc-second (30 m) DEM [67] and Copernicus Sentinel-1A and Sentinel-1B precise orbital files (AUX_POEORB products), obtained from the Copernicus Open Access Hub.

Satellite data available for mapping and monitoring flood coverage were obtained by passive (e.g., onboard Landsat, Aqua, Sentinel-2, Resourcesat-2) and active (e.g., Sentinel-1, RADARSAT, ENVISAT) sensors [68,69]. Sentinel-1 level-1 ground range detected (GRD) images were obtained from the Alaska Satellite Facility and used to identify flooded areas. Sentinel-1 level-1 (GRD) products consist of focused SAR data that has been detected, analyzed, and projected to the ground range using a ground ellipsoid model. The pixel values of the Sentinel-1 level-1 (GRD) image represent the detected amplitude of the backscattered signal, without phase information. Two pairs of GRD images with dual-polarization from descending orbital pass were acquired, covering the flood events in 2018 and 2019. Only VV polarization bands were used because, for flooded area detection, the co-polarization comparison gives better results than the cross-polarization one, while the use of VV polarization is recommended over the use of VH data [70]. The characteristics of the Sentinel-1 GRD products used in this study are presented in Table 2.

Table 2. Characteristics of the Sentinel-1 GRD data used to identify flooded areas.

Image	Date	Format-Mode	Polarization	Land Surface Condition
1	6 January 2018	GRD-IW	VV	Dry
2	5 February 2018	GRD-IW	VV	Wet
3	16 October 2019	GRD-IW	VV	Dry
4	21 November 2019	GRD-IW	VV	Wet

2.3. SAR Differential Interferometry Background

The fundamentals of the conventional DInSAR technique have been presented in many publications [29,31,70]. Therefore, only some aspects relevant to this study are briefly described.

In principle, SAR interferometry exploits the information in the interferometric phase, calculated as the phase difference between two coregistered SAR images acquired from slightly different orbit positions (spatial baseline) and different times (temporal baseline). The interferometric phase (ϕ_{int}) is the sum of contributions from several factors, and the following equation can express this:

$$\phi_{int} = \phi_{displ} + \phi_{topo} + \phi_E + \phi_{atm} + \phi_{noise} \quad (1)$$

where ϕ_{displ} represents the phase due to surface displacement, ϕ_{topo} refers to the phase caused by local topography (or topographic phase), ϕ_E is the phase produced by a surface of constant elevation on a spherical Earth (curved Earth), also known as the orbital phase, ϕ_{atm} denotes the phase components due to the variation of atmospheric conditions between the image acquisitions (the so-called atmospheric phase screen (APS)), and ϕ_{noise} includes all the phase noise contributions that corrupt the interferometric SAR signal.

All other contributors to the interferometric phase must be removed or diminished, to obtain the Earth surface displacement measurement. Using external DEM and precise orbital information, phase contributions caused by topography and the curved Earth can be estimated and removed from the interferometric phase. This is the basic concept of the Differential SAR Interferometry (DInSAR) approach. However, the differential interferometric phase can still contain some "unwanted" phase components. APS is one of the main sources of errors that influence the differential interferometric phase, and it can degrade the accuracy of surface displacement estimates using DInSAR. Topographic and orbital errors can also contribute to the differential interferometric phase.

The accuracy of surface displacement measurements from DInSAR greatly depends on the quality of the differential interferometric phase. The established criterion to measure the quality of the differential interferometric phase is the value of the complex correlation coefficient, the so-called coherence. The coherence ($\hat{\gamma}$) is a measure of phase correlation (or phase reliability) between two complex SAR images, M (master image) and S (slave image), and is defined as [71]:

$$\hat{\gamma} = \frac{\sum_{i=1}^{N} M_i \, S_i{}^*}{\sqrt{\sum_{i=1}^{N} \left| M_i \right|^2} \sqrt{\sum_{i=1}^{N} \left| S_i \right|^2}} \quad (2)$$

where S^* is the complex conjugate of the complex slave image (S), |M| and |S| are the amplitude of complex SAR master and slave image, respectively, and N indicates the spatial set of samples employed in the coherence estimation. The coherence values lie in the range $0 \leq \hat{\gamma} \leq 1$; a value of zero indicates complete incoherence and a differential interferometric phase value with no useful information, whereas a value of one indicates complete coherence and a differential interferometric phase value with no noise. DInSAR is effectively applied only in areas where the differential interferometric phase is characterized by high coherence. The main factors that affect the coherence are temporal and spatial decorrelation and low accuracy of image coregistration.

The temporal decorrelation phenomenon is caused by changes in the physical and geometric properties of the scatterers on the Earth's surface [72]. Some of the main sources of temporal decorrelation are erosion, vegetation growth, cultivation, snow, and near-surface moisture changes.

Spatial or geometric decorrelation may result from high variations in imaging geometry. Thus, images with a short spatial baseline must be selected for interferometric processing.

A practical way to overcome the conventional DInSAR limitations mentioned above is combining the information from multiple short-interval differential interferograms, to extract common information. The most basic procedure is to compute integer linear combinations of unwrapped differential interferograms [33] or perform their temporal averaging, the so-called differential interferograms stacking (DIS) approach. The main assumption of this method is that the deformation phase is highly correlated, and the error/noise terms (e.g., APS, signal noise, orbital errors, and nonlinear ground displacements) are uncorrelated between independent pairs. The application of this method increases the signal-to-noise ratio (SNR) and improves the reliability of the Earth surface displacement measurements [73].

2.4. DInSAR Processing

The GAMMA software package developed by GAMMA Remote Sensing and Consulting AG, Bern, Switzerland [74] was used for S1 Level-1 SLC product processing. The processing chain was divided into three stages: pre-processing, conventional repeat-pass DInSAR, and stacking. Pre-processing and conventional repeat-pass DInSAR were performed following the standard workflow used for processing S1 TOPS mode image pairs. This workflow is comprehensively explained in [75].

The pre-processing stage consisted of importing SLC data; updating of image metadata with precise orbital state vectors; S1 TOPS splitting, which included polarization selection; selection of sub-swaths and bursts covering the AOI; and selection of suitable S1 SLC image pairs and coregistration. Here, suitable S1 SLC image pairs were selected within the thresholds of the perpendicular and temporal baselines, which were set to 200 m and 18 days, respectively. The connection graph for ascending and descending datasets generated using these thresholds is presented in Figure 4a,b, respectively. A total of 321 ascending and 290 descending interferograms were generated and used in the stacking procedure. However, for coherence analysis, one additional interferogram from a descending orbital pass with a temporal baseline of 24 days was generated.

The conventional repeat-pass DInSAR stage included the formation of interferograms, multi-looking, simulation of topographic phases, differential interferogram generation, coherence calculation, phase filtering, phase unwrapping, orbital error correction, atmospheric correction, phase to displacement conversion, and interferometric product (i.e., coherence, differential interferograms, displacements map) geocoding (Latitude/Longitude WGS84 coordinate system).

Topographic phases were simulated using the precise orbits and an external DEM. Differential interferograms were formed at a default 2 looks in azimuth and 10 looks in range, to obtain a pixel size of ~40×40 m^2. To improve the quality of differential interferograms and optimize the phase unwrapping procedure, the differential interferograms were filtered using an adaptive Goldstein filter [76], with an optimal filter strength of 0.7 being employed in this study, after a number of trials. After phase filtering, a minimum-cost flow (MCF) algorithm [77,78] was used for phase unwrapping. Areas with a coherence smaller than 0.2 were masked out before unwrapping. The linear trend was estimated and subtracted from the unwrapped differential interferograms, to correct the residual linear ramp caused by orbital errors. Differential atmospheric delay in the interferometric phase, which is correlated with the topography, was reduced using the empirical phase-based method, for which the linear correlation between the unwrapped phase and the elevation of DEM was calculated [79,80].

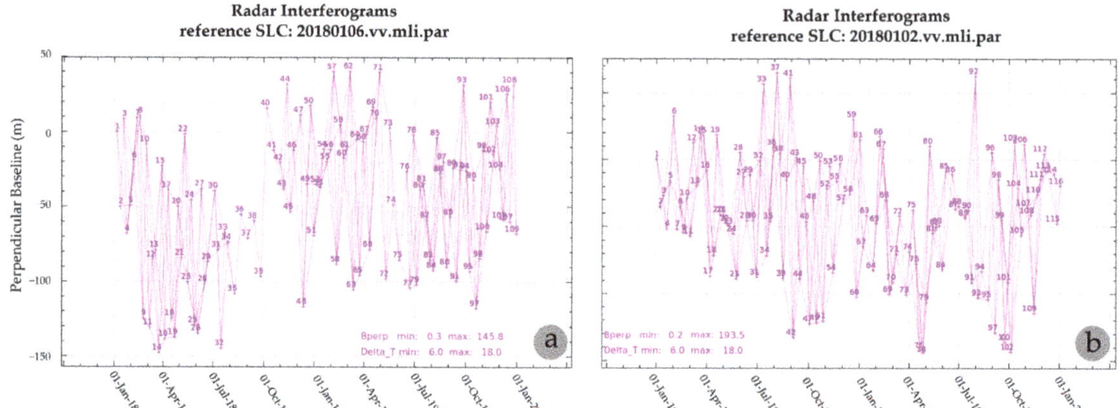

Figure 4. Temporal and spatial (perpendicular) baseline connection diagram of the Sentinel-1 SLC image pairs from (**a**) ascending and (**b**) descending orbital pass used in the DIS approach.

During the stacking stage, the unwrapped differential interferograms of each set were summed and divided by the total (cumulative) time interval of all interferograms of the set in years, to obtain an average annual LOS displacement rate. Before stacking, the interferograms were referenced to a common (32 × 32 pixels) area and were shifted accordingly, to set the reference phase to zero. The common reference area was in the center of Villahermosa.

After stacking, phase to displacement conversion was performed, and the resulting LOS displacement rate maps were geocoded.

2.5. *Identification of Flooded Areas*

Sentinel-1 GRD images were pre-processed by implementing radiometric calibration, spot filtering, and geometric correction of the data, to identify flooded areas. Radiometric calibration was initially applied, as it is an essential step in SAR GRD image pre-processing. The pixel values of the images could directly represent the radar backscatter [81], achieving results in dB. Image pixels representing bodies of water have a lower radar backscatter coefficient than other features [82], such as land or vegetation. The effects of thermal noises were also removed, and a precise orbit file was applied to the images. Lee Sigma filtering was applied to reduce the speckle noise caused by random effects of multiple backscattering within each resolution cell, which is best suited for this processing [81], leading to better results, with a filter size of 7 × 7. Finally, atmospheric correction was performed, to compensate for topography variations caused by the satellite sensor's viewing angle [81,82].

To obtain the flooded areas, the thresholding method was used, which is the simplest method of image segmentation [83]. Here, the areas affected by flooding were identified for two flood events: February 2018 and November 2019. The binary images (water/non-water) were created using thresholds estimated from intensity (in dB) histograms of pre-processed Sentinel-1 GRD images. The used threshold values varied for the analyzed images between −12 and −10 dB; the water areas being those with an intensity below the applied threshold's value. To separate the permanent water bodies from the areas affected by floods, the permanent water bodies were identified using pre-flood event images (dry conditions) and then masked out in co-flood event binary images, so that, as a result, binary images of areas affected/non-affected by floods were obtained.

A permanent water bodies mask was also used to exclude water bodies from interferometric coherence analysis, as water bodies generally have a low coherence (near zero).

SNAP software (Sentinel Application Platform) [84] was used for pre-processing, whereas GIS software was used to obtain the flooded areas from the pre-processed images.

3. Results

3.1. Coherence Analysis

As mentioned above, DInSAR can only be effectively applied in areas where the differential interferometric phase is characterized by high coherence. For a short spatial baseline interferometric pair, where two images are coregistered with high accuracy, the temporal decorrelation is the main factor of the coherence degradation.

To investigate the impact of the temporal baseline on the quality of interferometric results in the AOI, differential interferograms with a temporal baseline of 6, 12, 18, and 24 days and the common master image (12 February 2019) were processed, and their coherences were estimated (Figures 5 and 6). All image pairs had a short perpendicular baseline, to avoid the influence of spatial decorrelation on coherence degradation. All four analyzed image pairs covered relatively dry periods, without important or extreme precipitation (Figure 2) or floods. The parameters of interferometric pairs are presented in Table 3.

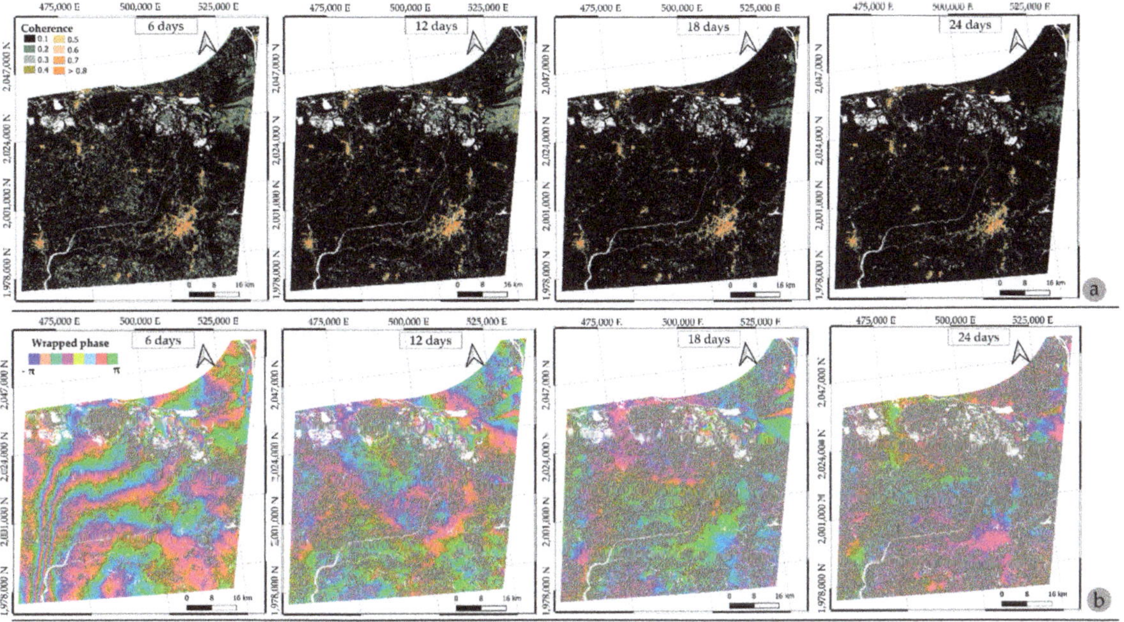

Figure 5. (**a**) Coherence and (**b**) wrapped differential interferograms for selected Sentinel-1 SLC image pairs (Table 3).

As it can be seen in Figure 5, the AOI is characterized by low coherence values, even for minimal possible temporal separation between images (6 days). The vegetation dominated areas, as for AOI, are especially likely to lose their coherence within a very short period. Moreover, an important loss of coherence is observed with a temporal baseline increase.

Figure 6 shows the coherence histograms for selected interferometric pairs (Figure 5; Table 3). As the temporal baseline increases, the frequency of pixels with low coherence also increases, whereas the mean coherence decreases from 0.13 (6 days) to 0.08 (24 days).

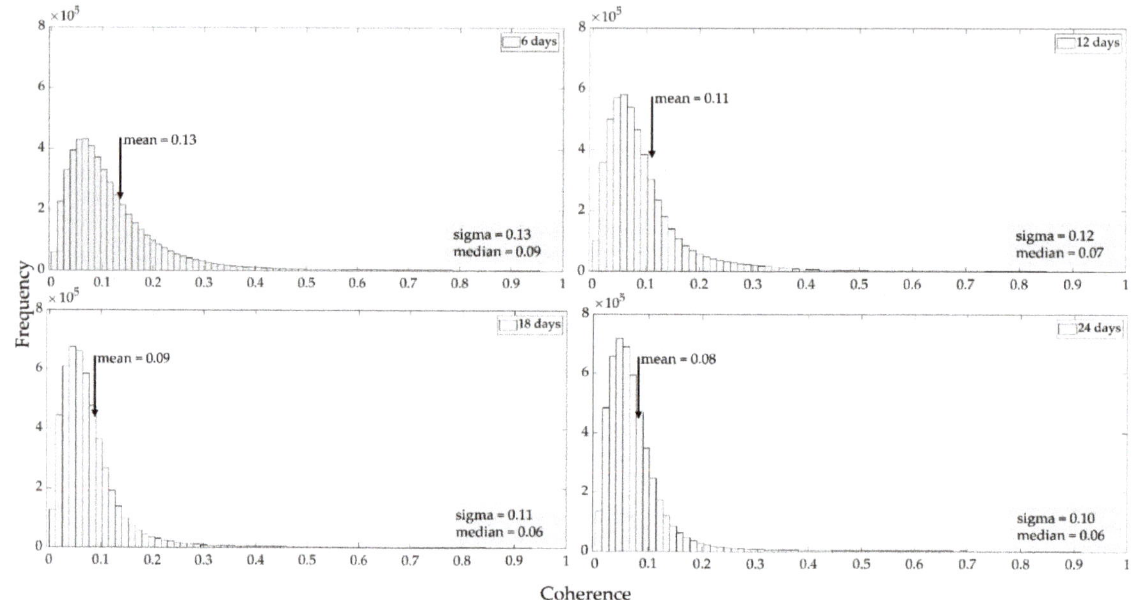

Figure 6. Coherence histograms for selected Sentinel-1 SLC image pairs (Table 3). Arrows indicate the mean coherence value.

Table 3. Parameters of image pairs selected for interferometric coherence analysis. Bp is the perpendicular baseline; Bt is the temporal baseline.

Interferometric Pair	Orbital Pass	Bp (m)	Bt (Days)	Mean Coherence
12 February 2019–18 February 2019	Descending	93	6	0.13
12 February 2019–24 February 2019	Descending	70	12	0.11
12 February 2019–02 March 2019	Descending	76	18	0.09
12 February 2019–08 March 2019	Descending	80	24	0.08

Figure 7 shows the average images for the ascending and descending orbital pass and associated coherence histograms. For each orbital direction, the average coherence image was obtained by averaging the coherence of all image pairs processed. These interferometric pairs correspond to the 2018–2019 period and meet the established baseline thresholds. The histograms (Figure 7c) show that the study area is dominated by low coherence (≤0.2), due to the land cover type (different types of vegetation). The highest values of average coherence (>0.3) correspond to urban areas and bare soil, reaching up to 0.99.

The average coherence values for different land cover classes are presented in Figure 8. In this case, the average coherence per class (ACC) value was calculated for each processed image pair.

As can be seen in Figure 8, the point cloud of ACC values is separated into two groups. The group with the highest ACC values corresponds to urban areas and bare soil, reaching a value of 0.5. The rest of the land cover classes belong to a group with lower ACC values, ranging from 0.05 to 0.2. The grassland, agriculture, and tule vegetation classes have the largest ACC values of this group (up to 0.2), whereas the lowest ACC values (≤0.05) were obtained for the mangrove class.

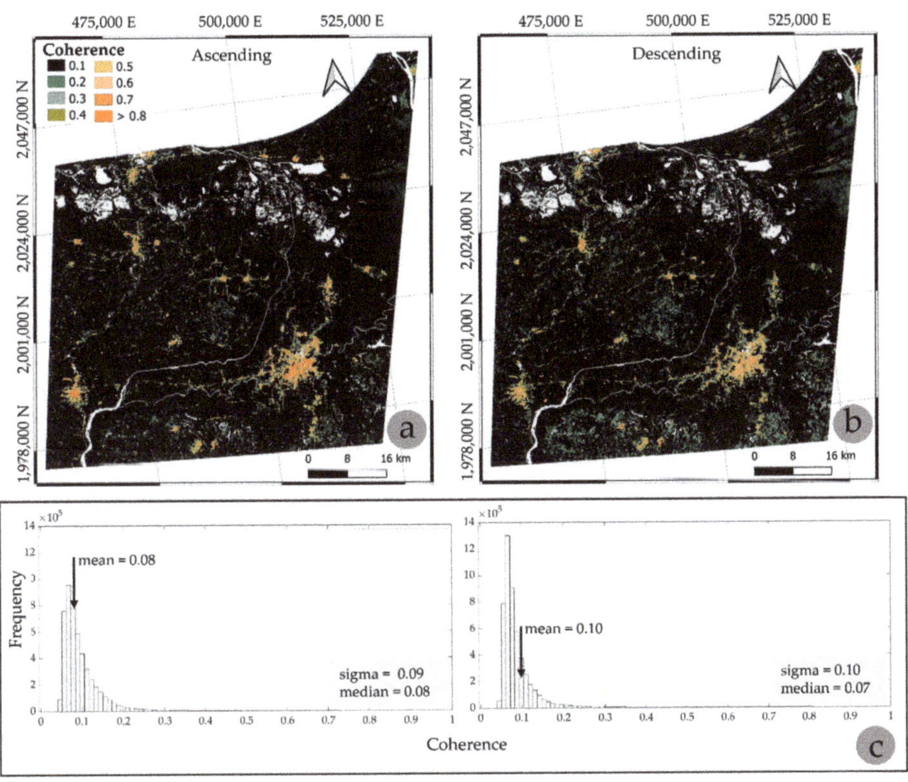

Figure 7. Average coherence estimated for the 2018–2019 period using image pairs from the (**a**) ascending and (**b**) descending pass; (**c**) coherence histograms for each average coherence. Arrows indicate the mean coherence value.

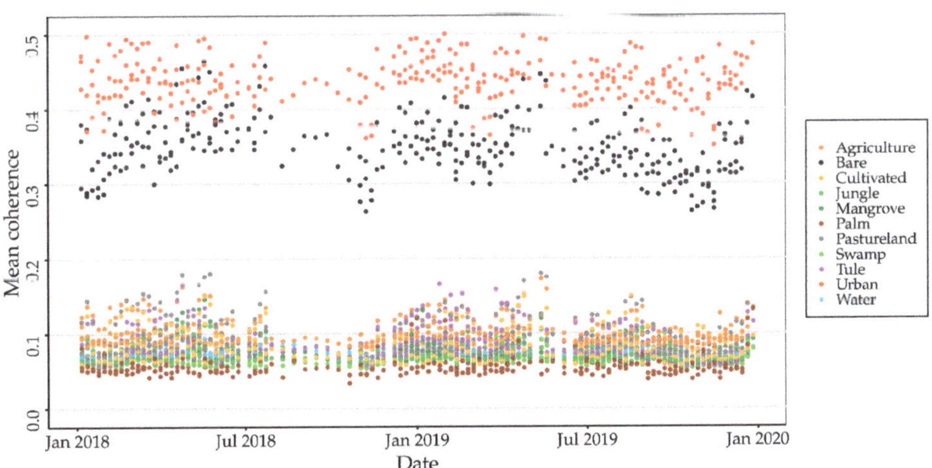

Figure 8. Average coherence for different land cover classes, estimated for image pairs obtained from the descending orbital pass during the 2018–2019 period.

3.2. Flooded Areas and Interferometric Coherence

As the AOI is recurrently affected by floods, their influence on coherence degradation was also investigated. The coherence of the two temporary closed short-baselines interferometric pairs (Table 4) is compared in Figure 9. The 14 November 2018–20 November 2018 interferometric pair spans the flood events caused by strong precipitations (Figure 2), whereas the 14 December 2018–20 December 2018 interferometric pair spans the period with relatively dry climatic conditions. Figure 9 shows the important coherence loss due to flood occurrence. Coherence degradation was observed even for urban (e.g., Villahermosa) and bare soil areas. The mean coherence for the image spanning the flood event was 0.08, while the image pair with relatively dry conditions has a mean coherence of 0.12.

Table 4. Parameters of interferometric pairs selected for the flood impact on the coherence degradation analysis. Bp is the perpendicular baseline; Bt is the temporal baseline.

Interferometric Pair	Orbital Pass	Bp (m)	Bt (Days)	Mean Coherence	Conditions
14 November 2018–20 November 2018	Descending	−85	6	0.08	Wet (flood)
14 December 2018–20 December 2018	Descending	82	6	0.12	Dry

As shown above, the floods had a significant negative impact on the interferometric product quality, degrading considerably the coherence, even in short temporal baseline pairs (Figure 9). Floods are recurrent in the AOI, so the areas repeatedly affected by floods are very challenging for DInSAR applications. To identify the recurrently flooded areas, analysis of Sentinel-1 GRD images was performed.

Figure 10 shows the intensity data (dB) from Sentinel-1 GRD images acquired before (Figure 10a,c) and during flood events (Figure 10b,d). Dark areas (low negative intensity) correspond to the areas covered by water.

Recurrently flooded areas obtained using the Sentinel-1 GRD images and the methodology described in Section 2.5 are shown in Figure 11. The recurrently flooded areas are located south-southeast of the city of Villahermosa, in the towns of Gaviota del Sur, Parrilla, and Huapinol. These regions have recently been reported as vulnerable to flooding. Large recurrently flooded areas are also observed northwest of Comalcalco and north of Paraíso, where the Dos Bocas refinery is located. The analyzed flood events of February 2018 and November 2019 had an affected area of 6.92 ha and 11.37 ha in 2018 and 2019, respectively.

3.3. DInSAR Results Analysis

As seen from the coherence analysis, temporal decorrelation is the major challenge for conventional DInSAR applications in the AOI. A considerable coherence loss was observed, even in short-temporal baseline interferometric pairs. Concerning the interferometric pairs where sufficient coherence remains (Bt $\leq$ 18 days, no-flood period), the main source of error that influences their differential interferometric phase and degrades the accuracy of LOS displacement estimates is the APS effect.

According to [85–87], a phase error of $\frac{\pi}{2}$ is considered relatively strong atmospheric distortion. At C-band, a phase error of $\frac{\pi}{2}$ results in an error of 0.7 cm in the LOS displacement estimate. To obtain reliable displacement values with DInSAR the displacement signal should dominate over the error terms. In this study, the DIS approach was applied, to improve the ratio between the displacement signal and the APS error.

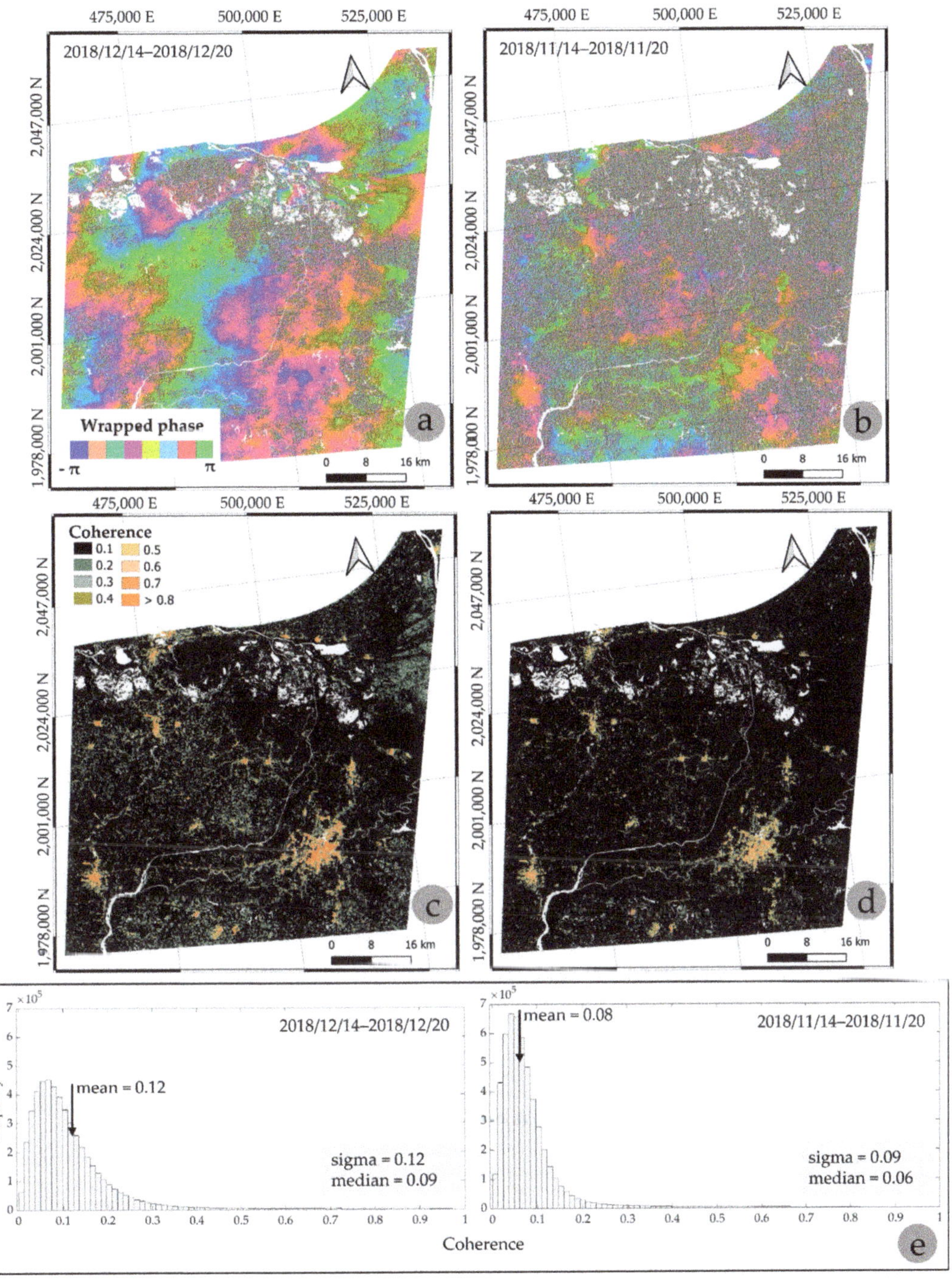

Figure 9. Wrapped differential interferograms: (**a**) 14 December 2018–20 December 2018 and (**b**) 14 November 2018–20 November 2018; their respective coherence images (**c**,**d**), and histograms (**e**).

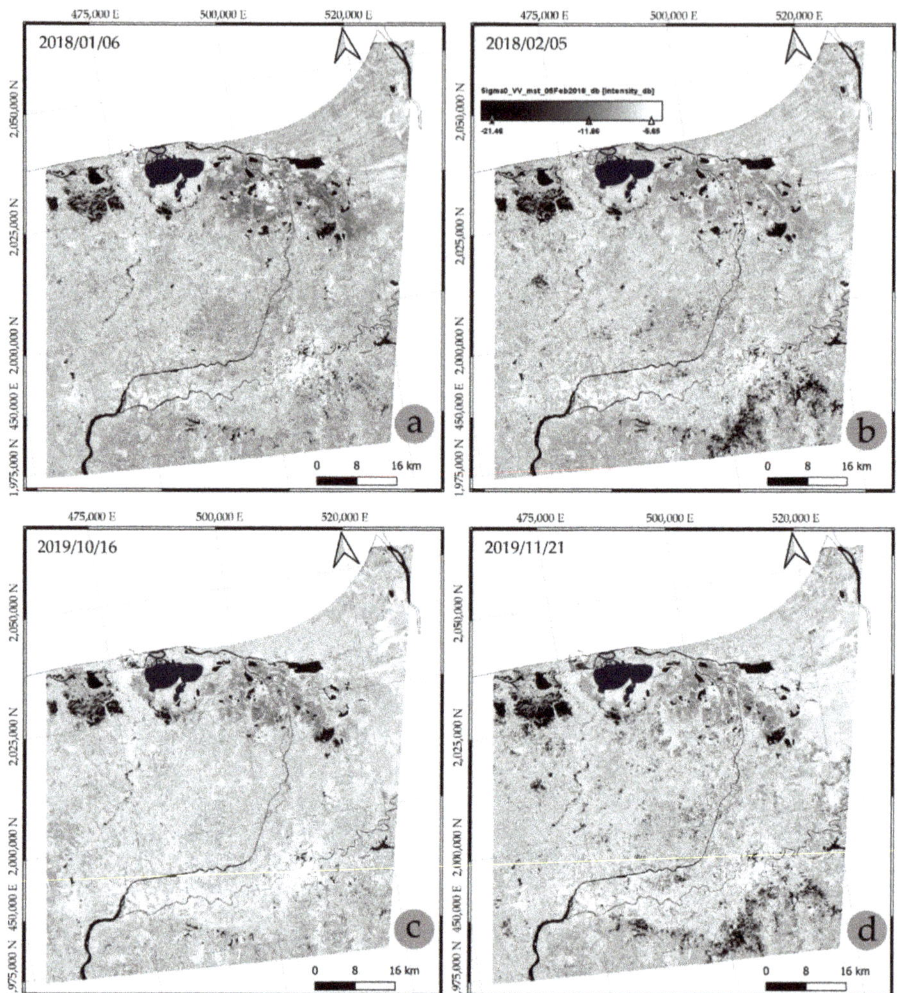

Figure 10. Sentinel-1 GRD intensity (in dB) of (**a**) 6 January 2018, (**b**) 5 February 2018, (**c**) 16 October 2019, and (**d**) 21 November 2019. (**a**,**c**) Correspond to the images before the flood event, and (**b**,**d**) to the images during the floods.

The DIS approach allowed obtaining the average LOS displacement rate estimation for the 2018–2019 period (2 years). The average LOS displacement rates obtained using Sentinel-1 SLC images from ascending and descending orbital pass are presented in Figures 12 and 13, respectively. In the average LOS displacement maps from both orbital passes, four zones with a higher average LOS displacement rate (magnitude) can be identified. These zones (b–e) are framed in Figures 12 and 13; the close-up to these zones is also presented. The LOS displacement obtained for these zones indicates the deformation of the Earth's surface away from the satellite in the ascending and descending pass results; this similarity suggests that the observed LOS displacements may be interpreted as mostly reflecting land subsidence. In zone b, the maximum average LOS displacement rates (-6 cm/yr) were obtained in the town of Paraíso, especially north of this urban center, where the Dos Bocas oil refinery is located. In Comalcalco and the surrounding areas, maximum average LOS displacement rates of -3 cm/yr were obtained.

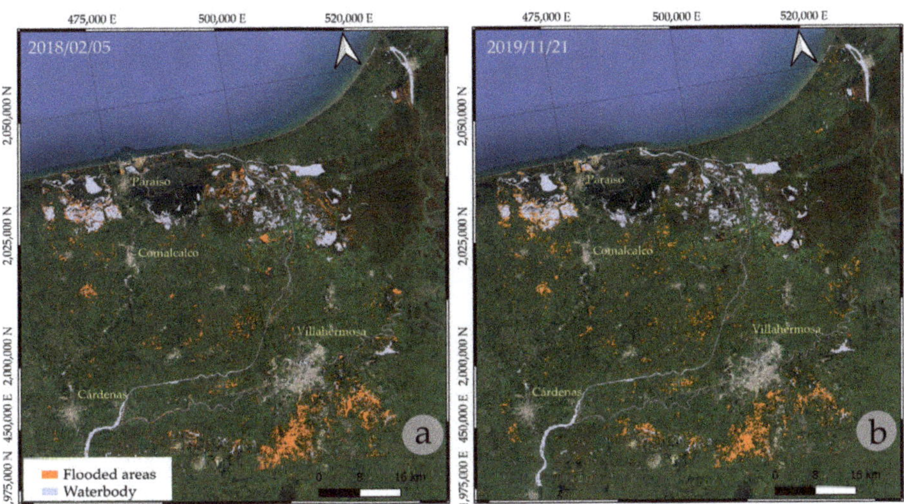

Figure 11. Areas affected by the flood events of (**a**) 5 February 2018 and (**b**) 21 November 2019. Permanent water bodies are shown. A digital globe image is used as a background, the background image was taken from QGIS XYZ Tiles (https://mt1.google.com/vt/, accessed on 8 August 2022).

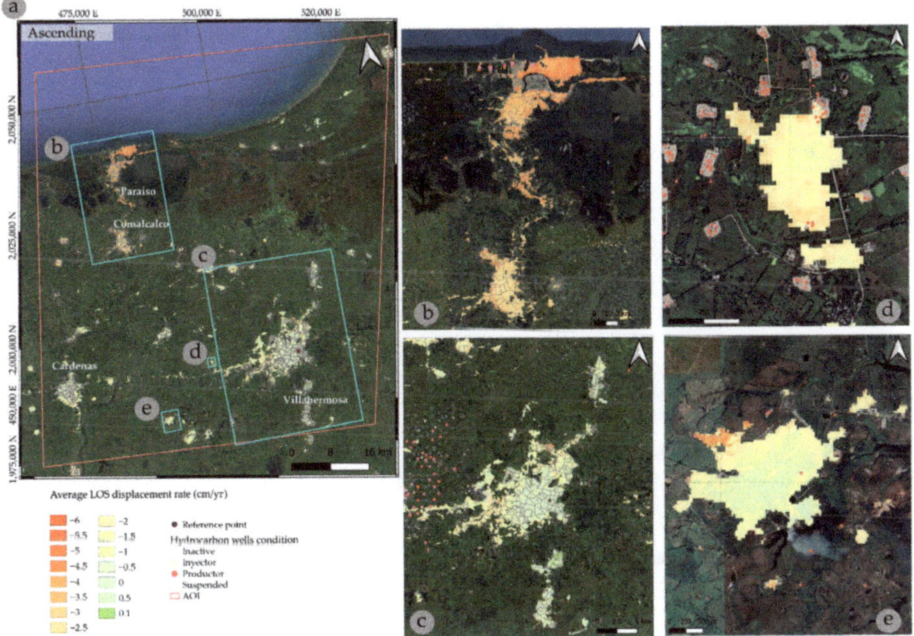

Figure 12. Map of average LOS displacement rates (cm/yr) obtained through the DIS approach, using the Sentinel-1 SLC images of ascending orbital pass acquired in the 2018–2019 period. Negative values indicate a movement away the satellite, (**a**) full AOI map. Blue rectangles enclose the zones with a higher average LOS displacement rate. Close-ups of the (**b**) Paraíso and Comalcalco; (**c**) Villahermosa; (**d**) Batería Samaria and (**e**) Batería Cactus zones are presented. The location of hydrocarbon wells is also shown.

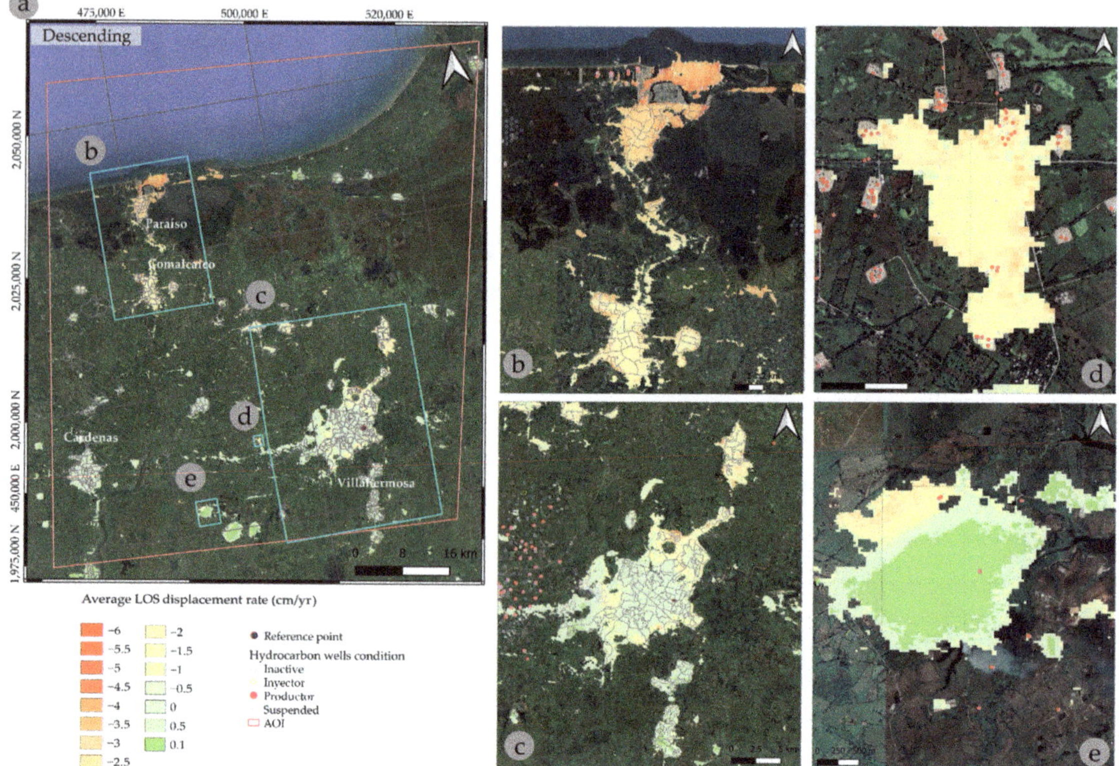

Figure 13. Map of the average LOS displacement rate (cm/yr) obtained through the DIS approach using the Sentinel-1 SLC images of the descending orbital pass acquired in 2018–2019 period. Negative values indicate a movement away the satellite. (**a**) Full AOI map. Blue rectangles enclose the zones with a higher average LOS displacement rate. Close-ups of the (**b**) Paraíso and Comalcalco, (**c**) Villahermosa, (**d**) Batería Samaria and (**e**) Batería Cactus zones are presented. The location of hydrocarbon wells is also shown.

In zone c, average LOS displacement rates of up to −4 cm/yr were obtained in the south limits of the Villahermosa urban area, with up to −6 cm/yr in Pomaca and Saloya 2nd located to the north of Villahermosa. In the west limits of the Villahermosa urban area, average LOS displacement rates of up to −3 cm/yr were obtained. This region is the closest urban area to oil-producing wells (Figures 12c and 13c). The center of Villahermosa city could be considered stable (±0.5 cm/yr).

In localities between Comalcalco and Villahermosa (such as Nacajuca, Soyateco, and Jalpa de Méndez), average LOS displacement rates of up to −2 cm/yr were obtained. The city of Cárdenas, one of the main urban centers of the area, did not present a displacement signal, except for a small region southwest of this town, with an average LOS displacement up to −1 cm/yr.

In zones d and e, the obtained average LOS displacement rates reached −3 cm/yr. There is a large number of hydrocarbon-producing wells, i.e., the Batería Samaría II and Batería Captus extraction zones, close to zones d and e, respectively, suggesting a relationship between the hydrocarbon extraction and surface deformation.

4. Discussion

In this work, a differential interferometric analysis using Sentinel-1 SLC images acquired between January 2018 and January 2020 was conducted, to map the LS in TCP, as well as to investigate the potential and limitations of DInSAR for detecting and monitoring the LS in this region. The detailed interferometric coherence analysis revealed that temporal decorrelation is the major challenge for DInSAR application. The AOI is dominantly covered by different types of vegetation, which is a land surface cover with constantly and rapidly changing scattering properties. Moreover, the recurrent floods are an additional source of coherence degradation in TCP. On the other hand, for the short baseline interferometric pairs, where sufficient coherence remains, the APS effect degraded the accuracy of LOS displacement estimates. Therefore, the conventional DInSAR is not an appropriate approach for LS detection and monitoring in the AOI. However, the successful application of advanced multi-temporal DInSAR approaches is still possible. Here, the simplest of the advanced DInSAR approaches, the DIS approach, was applied.

The DIS results revealed that several zones within the AOI are subsiding. The maximum average LOS displacement rate detected in this study (−6 cm/yr) corresponded to the area located north of Villahermosa (Pomaca and Saloya 2nd) and the town of Paraíso. Subsiding zones at the west and south limits of the Villahermosa urban area, the major urban area of the AOI, were detected; whereas Villahermosa city center remained stable during the analyzed period. The zone located at the southern limits of the Villahermosa urban area presented an average LOS displacement rate of up to −4 cm/yr. This zone is also characterized by a location close to areas recurrently affected by floods (see Section 3.2). Thus, LS increases the flood vulnerability of this zone. It is estimated that the Villahermosa urban area will increase to about 15 km^2 by 2050, and one of the possible urban expansion scenarios assumes an expansion to south-southeast [88], which will further increase the flood vulnerability of the zone. LS in the Comalcalco urban area was also detected, as well as in localities between Comalcalco and Villahermosa.

Three subsiding zones were identified near hydrocarbon extraction zones: at the western limits of the Villahermosa urban area, and two zones to the west-southwest of Villahermosa, close to the Batería Samaria II and Batería Cactus hydrocarbons extraction zones, suggesting a possible relationship between the hydrocarbon extraction and surface deformation. However, the possible subsidence caused in the rest of the identified subsidence zones is unclear. LS can be the result of natural processes and anthropic activities. Natural causes such as tectonics (except co-seismic displacement) and soil compaction can cause subsidence of a few mm/yr [40,89]. However, the natural characteristics of TCP are not significant triggers of subsidence: the AOI is located far from any active tectonic plate boundaries (e.g., the Pacific margin), and the compaction of fluvial sediments is maintained only in active alluvial plain areas that have not been subjected to direct anthropic modification (mainly the eastern–southeastern part of the AOI). In these areas, floodable geoforms prevail that accumulate sediments in the rainy season and are characterized by the overflow of rivers [90]. The central and west parts of the AOI belong to an inactive fluvio-deltaic plain, which currently does not receive alluvial sediments, due to the dam system in the middle basin of the Grijalva River, protection boards, and drainage systems [89] that control the river and rain water flows. On the other hand, hydrocarbon production is expected to be the main cause of anthropogenic subsidence in the AOI, as it is the main economic activity. The anthropogenic subsidence caused by gas and oil extraction can reach up to tens of cm/yr [44]. Therefore, anthropic activities could be responsible for the detected LOS displacement rates in the AOI. However, for the zones where there is not a direct spatial correlation between the subsiding zones and hydrocarbon extraction zones, it is impossible to draw conclusions about the origin of subsidence, and more investigations are required.

5. Conclusions

The present study evaluated the potential of DInSAR techniques for detecting LS in the TCP. Coherence degradation, due to temporal decorrelation caused by vegetation, which is the dominant land cover in the AOI, and due to recurrent floods, as well the degradation of the precision of DInSAR measurements due to APS effects, affected the effectiveness of conventional DInSAR application. However, advanced differential SAR interferometry, e.g., the DIS approach tested in this study, could be efficiently used to investigate the LS in the AOI. Using the DIS approach, average LOS displacement rates were obtained for the 2018–2019 period and several subsiding zones were identified. The subsiding zones are located in Paraíso and Comalcalco, at the limits of the Villahermosa urban area and its outskirts, such as Pomaca and Saloya 2nd, as well as in other localities, such as Nacajuca, Soyateco, and Jalpa de Méndez. The subsiding zone at the south limits of the Villahermosa urban area has a spatial correlation with the area recurrently affected by floods, indicating the possible influence of LS on the flood vulnerability of this zone. Three of the detected subsiding areas have a spatial correlation with hydrocarbon extraction areas, suggesting a possible relationship between the hydrocarbon extraction and surface deformation. However, more detailed investigations are required for more precise determination of the origin of subsidence in these, and the other subsiding zones identified in this study. This work represents the first effort to address the topic of subsidence in the TCP and could be used as a reference in future investigations.

Author Contributions: Conceptualization, G.M.-F. and O.S.; formal analysis, Z.P.-F.; funding acquisition, G.M.-F.; investigation, O.S. and G.M.-F.; methodology, O.S.; resources, O.S. and G.M.-F.; software, O.S., Z.P.-F. and G.M.-F.; supervision, G.M.-F. and O.S.; visualization, Z.P.-F.; writing—original draft, Z.P.-F.; writing—review and editing, G.M.-F. and O.S. All authors have read and agreed to the published version of the manuscript.

Funding: The Consejo Nacional de Ciencia y Tecnología de México awarded a doctoral scholarship to Z.P.-F. This work has been supported by the Centro Interdisciplinario de Ciencias Marinas, Instituto Politécnico Nacional (CICIMAR-IPN) within the framework of the project SIP-20195187. The APC was funded by Secretaría de Investigación y Posgrado (SIP-IPN).

Institutional Review Board Statement: Not applicable.

Informed Consent Statement: Not applicable.

Data Availability Statement: Sentinel-1 data are available at Copernicus Sentinel data 2018–2019, processed by the ESA.

Acknowledgments: The authors would like to thank the CICIMAR-IPN and Centro de Estudios Superiores de Ensenada (CICESE) for the support to carry out this work. We thank the anonymous reviewers for their insightful comments and suggestions.

Conflicts of Interest: The authors declare no conflict of interest.

References

1. Prokopovich, N.P. *Classification of Land Subsidence by Origin*; IAHS-AISH Publication; International Association of Hydrological Sciences: Wallingford, UK, 1986; pp. 281–290.
2. Marker, B.R. Land Subsidence. In *Encyclopedia of Natural Hazards*; Bobrowsky, P.T., Ed.; Springer Netherlands: Dordrecht, The Netherlands, 2013; pp. 583–590.
3. Peltier, W.R. Global glacial isostasy and the surface of the ice-age Earth: The ICE-5G (VM2) Model and GRACE. *Annu. Rev. Earth Planet. Sci.* **2004**, *32*, 111–149. [CrossRef]
4. Karegar, M.A.; Dixon, T.H.; Engelhart, S.E. Subsidence along the Atlantic Coast of North America: Insights from GPS and late Holocene relative sea level data. *Geophys. Res. Lett.* **2016**, *43*, 3126–3133. [CrossRef]
5. Dokka, R.K. Modern-day tectonic subsidence in coastal Louisiana. *Geology* **2006**, *34*, 281–284. [CrossRef]
6. Howell, S.; Smith-Konter, B.; Frazer, N.; Tong, X.; Sandwell, D. The vertical fingerprint of earthquake cycle loading in southern California. *Nat. Geosci.* **2016**, *9*, 611–614. [CrossRef]
7. Sarah, D.; Hutasoit, L.M.; Delinom, R.M.; Sadisun, I.A. Natural Compaction of Semarang-Demak Alluvial Plain and Its Relationship to the Present Land Subsidence. *Indones. J. Geosci.* **2020**, *7*, 273–289. [CrossRef]
8. Zhou, X.; Wang, G.; Wang, K.; Liu, H.; Lyu, H.; Turco, M.J. Rates of Natural Subsidence along the Texas Coast Derived from GPS and Tide Gauge Measurements (1904–2020). *J. Surv. Eng.* **2021**, *147*, 04021020. [CrossRef]
9. Galloway, D.; Riley, F.S. San Joaquin Valley, California: Largest human alteration of the Earth's surface. *U.S. Geol. Surv. Circ.* **1999**, *1182*, 23–34.

10. Motagh, M.; Walter, T.R.; Sharifi, M.A.; Fielding, E.; Schenk, A.; Anderssohn, J.; Zschau, J. Land subsidence in Iran caused by widespread water reservoir overexploitation. *Geophys. Res. Lett.* **2008**, *35*. [CrossRef]
11. Zhang, Y.; Gong, H.; Gu, Z.; Wang, R.; Li, X.; Zhao, W. Characterization of land subsidence induced by groundwater withdrawals in the plain of Beijing city, China. *Hydrogeol. J.* **2014**, *22*, 397–409. [CrossRef]
12. Figueroa-Miranda, S.; Tuxpan-Vargas, J.; Ramos-Leal, J.A.; Hernández-Madrigal, V.M.; Villaseñor-Reyes, C.I. Land subsidence by groundwater over-exploitation from aquifers in tectonic valleys of Central Mexico: A review. *Eng. Geol.* **2018**, *246*, 91–106. [CrossRef]
13. Martin, J.C.; Serdengecti, S.; Holzer, T.L. Subsidence over oil and gas fields. In *Man-Induced Land Subsidence*; Geological Society of America: Boulder, CO, USA, 1984; Volume 6, p. 23.
14. Fielding, E.J.; Blom, R.G.; Goldstein, R.M. Rapid subsidence over oil fields measured by SAR interferometry. *Geophys. Res. Lett.* **1998**, *25*, 3215–3218. [CrossRef]
15. Bayramov, E.; Buchroithner, M.; Kada, M.; Zhuniskenov, Y. Quantitative Assessment of Vertical and Horizontal Deformations Derived by 3D and 2D Decompositions of InSAR Line-of-Sight Measurements to Supplement Industry Surveillance Programs in the Tengiz Oilfield (Kazakhstan). *Remote Sens.* **2021**, *13*, 2579. [CrossRef]
16. Allis, R.G. Review of subsidence at Wairakei field, New Zealand. *Geothermics* **2000**, *29*, 455–478. [CrossRef]
17. Fialko, Y.; Simons, M. Deformation and seismicity in the Coso geothermal area, Inyo County, California: Observations and modeling using satellite radar interferometry. *J. Geophys. Res. Solid Earth* **2000**, *105*, 21781–21793. [CrossRef]
18. Sarychikhina, O.; Glowacka, E.; Robles, B. Multi-sensor DInSAR applied to the spatiotemporal evolution analysis of ground surface deformation in Cerro Prieto basin, Baja California, Mexico, for the 1993–2014 period. *Nat. Hazards* **2018**, *92*, 225–255. [CrossRef]
19. Samsonov, S.; d'Oreye, N.; Smets, B. Ground deformation associated with post-mining activity at the French–German border revealed by novel InSAR time series method. *Int. J. Appl. Earth Obs. Geoinf.* **2013**, *23*, 142–154. [CrossRef]
20. Grzovic, M.; Ghulam, A. Evaluation of land subsidence from underground coal mining using TimeSAR (SBAS and PSI) in Springfield, Illinois, USA. *Nat. Hazards* **2015**, *79*, 1739–1751. [CrossRef]
21. Strozik, G.; Jendruś, R.; Manowska, A.; Popczyk, M. Mine Subsidence as a Post-Mining Effect in the Upper Silesia Coal Basin. *Pol. J. Environ. Stud.* **2016**, *25*, 777–785. [CrossRef]
22. Demin, D.; Fengshan, M.; Zhang, Y.; Wang, J.; Guo, J. Characteristics of Land Subsidence Due to Both High-Rise Building and Exploitation of Groundwater in Urban Are. *J. Eng. Geol.* **2011**, *19*, 433–439.
23. Yuan, L.; Cui, Z.-D.; Yang, J.-Q.; Jia, Y.-J. Land subsidence induced by the engineering-environmental effect in Shanghai, China. *Arab. J. Geosci.* **2020**, *13*, 251. [CrossRef]
24. Turner, E.R. Coastal wetland subsidence arising from local hydrologic manipulations. *Estuaries* **2004**, *27*, 265–272. [CrossRef]
25. Ma, F.; Sui, L. Investigation on Mining Subsidence Based on Sentinel-1A Data by SBAS-InSAR technology-Case Study of Ningdong Coalfield, China. *Earth Sci. Res. J.* **2020**, *24*, 373–386. [CrossRef]
26. Erkens, G.; Bucx, T.; Dam, R.; de Lange, G.; Lambert, J. Sinking coastal cities. *Proc. IAHS* **2015**, *372*, 189–198. [CrossRef]
27. Liu, Y.; Li, J.; Fasullo, J.; Galloway, D.L. Land subsidence contributions to relative sea level rise at tide gauge Galveston Pier 21, Texas. *Sci. Rep.* **2020**, *10*, 17905. [CrossRef]
28. Nicholls, R.J.; Lincke, D.; Hinkel, J.; Brown, S.; Vafeidis, A.T.; Meyssignac, B.; Hanson, S.E.; Merkens, J.-L.; Fang, J. A global analysis of subsidence, relative sea-level change and coastal flood exposure. *Nat. Clim. Chang.* **2021**, *11*, 338–342. [CrossRef]
29. Zhou, X.; Chang, N.-B.; Li, S. Applications of SAR Interferometry in Earth and Environmental Science Research. *Sensors* **2009**, *9*, 1876–1912. [CrossRef]
30. Gabriel, A.K.; Goldstein, R.M.; Zebker, H.A. Mapping small elevation changes over large areas: Differential radar interferometry. *J. Geophys. Res. Solid Earth* **1989**, *94*, 9183–9191. [CrossRef]
31. Massonnet, D.; Feigl, K.L. Radar interferometry and its application to changes in the Earth's surface. *Rev. Geophys.* **1998**, *36*, 441–500. [CrossRef]
32. Bürgmann, R.; Rosen, P.A.; Fielding, E.J. Synthetic Aperture Radar Interferometry to Measure Earth's Surface Topography and Its Deformation. *Annu. Rev. Earth Planet. Sci.* **2000**, *28*, 169–209. [CrossRef]
33. Hanssen, R.F. *Radar Interferometry: Data Interpretation and Error Analysis*; Kluwer Academic Publishers: Dordrecht, The Netherlands, 2001; p. 308.
34. Zebker, H.A.; Villasenor, J. Decorrelation in interferometric radar echoes. *IEEE Trans. Geosci. Remote Sens.* **1992**, *30*, 950–959. [CrossRef]
35. Ferretti, A.; Prati, C.; Rocca, F. Permanent scatterers in SAR interferometry. *IEEE Trans. Geosci. Remote Sens.* **2001**, *39*, 8–20. [CrossRef]
36. Berardino, P.; Fornaro, G.; Lanari, R.; Sansosti, E. A new algorithm for surface deformation monitoring based on small baseline differential SAR interferograms. *IEEE Trans. Geosci. Remote Sens.* **2002**, *40*, 2375–2383. [CrossRef]
37. Hooper, A. A multi-temporal InSAR method incorporating both persistent scatterer and small baseline approaches. *Geophys. Res. Lett.* **2008**, *35*, 16. [CrossRef]
38. Aparicio, J.; Martínez-Austria, P.F.; Güitrón, A.; Ramírez, A.I. Floods in Tabasco, Mexico: A diagnosis and proposal for courses of action. *J. Flood Risk Manag.* **2009**, *2*, 132–138. [CrossRef]

39. Gama, L.; Ordoñez, E.M.; Villanueva-García, C.; Ortiz-Pérez, M.A.; Lopez, H.D.; Torres, R.C.; Valadez, M.E.M. Floods In Tabasco Mexico History And Perspectives. *WIT Trans. Ecol. Environ.* **2010**, *133*, 25–33.
40. Da Lio, C.; Tosi, L. Land subsidence in the Friuli Venezia Giulia coastal plain, Italy: 1992–2010 results from SAR-based interferometry. *Sci. Total Environ.* **2018**, *633*, 752–764. [CrossRef]
41. Di Paola, G.; Alberico, I.; Aucelli, P.P.C.; Matano, F.; Rizzo, A.; Vilardo, G. Coastal subsidence detected by Synthetic Aperture Radar interferometry and its effects coupled with future sea-level rise: The case of the Sele Plain (Southern Italy). *J. Flood Risk Manag.* **2018**, *11*, 191–206. [CrossRef]
42. Hung, W.-C.; Hwang, C.; Chen, Y.-A.; Zhang, L.; Chen, K.-H.; Wei, S.-H.; Huang, D.-R.; Lin, S.-H. Land Subsidence in Chiayi, Taiwan, from Compaction Well, Leveling and ALOS/PALSAR: Aquaculture-Induced Relative Sea Level Rise. *Remote Sens.* **2018**, *10*, 40. [CrossRef]
43. Nguyen Hao, Q.; Takewaka, S. Detection of Land Subsidence in Nam Dinh Coast by Dinsar Analyses. In Proceedings of the International Conference on Asian and Pacific Coasts, Singapore, 25–28 September 2020; pp. 1287–1294.
44. Bayramov, E.; Buchroithner, M.; Kada, M.; Duisenbiyev, A.; Zhuniskenov, Y. Multi-Temporal SAR Interferometry for Vertical Displacement Monitoring from Space of Tengiz Oil Reservoir Using SENTINEL-1 and COSMO-SKYMED Satellite Missions. *Front. Environ. Sci.* **2022**, *10*, 783351. [CrossRef]
45. Haley, M.; Ahmed, M.; Gebremichael, E.; Murgulet, D.; Starek, M. Land Subsidence in the Texas Coastal Bend: Locations, Rates, Triggers, and Consequences. *Remote Sens.* **2022**, *14*, 192. [CrossRef]
46. Pérez-Falls, Z.; Martínez-Flores, G. *Land Subsidence in Villahermosa Tabasco Mexico, Using Radar Interferometry*; Springer Nature: Cham, Switzerland, 2020; pp. 18–29.
47. Zavala-Cruz, J.; Palma-López, D.J.; Fernández-Cabrera, C.R.; López-Castañeda, A.; Shirma-Torres, E. *Degradación y Conservación de Suelos en la Cuenca del río Grijalva, Tabasco*; Secretaría de Recursos Naturales y Protección Ambiental, Colegio de Postgraduados, Campus Tabasco: Villahermosa, Tabasco, México, 2011.
48. Palma-López, D.J.; Cisneros-Domínguez, J.; Moreno-Cáliz, E.; Rincón-Ramírez, J.A. *Suelos de Tabasco: Su uso y Manejo Sustentable*; Colegio de Postgraduados-Campus Tabasco, Instituto para el Desarrollo de Sistemas de Producción del Territorio Húmedo de Tabasco, Fundación Produce Tabasca: Villahermosa, Tabasco, México, 2007.
49. Vazquez-Meneses, M.E.; España-Pinto, A.; Rosales-Contreras, E.; Rosales-Rodriguez, J.; Ruiz-Violante, A.; del Valle-Reyes, A. Structural evolution in the Tabasco Coastal Plain, Mexico. *Gulf Coast Assoc. Geol. Soc. Trans.* **2011**, *61*, 671–674.
50. Krasilnikov, P.; Gutiérrez-Castorena, M.D.C.; Ahrens, R.; Cruz-Gaistardo, C.; Sedov, S.; Solleiro-Rebolledo, E. *The Soils of Mexico*; Springer Dordrecht: Madison, WI, USA, 2013.
51. Zavala-Cruz, J. Actualización de la clasificación de suelos de Tabasco, México. *Agro Product.* **2018**, *10*, 29–35.
52. García-Amaro, E. *Modificaciones al Sistema de Clasificación Climática de Köppen*, 5th ed.; Instituto de Geografía, UNAM: Mexico City, México, 2004.
53. Perevochtchikova, M.; Lezama de la Torre, J.L. Causas de un desastre: Inundaciones del 2007 en Tabasco, México. *J. Lat. Am. Geogr.* **2010**, *9*, 73–98. [CrossRef]
54. Copernicus. EMSR479: Flood in Tabasco, Mexico. Available online: https://emergency.copernicus.eu/mapping/list-of-components/EMSR479 (accessed on 20 April 2022).
55. CONAGUA. Resúmenes Mensuales de Temperaturas y Lluvia. Available online: https://smn.conagua.gob.mx/es/climatologia/temperaturas-y-lluvias/resumenes-mensuales-de-temperaturas-y-lluvias (accessed on 25 April 2022).
56. CNH. Atlas Geológico Cuencas del Sureste-Cinturón Plegado de la Sierra de Chiapas. Available online: https://hidrocarburos.gob.mx/media/3094/atlas_geologico_cuencas_sureste_v3.pdf (accessed on 1 August 2022).
57. Chavez Valois, V.M.; Valdés, M.d.L.C.; Juárez Placencia, J.I.; Ortiz, I.A.; Jurado, M.M.; Yánez, R.V.; Tristán, M.G.; Ghosh, S.; Bartolini, C.; Ramos, J.R.R. A New Multidisciplinary Focus in the Study of the Tertiary Plays in the Sureste Basin, Mexico. In *Petroleum Systems in the Southern Gulf of Mexico*; American Association of Petroleum Geologists: Tulsa, OK, USA, 2009; Volume 90. Available online: https://pubs.geoscienceworld.org/books/book/1892/chapter-abstract/107094861/A-New-Multidisciplinary-Focus-in-the-Study-of-the?redirectedFrom=fulltext (accessed on 2 July 2022).
58. Ketelaar, V.B.H. Subsidence Due to Hydrocarbon Production in the Netherlands. In *Satellite Radar Interferometry: Subsidence Monitoring Techniques*; Ketelaar, V.B.H., Ed.; Springer Netherlands: Dordrecht, The Netherlands, 2009; Volume 14, pp. 7–26.
59. Fatholahi, S.N.; He, H.; Wang, L.; Syed, A.; Li, J. Monitoring Surface Deformation Over Oilfield Using MT-Insar and Production Well Data. In Proceedings of the 2021 IEEE International Geoscience and Remote Sensing Symposium IGARSS, Virtual, 11–16 July 2021; pp. 2298–2301.
60. Allen, D.R.; Mayuga, M.N. The mechanics of compaction and rebound, Wilmington oil field, Long Beach, California, USA. In *Land Subsidence*; Tison, L.J., Ed.; International Association of Scientific Hydrology, UNESCO: Wallingford, UK, 1970; pp. 66–79.
61. INEGI. Uso del Suelo y Vegetación, Escala 1:250,000. Available online: http://www.conabio.gob.mx/informacion/gis/?vns=gis_root/usv/inegi/usv250s6gw (accessed on 9 May 2022).
62. IICNIH. Ubicación de Exploración, Perforación y Producción Petrolera. Available online: https://mapa.hidrocarburos.gob.mx/ (accessed on 9 May 2022).
63. ESA. Sentinel-1 SAR Technical Guide. Available online: https://sentinels.copernicus.eu/web/sentinel/technical-guides/sentinel-1-sar (accessed on 26 May 2022).

64. Zan, F.D.; Guarnieri, A.M. TOPSAR: Terrain Observation by Progressive Scans. *IEEE Trans. Geosci. Remote Sens.* **2006**, *44*, 2352–2360. [CrossRef]
65. Yagüe-Martínez, N.; Prats-Iraola, P.; González, F.R.; Brcic, R.; Shau, R.; Geudtner, D.; Eineder, M.; Bamler, R. Interferometric Processing of Sentinel-1 TOPS Data. *IEEE Trans. Geosci. Remote Sens.* **2016**, *54*, 2220–2234. [CrossRef]
66. Torres, R.; Snoeij, P.; Geudtner, D.; Bibby, D.; Davidson, M.; Attema, E.; Potin, P.; Rommen, B.; Floury, N.; Brown, M.; et al. GMES Sentinel-1 mission. *Remote Sens. Environ.* **2012**, *120*, 9–24. [CrossRef]
67. Farr, T.G.; Rosen, P.A.; Caro, E.; Crippen, R.; Duren, R.; Hensley, S.; Kobrick, M.; Paller, M.; Rodriguez, E.; Roth, L.; et al. The Shuttle Radar Topography Mission. *Rev. Geophys.* **2007**, *45*, RG2004. [CrossRef]
68. Anusha, N.; Bharathi, B. Flood detection and flood mapping using multi-temporal synthetic aperture radar and optical data. *Egypt. J. Remote Sens. Space Sci.* **2020**, *23*, 207–219. [CrossRef]
69. Dao, P.D.; Liou, Y.-A. Object-Based Flood Mapping and Affected Rice Field Estimation with Landsat 8 OLI and MODIS Data. *Remote Sens.* **2015**, *7*, 5077–5097. [CrossRef]
70. Pepe, A.; Calò, F. A Review of Interferometric Synthetic Aperture RADAR (InSAR) Multi-Track Approaches for the Retrieval of Earth's Surface Displacements. *Appl. Sci.* **2017**, *7*, 1264. [CrossRef]
71. Zhang, Y.; Prinet, V. InSAR coherence estimation. In Proceedings of the IGARSS 2004, 2004 IEEE International Geoscience and Remote Sensing Symposium, Anchorage, AK, USA, 20–24 September 2004; Volume 3355, pp. 3353–3355.
72. Ma, G.; Zhao, Q.; Wang, Q.; Liu, M. On the Effects of InSAR Temporal Decorrelation and Its Implications for Land Cover Classification: The Case of the Ocean-Reclaimed Lands of the Shanghai Megacity. *Sensors* **2018**, *18*, 2939. [CrossRef] [PubMed]
73. Wegnüller, U.; Werner, C.; Strozzi, T.; Wiesmann, A.; Frey, O.; Santoro, M. Sentinel-1 Support in the GAMMA Software. *Procedia Comput. Sci.* **2016**, *100*, 1305–1312. [CrossRef]
74. Wegnüller, U.; Werner, C. Gamma SAR processor and Interferometry Software. In Proceedings of the Proceedings 3rd ERS Scientific Symposium, Florence, Italy, 17–20 March 1997.
75. GAMMA Remote Sensing AG. *Sentinel-1 Processing with GAMMA Software: Documentation User's Guide*; GAMMA Remote Sensing AG: Gümligen, Switzerland, 2015; p. 38.
76. Goldstein, R.M.; Werner, C.L. Radar interferogram filtering for geophysical applications. *Geophys. Res. Lett.* **1998**, *25*, 4035–4038. [CrossRef]
77. Costantini, M. A novel phase unwrapping method based on network programming. *IEEE Trans. Geosci. Remote Sens.* **1998**, *36*, 813–821. [CrossRef]
78. Costantini, M.; Rosen, P.A. A generalized phase unwrapping approach for sparse data. In Proceedings of the IEEE 1999 International Geoscience and Remote Sensing Symposium, IGARSS'99 (Cat. No.99CH36293), Hamburg, Germany, 28 June–2 July 1999; Volume 261, pp. 267–269.
79. Beauducel, F.; Briole, P.; Froger, J.-L. Volcano-wide fringes in ERS synthetic aperture radar interferograms of Etna (1992–1998): Deformation or tropospheric effect? *J. Geophys. Res. Solid Earth* **2000**, *105*, 16391–16402. [CrossRef]
80. Bekaert, D.P.S.; Walters, R.J.; Wright, T.J.; Hooper, A.J.; Parker, D.J. Statistical comparison of InSAR tropospheric correction techniques. *Remote Sens. Environ.* **2015**, *170*, 40–47. [CrossRef]
81. Psomiadis, E. *Flash Flood Area Mapping Utilising SENTINEL-1 Radar Data*; SPIE: Washington, DC, USA, 2016; Volume 10005.
82. Tavus, B.; Kocaman, S.; Nefeslioglu, H.A.; Gökçeoğlu, C. Flood Mapping Using Sentinel-1 SAR Data: A Case Study of Ordu 8 August 2018 Flood. *Int. J. Environ.* **2019**, *6*, 333–337. [CrossRef]
83. Ratna, R.D.; Neelima, G. Image Segmentation by using Histogram Thresholding. *Int. J. Comp. Sci. Eng. Technol.* **2012**, *2*, 776–779.
84. ESA. Sentinel Application Platform v.7.0. Available online: http://step.esa.int/main/toolboxes/snap/ (accessed on 18 November 2018).
85. Strozzi, T.; Wegmüller, U.; Tosi, L.; Bitelli, G.; Spreckels, V. Land subsidence monitoring with differential SAR Interferometry. *Photogramm. Eng. Remote Sens.* **2001**, *67*, 1261–1270.
86. Wegmüller, U.; Werner, C.; Strozzi, T.; Wiesmann, A. Application of SAR Interferometric techniques for surface deformation monitoring. In Proceedings of the 12th FIG Symposium, Baden, Austria, 22–24 May 2006.
87. Wegmüller, U.; Strozzi, T. Diferential SAR Interferometry for Land Subsidence Monitoring: Methodology and examples. In Proceedings of the International Symposium on Land Subsidence, Ravenna, Italy, 25–29 September 2000.
88. Areu-Rangel, O.S.; Cea, L.; Bonasia, R.; Espinosa-Echavarria, V.J. Impact of Urban Growth and Changes in Land Use on River Flood Hazard in Villahermosa, Tabasco (Mexico). *Water* **2019**, *11*, 304. [CrossRef]
89. Aslan, G.; Cakir, Z.; Lasserre, C.; Renard, F. Investigating Subsidence in the Bursa Plain, Turkey, Using Ascending and Descending Sentinel-1 Satellite Data. *Remote Sens.* **2019**, *11*, 85. [CrossRef]
90. Santoro, M.; Wegmuller, U.; Askne, J.I.H. Signatures of ERS–Envisat Interferometric SAR Coherence and Phase of Short Vegetation: An Analysis in the Case of Maize Fields. *IEEE Trans. Geosci. Remote Sens.* **2010**, *48*, 1702–1713. [CrossRef]

 land

Article

A Novel Spectral Index to Identify Cacti in the Sonoran Desert at Multiple Scales Using Multi-Sensor Hyperspectral Data Acquisitions

Kyle Hartfield [1,*], Jeffrey K. Gillan [1], Cynthia L. Norton [1], Charles Conley [1] and Willem J. D. van Leeuwen [2]

[1] University of Arizona School of Natural Resources and the Environment, Arizona Remote Sensing Center, The University of Arizona, 1064 E. Lowell Street, Tucson, AZ 85721, USA; jgillan@email.arizona.edu (J.K.G.); cnorton1@email.arizona.edu (C.L.N.); charlesconley@email.arizona.edu (C.C.)

[2] University of Arizona School of Natural Resources and the Environment and School of Geography, Development & Environment, Arizona Remote Sensing Center, The University of Arizona, 1064 E. Lowell Street, Tucson, AZ 85721, USA; leeuw@email.arizona.edu

* Correspondence: kah7@email.arizona.edu

Abstract: Accurate identification of cacti, whether seen as an indicator of ecosystem health or an invasive menace, is important. Technological improvements in hyperspectral remote sensing systems with high spatial resolutions make it possible to now monitor cacti around the world. Cacti produce a unique spectral signature because of their morphological and anatomical characteristics. We demonstrate in this paper that we can leverage a reflectance dip around 972 nm, due to cacti's morphological structure, to distinguish cacti vegetation from non-cacti vegetation in a desert landscape. We also show the ability to calculate two normalized vegetation indices that highlight cacti. Furthermore, we explore the impacts of spatial resolution by presenting spectral signatures from cacti samples taken with a handheld field spectroradiometer, drone-based hyperspectral sensor, and aerial hyperspectral sensor. These cacti indices will help measure baseline levels of cacti around the world and examine changes due to climate, disturbance, and management influences.

Keywords: hyperspectral; cacti; drone

Citation: Hartfield, K.; Gillan, J.K.; Norton, C.L.; Conley, C.; van Leeuwen, W.J.D. A Novel Spectral Index to Identify Cacti in the Sonoran Desert at Multiple Scales Using Multi-Sensor Hyperspectral Data Acquisitions. *Land* **2022**, *11*, 786. https://doi.org/10.3390/land11060786

Academic Editors: Sara Venafra, Carmine Serio and Guido Masiello

Received: 21 April 2022
Accepted: 24 May 2022
Published: 26 May 2022

Publisher's Note: MDPI stays neutral with regard to jurisdictional claims in published maps and institutional affiliations.

1. Introduction

The Cactaceae (cacti) family is one of the most threatened plant families on the planet, while being some of the most important flora in arid regions of the North American continent [1,2]. A variety of mammals, birds, and insects rely on cacti for shelter and as a source of nutrients and hydration during the hot summer season [3–10]. People also use various cacti for ornamental horticulture, food, and medicinal purposes [2]. Facing pressure from land conversion for agriculture, horticulture collection, and urban development, nearly a third of cacti worldwide fall into the threatened category [2]. Cacti also face threats from climate change as arid regions become more arid [11,12]. Although cacti are adapted to survive in areas with limited precipitation, extended periods of drought and increases in summer temperatures harm the establishment of seedlings that require moist soils to flourish [11,13].

In other regions of the world, cacti are an invasive species that threatens native plants. In Kenya, degraded rangelands overrun with prickly pear impact forage for both wildlife and livestock [14,15]. Ranchers in the Edwards Plateau region of Texas work to determine the best strategy to control prickly pear encroachment on fire-disturbed rangelands [16]. Research shows clusters of invasive cacti in South Africa, Australia, and Spain due to ideal climates for cacti. Climate change trends also indicate that parts of China, eastern Asia, and central Africa are suitable for future cacti invasions [17].

Despite the recognized need to monitor cacti occurrence and density, information on cacti population trends are relatively unknown over large areas [2]. The limited geographic extent of field surveys (e.g., transects and plots) makes remote sensing a useful approach for mapping cacti. Other studies have demonstrated cacti mapping with the use of convolution neural networks from drone imagery [18] and random forest supervised classification with Sentinel 2 satellite imagery [15]. These methods may be challenging in the southwestern United States, where individual cacti plants are often much smaller than satellite imagery spatial resolution. Additionally, multispectral satellite imagery will have a difficult time separating cacti from other spectrally similar plant species.

A potential key to distinguishing cacti from non-cacti plants lies in their morphological differences and how they perform photosynthesis. Cacti contain cells designed to store water long term and exhibit a form of photosynthesis allowing their stomata to open at night to limit water loss through transpiration. Most cacti do not have leaves and perform photosynthesis with the tissue layer of their stems [19–21]. These structural differences influence the fate of solar radiation in nuanced ways, best observed with hyperspectral imaging sensors [22–25]. Previous research has shown that saguaro cacti (*Carnegiea gigantea*) exhibit a strong dip in near-infrared reflectance values around the 970 nm due to water absorption [24]. Other research showed the utility of using measurements at 970 nm to estimate plant water concentration [26].

This paper had two main objectives:

1. Using hyperspectral field spectroradiometer measurements to examine the spectral signatures of cacti and non-cacti desert adapted plants to find distinguishing characteristics that would allow for the development of a spectral 'cacti index'.
2. Examine the efficacy of cacti signatures and the index to identify cacti from a drone-mounted hyperspectral sensor (3 cm resolution) and an airplane-mounted hyperspectral sensor (1 m resolution).

Accurate identification of cacti will allow us to monitor extent baselines and changes due to climate variability, climate change, disturbance events, and other human-related modifications of the environment.

2. Materials and Methods

2.1. Study Area

Our study used data collected in the Santa Rita Experimental Range (SRER) located about 30 km south of Tucson, AZ (Figure 1). SRER, founded in 1903, is the longest continuously active rangeland research facility in the United States [27]. The National Science Foundation (NSF) National Ecological Observation Network (NEON) has designated SRER as a terrestrial core site [28].

SRER is located in a semi-arid ecosystem with a bimodal precipitation distribution. The area receives 28 to 50 cm of rain annually with the majority of events occurring in the winter and summer. The average annual temperature is 20 °C. Cacti are common in the area due to a lack of extended freezing temperatures during the winter months [27].

Elevation levels range from 900 m in the northwest to 1400 m in the southeast. The vegetation composition consists of creosote (*Larrea tridentata*), prickly pear (*Opuntia engelmannii*), cholla (*Cylindropuntia fulgida*), barrel cactus (*Ferocactus wislizeni*), mesquite (*Prosopis velutina*), palo verde (*genus Parkinsonia*), whitethorn acacia (*Acacia constricta*), yucca (*Yucca rostrata*), lotebush (*Ziziphus obtusifolia*), and various grasses [27].

The airplane-mounted hyperspectral AVIRIS (Airborne Visible-Infrared Imaging Spectrometer) exploration occurred in the northwest part of SRER, an ideal environment for cacti observed through a high occurrence of prickly pear and cholla cacti (Figure 1). The drone-based hyperspectral analysis occurred closer to the center of the SRER at an elevation of 1130 m. This area had plentiful prickly pear and barrel cacti. Due to the higher elevation and monsoon rains (collected in August 2021), the area had a substantial herbaceous cover of grasses and forbs.

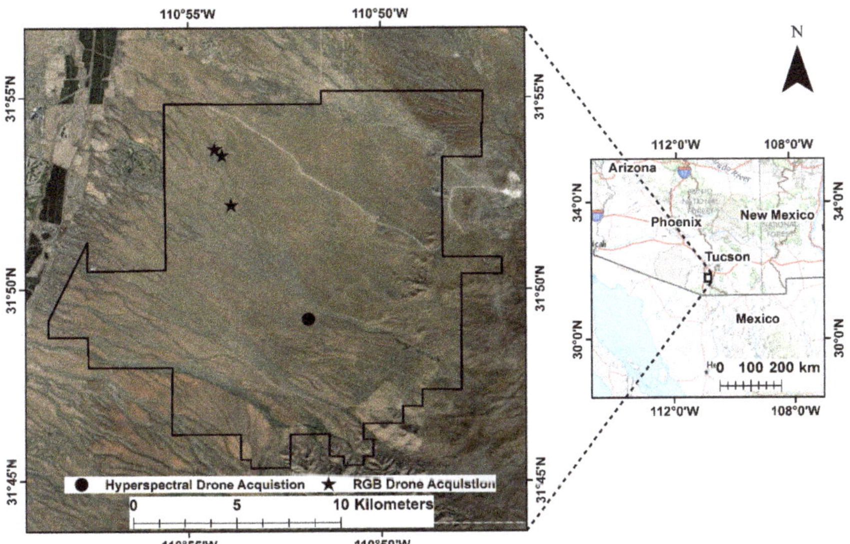

Figure 1. We conducted this study at the Santa Rita Experimental Range (SRER), located about 30 km south of Tucson, Arizona. We identified cacti samples for the NEON AVIRIS imagery analysis using photos acquired with a DJI Mavic Pro multirotor in the northwest portion of the study area (black stars), an area with a high density of cacti. We collected drone-mounted hyperspectral imagery in the center of the SRER, where prickly pear and barrel cacti are plentiful (black circle).

2.2. Data and Methods

2.2.1. Field Spectroradiometer

We acquired outdoor clear-sky hyperspectral nadir measurements with a full-range (350–2500 nm) spectroradiometer, FieldSpec®3 by Analytical Spectral Devices Inc. (ASD) Boulder, CO, USA, with a 15-degree field of view during the summer of 2017. We used a calibrated Spectralon panel to compute component/surface reflectance data [24]. We took measurements from 5 chollas, 11 prickly pear cacti, 9 barrel cacti, 32 mesquite, 9 creosote, 26 patches of bare soil, and 28 plots of grass. We averaged the measurements for each type of sample.

From the field spectroradiometer data, we identified the same reflectance dip at 972 nm first reported by van Leeuwen. This reflectance dip is due to water absorption in succulent cacti and can serve as a distinguishing characteristic from non-cacti vegetation [24].

We measured the magnitude of the dip with a normalized difference approach between the reflectance values at the bottom of the dip and reflectance values immediately outside of the dip. This is similar to the concept of the normalized difference vegetation index (NDVI) [29].

A priori, it was unknown which specific bands outside of the dip would provide a robust index capable of distinguishing cacti from non-cacti vegetation. We tested two different spectral indices.

Cacti Index 1 (CI1) uses reflectance at 862 nm, which occurs immediately before the dip at 972 nm:

$$\text{Cacti Index 1} = \frac{862\ \text{nm} - 972\ \text{nm}}{862\ \text{nm} + 972\ \text{nm}}$$

The second equation, Cacti Index 2 (CI2), uses reflectance at 1072 nm, which occurs immediately after the dip at 972 nm:

$$\text{Cacti Index 2} = \frac{1072\ \text{nm} - 972\ \text{nm}}{1072\ \text{nm} + 972\ \text{nm}}$$

2.2.2. Taking the Cacti Index Airborne

Using the cacti index with aerial and drone-acquired imagery would greatly improve its utility for mapping and monitoring cacti populations across large tracts of land. We also wanted to examine the impact of spatial resolution on the uniqueness and clarity of the cacti signature. We tested the efficacy of the cacti index using both drone- (3 cm spatial resolution) and airplane-mounted (1 m spatial resolution) hyperspectral sensors.

To investigate the utility of the cacti index with drone scale imagery, we used a Nano Hyperspectral (https://www.headwallphotonics.com/products/vnir-400-1000nm (accessed on 23 May 2022)) Visual & Near Infrared (VNIR) sensor by Headwall Photonics gimbal mounted on a DJI Matrice 600 Pro 6 rotor copter, which is a push broom slit sensor with 640 linear array detectors. The data contain 270 bands ranging from 400 nm (blue) to 1000 nm (near infrared) at ~2.2 nm slices. In August 2021, we collected imagery over a 2 ha plot in SRER known to have a mix of cacti and non-cacti vegetation. We flew the drone ~65 m above ground level, yielding a spatial resolution of ~3 cm. Using Headwall software, we converted the raw imagery digital numbers to radiance and then to reflectance using a tarp with known reflectance values. Individual frames were then orthorectified and mosaicked into a stacked imagery product. On the stacked orthomosaic, we identified and extracted spectral signatures from 11 prickly pear, 7 mesquite, 10 barrel cacti, 10 patches of bare ground, and 10 samples of grasses. Due to known sensor noise in the near-infrared region of the spectrum, we employed a three-band moving average to smooth the spectral signatures. We calculated CI1 using bands 212 (864 nm) and 260 (970 nm) and extracted these values from the same vegetation samples. The sensor is not sensitive to radiation beyond 1000 nm, so we were unable to calculate CI2.

To investigate the utility of the Cacti Indices at an airplane imagery scale, we used NEON Airborne Observation Platform hyperspectral data collected across the SRER between August 24 and 29 of 2018. The sensor was a next-gen version of the Airborne Visible/Infrared Imaging Spectrometer (AVIRIS-NG) (https://avirisng.jpl.nasa.gov/aviris-ng.html (accessed on 23 May 2022)). The data contain 426 bands ranging from 380 nm (blue) to 2500 nm (near infrared) at ~5.5 nm slices. The aircraft flew ~1000 m above ground level with a nominal spatial resolution of 1 m. NEON used ATCOR4r [30] to atmospherically correct the data and serve it as unitless surface reflectance values scaled by 10,000.

To facilitate the identification of individual cacti and non-cacti vegetation samples, we collected 35 ha of high-resolution (1.6 cm) RGB drone imagery using a DJI Mavic Pro multirotor. We found this method to be more efficient than locating samples on foot. We co-registered orthomosaics created from the drone imagery with the NEON AVIRIS imagery using ArcGIS Pro. We identified and extracted the spectral signature from many samples of cholla (N = 418), prickly pear (N = 325), mesquite (N = 105), palo verde (N = 100), creosote (N = 100), and bare ground (N = 300). We calculated CI1 and CI2 using AVIRIS bands 97 (862 nm), 119 (972 nm), and 139 (1072 nm).

3. Results

3.1. Spectral Signatures from Ground-, Drone-, and Airplane-Based Sensors

Spectral signatures for all three sensors showed similar characteristics but nuanced differences. The prickly pear, cholla, and barrel cactus spectral signatures, as measured from the ASD spectroradiometer, all exhibit a dip in reflectance centered at 972 nm and a peak around 1072 nm (Figure 2). The non-cacti classes do not have this dip. The prickly pear and cholla signatures show a similar dip at 972 nm and a peak at 1072 nm when extracted from the NEON AVIRIS data. However, the absolute reflectance values are lower and the dip is shallower. From the drone-mounted Nano Hyperspectral sensor, the same water absorption dip at 972 nm is present in the barrel and prickly pear cacti signatures, but the absolute reflectance values are also lower than the ASD.

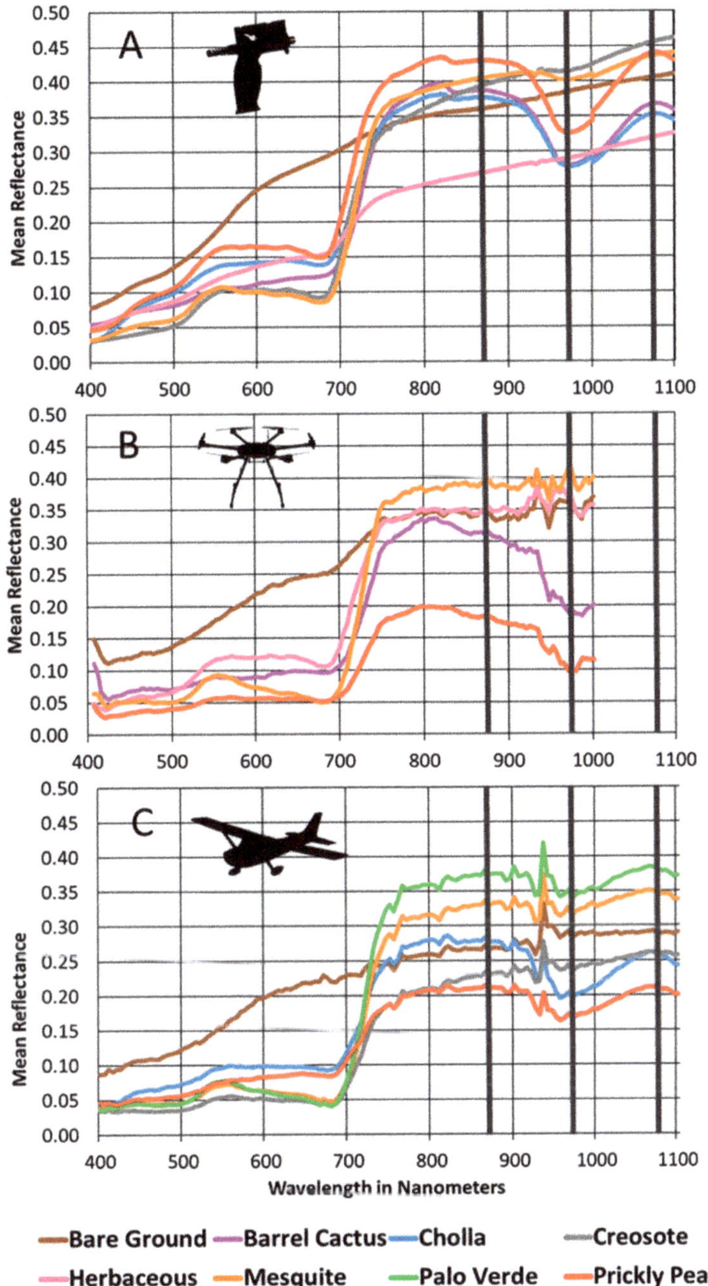

Figure 2. Average spectral signatures of mesquite (orange line), creosote (grey line), bare ground (brown line), herbaceous (pink line), barrel cactus (purple line), cholla (blue line), and prickly pear (red line), as measured from by the ASD spectroradiometer (**A**), drone-mounted Nano Hyperspectral sensor (**B**), and airplane-mounted NEON AVIRIS (**C**). The three dark grey lines represent the three portions of the electromagnetic spectrum used to calculate the two cacti indices.

3.2. Range of Cacti Indices by Plant Type

Using the field spectroradiometer, we sampled bare ground (N = 26), barrel cactus (N = 9), cholla (N = 5), creosote (N = 9), herbaceous (N = 28), mesquite (N = 32), and prickly pear (N = 11). CI1 values calculated from the field spectroradiometer show a separation of the cacti from non-cacti vegetation and bare ground (Figure 3). The majority of CI1 values for the cholla samples range from 0.173 to 0.186, with a mean value of 0.133. The majority of CI1 values for prickly pear samples range from 0.095 to 0.177, with a mean value of 0.103. The majority of CI1 values for the barrel cactus samples range from 0.101 to 0.257 with a mean value of 0.125.

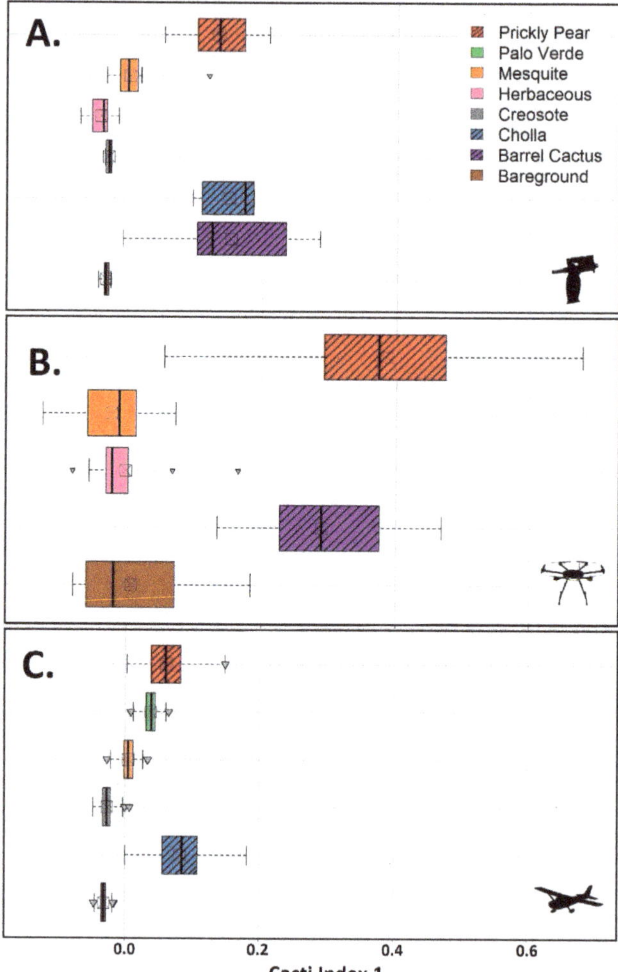

Figure 3. This graph shows the range of Cacti Index 1 (862 and 972 nm) values for cacti, non-cacti vegetation, and bare ground computed from the Field Spectrometer (**A**), Nano Hyperspectral (**B**), and NEON AVIRIS (**C**) collections. The red hatched boxes represent prickly pear, the blue hatched boxes represent cholla, and the purple hatched boxes represent barrel cactus. The square in the middle of each box is the mean for that series of samples. The whiskers represent the minimum and maximum values within the interquartile range. Any triangles outside the whiskers are outlier values.

We identified bare ground (N = 10), barrel cactus (N = 10), herbaceous (N = 10), mesquite (N = 7), and prickly pear (N = 11) using the drone-mounted Nano hyperspectral sensor. CI1 values calculated using the Nano hyperspectral data for the selected samples show a separation of the barrel cactus and prickly pear (cacti) from non-cacti vegetation and bare ground. The majority of values for the barrel cactus samples range from 0.205 to 0.390, while the majority of values for the prickly pear samples range from 0.264 to 0.482. The mean value for the barrel cactus samples is 0.289 and the mean value for the prickly pear samples is 0.385 (Figure 3).

The greater coverage of the NEON AVIRIS data made it possible for us to recognize bare ground (N = 300), creosote (N = 418), herbaceous (N = 100), mesquite (N = 105), palo verde (N = 100), and prickly pear (N = 325). CI1 values calculated using the NEON AVIRIS data for the selected pixels show a separation of the cholla and prickly pear (cacti) from non-cacti vegetation and bare ground. The majority of values for the cholla samples range from 0.056 to 0.109, while the majority of values for the prickly pear samples range from 0.039 to 0.083 on the CI1. The mean value for the cholla samples is 0.083 and the mean value for the prickly pear samples is 0.062. The only other series of samples that overlaps the two cacti boxes are those for palo verde with a range from 0.032 to 0.045 and a mean of 0.040 (Figure 3).

We used the same bare ground (N = 26), barrel cactus (N = 9), cholla (N = 5), creosote (N = 9), herbaceous (N = 28), mesquite (N = 32), and prickly pear (N = 11) samples pulled from the field spectroradiometer to investigate CI2. The CI2 values show a separation of cacti from the other land cover types (Figure 4). The mean value for the cholla samples is 0.083 with the majority of values falling between 0.173 and 0.186. The mean value for the prickly pear samples is 0.143 with the majority of values falling between 0.948 and 0.177. The majority of barrel cactus samples have values between 0.101 and 0.257, with a mean value of 0.134.

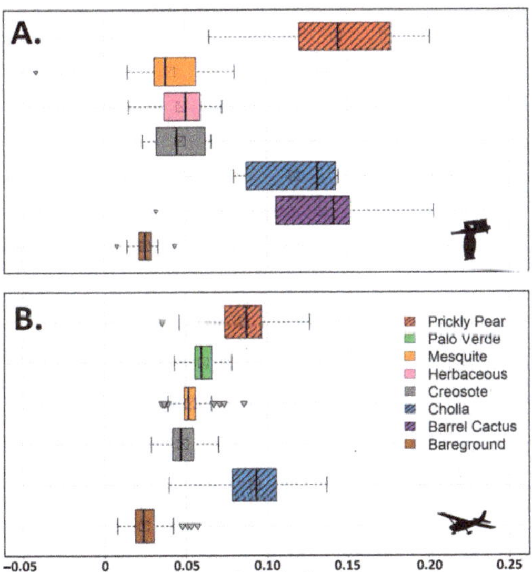

Figure 4. This graph shows the range of Cacti Index 2 (972 and 1072 nm) values for cacti, non-cacti vegetation, and bare ground computed from the Field Spectrometer (**A**) and NEON AVIRIS (**B**) collections. The red hatched boxes represent prickly pear, the blue hatched boxes represent cholla, and the purple hatched boxes represent barrel cactus. The square in the middle of each box is the mean for that series of samples. The whiskers represent the minimum and maximum values within the interquartile range. Any triangles outside the whiskers are outlier values.

Using the same examples of bare ground (N = 300), creosote (N = 418), herbaceous (N = 100), mesquite (N = 105), palo verde (N = 100), and prickly pear (N = 325) extracted from the NEON AVIRIS data we computed CI2. The calculated CI2 values for the selected pixels show a separation of the cholla and prickly pear (cacti) from non-cacti vegetation and bare ground. The majority of values for the cholla samples range from 0.078 to 0.106, while the majority of values for the prickly pear samples range from 0.074 to 0.097. The mean value for the cholla samples is 0.092 and the mean value for the prickly pear samples is 0.086 (Figure 4).

4. Discussion

The spectral signatures created from the direct measurements performed with the hand-held spectroradiometer demonstrate the unique features of cacti. It is clear that a water absorption dip occurs around 972 nm and reflectance peaks occur around 862 nm and 1072 nm for barrel, cholla, and prickly pear cacti. The dip is also present in the data of drone-mounted and airplane-mounted hyperspectral sensors. Nuanced differences between the aerial spectra and the hand-held spectra could be caused by a variety of factors, including 1. spatial resolution, 2. radiometric sensitivity, 3. The reflectance calculation method, and 4. atmospheric water absorption of spectra around 972 nm [31].

We observed diminished sensitivity of the Cacti Indices within the NEON AVIRIS data. This was probably a function of the coarser spatial resolution of that imagery. Within 1 m pixels, instead of being pure cacti reflectance, the features are often a mix of vegetation, bare ground, and shadow. Prickly pear samples were especially prone to mixed signals due to their spreading structural form (Figure 5). However, prickly pears tended to be large, 2–4 m in diameter, making up for shadowing created by their structural composition. Barrel cactus and cholla usually had less mixed signals due to their more compact morphology with diameters as small as 50 cm. Despite reduced sensitivity from AVIRIS NEON data, the cacti indices still demonstrated the ability to separate cacti from non-cacti vegetation. Though both CI1 and CI2 show separability between cacti and non-cacti vegetation, CI1 shows a broader separation, making it the preferred index in our study area.

Figure 5. On the **left**: Example of a prickly pear cactus and a cholla cactus as captured by the DJI Mavic Pro Quadcopter. The nominal spatial resolution of the image is 1.6 cm. The difference in structure between the cholla and prickly pear is clear as seen by the influence of shadowing on the prickly pear cactus. On the **right**: The Cacti Index 1 visualized using the 1 m NEON AVIRIS reflectance data. It can be seen that cholla and prickly pear have higher cacti index values.

The Nano Hyperspectral sensor has diminished radiometric sensitivity near the edges of its range (near 400 nm and 1000 nm), which leads to a lower signal to noise ratio than other spectral bands in the sensor. This hardware limitation impacts CI1 produced from the Nano Hyperspectral sensor because 972 nm is near the edge of the silicon-based sensitivity.

The implication is that in addition to identifying cacti, the index produces many false high values for low light shadowed areas (Figure 6). Mitigation strategies could include identifying and removing low radiance pixels found in shadowed areas, or collecting imagery with longer exposure time. Ideally, drone-based mapping of cacti should use a hyperspectral sensor with a wider range (i.e., >1000 nm) than the Nano Hyperspectral sensor provides.

Figure 6. Drone-mounted Nano hyperspectral imagery shown as true color (**left** panel) and Cacti Index 1 (**right** panel). Cacti were easily identifiable with the index; however, sensitivity problems near the edge of the sensor range created false high index values in low light shadowed areas of the study area.

The spectral slices captured by the AVIRIS NEON sensor (5.5 nm) and Nano Hyperspectral sensor (2.4 nm) played a role in cacti detection. The smaller the slices the better the sensor is at capturing the difference between the peak at 862 nm and the dip at 972 nm. This led to a smaller dynamic range in the CI1 values calculated with the AVIRIS NEON data compared to the CI1 values calculated with the Nano Hyperspectral data.

The water absorption dip (972 nm) observed in the drone-based and airplane-based imagery demonstrate that the Cacti Indices can be used for the identification and mapping of cacti species across larger extents. Depending on the application, the Cacti Indices could be combined with other sensor data (e.g., LiDAR height information), used within a supervised classification framework, or implemented with a user-defined threshold.

The Cacti Indices should be effective in identifying succulent cacti in other regions of the world if the plant species store water in their tissue similar to barrel, cholla, and prickly pear cacti. However, localized research should investigate the extent to which non-target species might exhibit high index values and confuse cacti identification. In our study area for example, we discovered that Palo Verde, with their photosynthesizing stems, exhibited CI1 values nearly as high as prickly pear samples.

Using the Cacti Indices with satellite imagery may be possible but will be challenging. Our methods require narrow spectral bands and high spatial resolution to identify individual cacti. Additionally, the spectral area of interest near 972 nm is greatly impacted by water vapor in the atmosphere [31]. As a result, many satellite sensors do not have bands sensitive to this spectral region, and if they do, the signal will be quite weak. The Earth Observing-1 (EO-1) Hyperion (2000–2017) and a few other orbital hyperspectral reflectance sensors (HISUI-Hyperspectral Imager Suite-onboard the International Space Station; EnMAP-Environmental Mapping and Analysis Program; PRISMA-Hyperspectral Precursor and Application Mission) could possibly leverage the Cacti Indices [32]. These hyperspectral sensors all have moderate spatial resolutions (e.g., 30 m) that are unable to detect individual cacti, but could be used to estimate percent cover of cacti per pixel. More research on these sensors and platforms is needed.

5. Conclusions

In this paper, we demonstrated the ability to create an index from hyperspectral data to accurately identify cacti. We used a hand-held spectroradiometer and two imaging spectrometers to collect spectral information from cholla, barrel, and prickly pear cacti plants and other prevalent landscape features. Based on those collections, we identified three unique aspects in the spectral reflectance signatures for cacti at 862 nm, 972 nm, and 1072 nm bands of the electromagnetic spectrum. Using those three portions of the cacti spectral signatures, we calculated two normalized difference indices: Cacti 1 (862 nm and 972 nm) and Cacti 2 (972 nm and 1072 nm). We then used hyperspectral data captured by drone and airplane to show the applicability of the Cacti indices at various spatial resolutions. Cacti samples showed spectral uniqueness in both the 3 cm drone hyperspectral imagery and the 1 m aerial hyperspectral imagery using the cacti indices.

Whether for conservation or control applications, the cacti indices derived from aerial platforms can help identify cacti across larger landscapes than is possible with field-based measurements. Hyperspectral data provide more precise spectral observations of plant characteristics than multi-spectral imagery sources. Though hyperspectral imagery availability is currently limited, interest in the technology is strong across land management, agriculture, and mining industries. Availability of hyperspectral imagery at multiple airborne and spaceborne scales is likely to proliferate in the near future.

Author Contributions: Conceptualization, W.J.D.v.L. and C.C.; methodology, W.J.D.v.L., C.C. and K.H.; software, K.H., C.C. and J.K.G.; validation, W.J.D.v.L., C.L.N., C.C., K.H. and J.K.G.; formal analysis, W.J.D.v.L., C.L.N., C.C., K.H. and J.K.G.; investigation, W.J.D.v.L., C.L.N., C.C., K.H. and J.K.G.; resources, W.J.D.v.L.; data curation, C.L.N., K.H. and J.K.G.; writing—original draft preparation, K.H.; writing—review and editing, W.J.D.v.L., K.H. and J.K.G.; visualization, K.H. and J.K.G.; supervision, W.J.D.v.L.; project administration, W.J.D.v.L.; funding acquisition, W.J.D.v.L. All authors have read and agreed to the published version of the manuscript.

Funding: This research received no external funding. Technology and Research Initiative Fund (TRIF) funding was provided to purchase the hyperspectral drone.

Data Availability Statement: Data can be obtained by contacting the lead author.

Acknowledgments: The authors appreciate the research support provided by the Arizona Remote Sensing Center. We are thankful to Angela Chambers, who helped with data acquisition and image interpretation. We downloaded the NEON AVIRIS hyperspectral data from the NSF NEON Data Portal.

Conflicts of Interest: The authors declare no conflict of interest. The funders had no role in the design of the study; in the collection, analyses, or interpretation of data; in the writing of the manuscript, or in the decision to publish the results.

References

1. Barthlott, W.; Burstedde, K.; Geffert, J.; Ibisch, P.; Korotkova, N.; Miebach, A.; Rafiqpoor, M.D.; Stein, A.; Mutke, J. Biogeography and biodiversity of cacti. *Schumannia* **2015**, *7*, 208.
2. Goettsch, B.; Hilton-Taylor, C.; Cruz-Piñón, G.; Duffy, J.; Frances, A.; Hernández, H.; Inger, R.; Pollock, C.; Schipper, J.; Superina, M.; et al. High proportion of cactus species threatened with extinction. *Nat. Plants* **2015**, *1*, 15142. [CrossRef] [PubMed]
3. Orr, T.J.; Newsome, S.D.; Wolf, B.O. Cacti supply limited nutrients to a desert rodent community. *Oecologia* **2015**, *178*, 1045–1062. [CrossRef] [PubMed]
4. Wolf, B.O.; Martinez del Rio, C. Use of saguaro fruit by white-winged doves: Isotopic evidence of a tight ecological association. *Oecologia* **2000**, *124*, 536–543. [CrossRef] [PubMed]
5. Wolf, B.O.; Rio, C.M.d. How important are columnar cacti as sources of water and nutrients for desert consumers? A review. *Isot. Environ. Health Stud.* **2003**, *39*, 53–67. [CrossRef]
6. Carter, P.; Rollins, D.; Scott, C. Initial effects of prescribed burning on survival and nesting success of northern bobwhites in west-central texas. *Natl. Quail Symp.* **2002**, *5*, 23.
7. Mader, W.J. A comparative nesting study of red-tailed hawks and harris' hawks in southern arizona. *Auk* **1978**, *95*, 327–337. [CrossRef]
8. Stutchbury, B.J. Coloniality and breeding biology of purple martins (*Progne subis hesperia*) in saguaro cacti. *Condor* **1991**, *93*, 666–675. [CrossRef]

9. Bravo-Avilez, D.; Navarrete-Heredia, J.L.; Rendón-Aguilar, B. New hosts of insects associated with the process of rot damage in edible columnar cacti of central mexico. *Southwest. Entomol.* **2019**, *44*, 637–646. [CrossRef]
10. Miller, T.E.X.; Louda, S.M.; Rose, K.A.; Eckberg, J.O. Impacts of insect herbivory on cactus population dynamics: Experimental demography across an environmental gradient. *Ecol. Monogr.* **2009**, *79*, 155–172. [CrossRef]
11. Aragón-Gastélum, J.; Badano, E.; Yáñez-Espinosa, L.; Ramírez-Tobías, H.; Rodas-Ortiz, J.; Gonzalez-salvatierra, C.; Flores, J. Seedling survival of three endemic and threatened mexican cacti under induced climate change. *Plant Species Biol.* **2017**, *32*, 92–99. [CrossRef]
12. Overpeck, J.; Udall, B. Climate change and the aridification of north america. *Proc. Natl. Acad. Sci. USA* **2020**, *117*, 202006323. [CrossRef] [PubMed]
13. Conver, J.L.; Foley, T.; Winkler, D.E.; Swann, D.E. Demographic changes over >70 yr in a population of saguaro cacti (*Carnegiea gigantea*) in the northern sonoran desert. *J. Arid Environ.* **2017**, *139*, 41–48. [CrossRef]
14. Moran, S. A Plague of Cactus. *bioGraphic*. 2019. Available online: https://www.biographic.com/a-plague-of-cactus/ (accessed on 20 April 2022).
15. Muthoka, J.M.; Salakpi, E.E.; Ouko, E.; Yi, Z.-F.; Antonarakis, A.S.; Rowhani, P. Mapping *Opuntia stricta* in the arid and semi-arid environment of kenya using sentinel-2 imagery and ensemble machine learning classifiers. *Remote Sens.* **2021**, *13*, 1494. [CrossRef]
16. Sosa, G. *Prescribed Fire as the Catalyst to Control Prickly Pear Cactus Encroachment and Restore the Ecological Intergrity of Texas Rangelands*; Texas A&M University: College Station, TX, USA, 2019.
17. Novoa, A.; Le Roux, J.J.; Robertson, M.P.; Wilson, J.R.U.; Richardson, D.M. Introduced and invasive cactus species: A global review. *AoB Plants* **2014**, *7*, plu078. [CrossRef]
18. Atitallah, S.B.; Driss, M.; Boulila, W.; Koubaa, A.; Atitallah, N.; Ben Ghezala, H. An Enhanced Randomly Initialized Convolutional Neural Network for Columnar Cactus Recognition in Unmanned Aerial Vehicle. *Imag. Procedia Comput. Sci.* **2021**, *192*, 573–581. [CrossRef]
19. Edwards, E.; Donoghue, M. Pereskia and the origin of the cactus life-form. *Am. Nat.* **2006**, *167*, 777–793. [CrossRef]
20. Gibson, A.C.; Nobel, P.S. *The Cactus Primer*; Harvard University Press: Cambridge, MA, USA, 1986.
21. Nobel, P.; Hartsock, T. Leaf and stem CO_2 uptake in the three subfamilies of the cactaceae. *Plant Physiol.* **1986**, *80*, 913–917. [CrossRef]
22. Everitt, I.H.; Richardson, A.J.; Nixon, P.R. Canopy reflectance characteristics of succulent and nonsucculent rangeland plant species. *Photogramm. Eng. Remote Sens.* **1986**, *52*, 1891–1897.
23. Kokaly, R.F.; Clark, R.N.; Swayze, G.A.; Livo, K.E.; Hoefen, T.M.; Pearson, N.C.; Wise, R.A.; Benzel, W.M.; Lowers, H.A.; Driscoll, R.L.; et al. *Usgs Spectral Library Version 7*; USGS Publications Warehouse: Reston, VA, USA, 2017; Volume 1035, p. 68.
24. van Leeuwen, W.J.D. Visible, near-ir, and shortwave ir spectral characteristics of terrestrial surfaces. In *The Sage Handbook of Remote Sensing*; Sage Publications Ltd.: London, UK, 2009; pp. 33–50.
25. Gates, D.M.; Keegan, H.J.; Schleter, J.C.; Weidner, V.R. Spectral properties of plants. *Appl. Opt.* **1965**, *4*, 11–20. [CrossRef]
26. Penuelas, J.; Pinol, J.; Ogaya, R.; Filella, I. Estimation of plant water concentration by the reflectance water index wi (r900/r970). *Int. J. Remote Sens.* **1997**, *18*, 2869–2875. [CrossRef]
27. McClaran, M.P. A Century of Vegetation Change on the Santa Rita Experimental Range. In *Santa Rita Experimental Range: 100 Years (1903 to 2003) of Accomplishments and Contributions, Conference Proceedings, Tucson, AZ, 30 October–1 November 2003*; McClaran, M.P., Ffolliott, P.F., Edminster, C.B., Eds.; U.S. Department of Agriculture, Forest Service: Tucson, AZ, USA, 2003; pp. 16–33.
28. National Ecological Observatory Network. About Field Sites and Domains. Available online: https://www.neonscience.org/field-sites/about-field-sites (accessed on 30 September 2021).
29. Tucker, C.J. Red and photographic infrared linear combinations for monitoring vegetation. *Remote Sens. Environ.* **1979**, *8*, 127–150. [CrossRef]
30. Richter, R. Status of Model Atcor4 on Atmospheric/Topographic Correction for Airborne Hyperspectral Imagery. In Proceedings of the 3rd EARSeL Workshop on Imaging Spectroscopy, Herrsching, Germany, 13–16 May 2003.
31. Jensen, J.R. *Remote Sensing of the Environment: An Earth Resource Perspective*; Pearson Prentice Hall: Hoboken, NJ, USA, 2007.
32. Ustin, S.L.; Middleton, E.M. Current and near-term advances in earth observation for ecological applications. *Ecol. Processes* **2021**, *10*, 1. [CrossRef] [PubMed]

Article

A New Earth Observation Service Based on Sentinel-1 and Sentinel-2 Time Series for the Monitoring of Redevelopment Sites in Wallonia, Belgium

Sophie Petit [1,*], Mattia Stasolla [2], Coraline Wyard [1], Gérard Swinnen [1], Xavier Neyt [2] and Eric Hallot [1]

1 Remote Sensing and Geodata Unit, Institut Scientifique de Service Public (ISSeP), 4000 Liège, Belgium; c.wyard@issep.be (C.W.); g.swinnen@issep.be (G.S.); e.hallot@issep.be (E.H.)
2 Signal and Image Centre, Royal Military Academy, 1000 Brussels, Belgium; mattia.stasolla@rma.ac.be (M.S.); xavier.neyt@rma.ac.be (X.N.)
* Correspondence: s.petit@issep.be

Abstract: Urban planning is a challenge, especially when it comes to limiting land take. In former industrial regions such as Wallonia, the presence of a large number of brownfields, here called "redevelopment sites", opens up new opportunities for sustainable urban planning through their revalorization. The Walloon authorities are currently managing an inventory of more than 2200 sites, which requires a significant amount of time and resources to update. In this context, the Sentinel satellites and the Terrascope platform, the Sentinel Collaborative Ground Segment for Belgium, enabled us to deploy SARSAR, an Earth observation service used for the automated monitoring of redevelopment sites that generates regular and automatic change reports that are directly usable by the Walloon authorities. In this paper, we present the methodological aspects and implementation details of the service, which combines two well-known and robust methods: the Pruned Exact Linear Time method for change point detection and threshold-based classification, which assigns the detected changes to three different classes (vegetation, building and soil). The overall accuracy of the system is in the range of 70–90%, depending on the different methods and classes considered. Some remarks on the advantages and possible drawbacks of this approach are also provided.

Keywords: automatic monitoring; time series; change detection; Sentinel-1; Sentinel-2; urban planning

Citation: Petit, S.; Stasolla, M.; Wyard, C.; Swinnen, G.; Neyt, X.; Hallot, E. A New Earth Observation Service Based on Sentinel-1 and Sentinel-2 Time Series for the Monitoring of Redevelopment Sites in Wallonia, Belgium. *Land* **2022**, *11*, 360. https://doi.org/10.3390/land11030360

Academic Editor: Sara Venafra

Received: 28 January 2022
Accepted: 25 February 2022
Published: 1 March 2022

1. Introduction

In former industrialized regions characterized by a large number of brownfields and a high population density, such as Wallonia (the southern region of Belgium), offering new living spaces while limiting land take has become a challenge. The management of vacant lands is then a key to urban planning, as monitoring abandoned sites can support policy and decision-making [1]. In Wallonia, many industrial sites were developed during three distinct periods between the end of the 18th century and the middle of the 20th century. However, since the middle of the 20th century, industrial sites have been increasingly abandoned, first due to the closure of coal mines, then of manufacturing and metallurgical industries. Moreover, a phenomenon of relentless de-urbanization has increasingly emptied the urban centers. This has led to the development of industrial and urban wastelands, which, depending on their origin, can vary in size from a few dozen square meters to a few dozen hectares (e.g., coal mines or blast furnaces), with 75% of them being less than one hectare. As the vast majority of these sites are located in urban areas, they negatively impact the urban fabric but also represent an opportunity for sustainable urban planning as they can be revalorized, with their reuse being a fundamental asset in land management [2]. Therefore, the Walloon authorities have proposed a detailed definition for those sites and have catalogued them into an exhaustive inventory [3,4]. The redevelopment sites (RDSs) are thus defined as "property or group of properties that have been or are intended to be used for an activity, excluding housing, and whose current state is against land management best practices, or constitutes a deconstruction of the urban fabric" [5]. The RDS inventory, which enables potential investors and public authorities to find out about vacant land and its condition, currently contains more than 2200 sites and is available online [6]. Updating it is essential to keep a record of all the sites that have already been enhanced and provide reliable information to the actors consulting the database. Currently, this update is performed, on the one hand, by the visual analysis of orthophotos annually provided, as open data, over the entire Walloon territory and, on the other hand, by systematic field visits. These methodologies are time-consuming and costly. Indeed, the first solution requires several months of work for the analysis of all the RDSs included in the inventory; moreover, the results can only be provided once a year, and there is also a delay between the moment of data acquisition and their availability. As for the second solution, the systematic field visits, the analysis is spread over several years. However, the Walloon authorities estimate that less than 10% of the RDSs are likely to be redeveloped from one year to the other and show major changes (the three classes of interest for the administration are buildings, vegetation and soil). It is, therefore, necessary to find a way to reduce the time spent on the inventory update by providing operators with a list of sites presenting indications of significant changes that would enable them to concentrate their efforts on these sites. The problem of how to efficiently monitor redevelopment areas (usually called brownfield sites or more generally, vacant lands, although with a slightly different meaning than ours) has been examined in many studies that mostly focus on either their potential for policy-makers by using GIS data [7] or the detection of new vacant lands. In particular, remote sensing data have been used in several studies for the detection of new brownfields: Ref. [8] investigated the potential of IKONOS data in the object-oriented classification approach and Ref. [9] investigated IKONOS, QuickBird and hyperspectral data. In a recent study [10], the fusion of remote sensing images thermal data, GIS layers and citizen science data is proposed for the identification of urban vacant land. Remote sensing is also used, at a fine scale, for the detection and monitoring of hazardous substances and materials, as shown in [11].

Change detection is one of the major applications of satellite-based remote sensing data [12], and many different satellite-based change detection methods have been developed and used in recent decades. Among the most commonly used methods are algebra methods (e.g., Image Differencing, Ratioing or Change Vector Analysis), transformation-based methods (e.g., Principal Component Analysis), classification-based methods [13] and time series analysis. In [14], the authors provide a review of the different techniques, a guide to compare them by placing a clear separation of variables between the analysis unit and classification method and report that pixel and post-classification change methods remain the most popular choices. The review also presents some advantages and limitations of the different techniques. These limitations and how to overcome them have been widely studied and have led to more refined methodologies, e.g., super-resolution mapping and the analysis of mixed pixels for the improvement of land-cover class maps [15]. In addition, many other methods have recently been developed, notably based on artificial intelligence [16,17]. However, in [16], it was highlighted that supervised AI methods require massive training samples to obtain a robust model and that processing remote sensing big data requires a large amount of computational resources, which limits the implementation of the AI model. It is, therefore, crucial to choose the methodologies based, on the one hand, on needs such as the scale of the application and the thematic objectives and, on the other hand, on aspects such as the resolution of the available images and their ability to provide the required comparison features [14]. In the framework of this project, we opted for a time series analysis approach as, depending on the method, it offers a number of advantages, e.g., being able to detect abrupt and gradual changes (BFAST) or to capture subtle but consistent trends (LandTrendR), Continuous Change Detection and Classification (CCDC) being able to detect a variety of LULC changes continuously with high spatial and temporal accuracies [18]. However, in [18], the limitations of these methods are also presented, e.g., time-consuming, requiring many resources, unsuitable for irregular observations, and some are unable to identify types of changes. It is, therefore, crucial that the choice of time series analysis method takes into account the objective of the research, and considers the need to find the change points as soon as possible in real-world applications and that there is a detection delay for many existing approaches [19,20].

Within this context, the European Copernicus program has opened, with the launch of Sentinel-1 and Sentinel-2 satellites, new opportunities thanks to their high spatial and temporal resolution. The Sentinel-1 mission consists of a constellation of two polar-orbiting satellites mounting a C-band synthetic aperture radar (SAR) imaging system. They offer a repeat cycle of six days and all-weather and day-and-night monitoring capabilities [21]. The two Sentinel-2 satellites A and B are characterized by a sun-synchronous orbit, phased at 180 to each other, and a repeat cycle of 5 days [22]. The temporal resolution of the Sentinel satellites ensures enough data to create time series [23–25], and their spatial resolution allows for the identification of landscape features [26] and monitoring urban areas [27], whereas the Sentinel-2 spectral resolution facilitates the thematic identification of land cover [28–30].

In addition to the use of SAR and optical data separately, the combination of SAR and optical data has been highlighted in domains such as vegetation monitoring [31] and urban mapping [32,33]. Combining the two types of data has the advantage of coupling features and thus overcoming some limitations, such as clouds, shadows and snow cover for the optical data. Regarding the Sentinel images, the combination has been investigated in various domains, such as forest disturbance [34], soil tillage [35] and urban mapping [36]. In [37], the use of Sentinel-1 data alone, Sentinel-2 data alone and their combined use for forest–agriculture mapping are compared.

The demand for automated operational services providing near-real time information for environmental monitoring has increased substantially in recent years, and several studies have investigated their feasibility and proposed possible implementations, mainly for natural events monitoring. In [20], the Thresholding Rewards and Penances TRP concept was applied for a near-real time forest disturbance alert system based on PlanetScope imagery, producing new forest change maps when a new image is made available. They proposed a robust statistical method to estimate forest clear-cuts, but the use of PlanetScope images makes the service costly as they need to acquire raw imagery. In [38], a near-real time automatic avalanche monitoring system based on Sentinel-1 data was presented, and an age tracking algorithm was developed, while, in [39], the focus was on burned forest areas using Sentinel-2 data. For mapping burned areas, the latter used a selection of spectral indices to compare the pre-fire and post-fire values. In [40], an automatic and repeatable plot-based change detection method, based on pre and post event Sentinel-1 and Sentinel-2 data, was designed and tested to map extreme storm-related damages. Most of the services are in the test or pre-operational phase and focus on localizing one type of change, with hindsight of events and/or using one type of remote sensing data being sometimes costly.

The goal of this paper is to present the methodological aspects and implementation details of SARSAR, a new Earth observation service for the monitoring of redevelopment sites in southern Belgium. For its deployment, a number of requirements made by the Walloon administration had to be met, namely: (i) the implementation of a straightforward automatic operational tool providing results on a regular basis (once every two months); (ii) the ability to detect changes in vegetation, buildings and soil, on a set of sites spread throughout the region's territory; (iii) the use of open-source data.

Differently from other methodologies and services mentioned above, the focus is, therefore, on providing a response to the administration need of monitoring RDSs on a regional scale and identifying the time and type of change at the site level using free and open-source technology. In brief, by exploiting Sentinel-1 and Sentinel-2 data, the service automatically detects and characterizes changes in user-defined sites of interest and provides a final change list that can be directly used by the Walloon authorities to prioritize their daily work and reduce the time needed for the inventory update.

To fulfil the free and open-source technology requirement, we exploited Terrascope, the Belgian contribution to the Sentinel Collaborative Ground Segment (CollGS), which provides access to pre-processed Sentinel data [41] and computer capacity for the execution of the process and its automation. The Sentinel Collaborative Ground Segments were created by ESA and its Member States to facilitate the access to the Sentinel data and the data exploitation. CollGS can be used for various applications, as shown by Ref. [42], who used Terrascope for geohazard monitoring.

To be able to provide a list of the RDSs that are likely to change, several steps were implemented. Considering the number of sites to be processed and the fact that aggregate information is needed for each RDS, we opted for an object-based approach. Moreover, since the number of training samples required to implement a solution based on AI would have been prohibitive, our final choice was a combination of unsupervised methodologies.

After data preparation, where the extraction of temporal features from the Sentinel time series was performed, two processes were run: first, the change point detection analysis based on the Pruned Exact Linear Time (PELT) [43], whose goal is to flag each site as changed/unchanged and to provide an estimate of the change date(s) [44] and then a rule-based classification based on threshold selection to characterize the types of changes.

Changepoint analysis is largely employed for the study of time series in many application domains, yet it is still underexploited within the remote sensing community, due to the fact that high resolution images were not easily accessible until a few years ago. In regard to our service, changepoint detection was chosen because it serves a twofold purpose: it directly provides an estimate of the date of change, which alone constitutes valuable information for the administration, and allows us to restrict the time window within which the change classification should be performed. As regards threshold selection, it is a common procedure in algebra-based change detection [45]. The selection of the best threshold could be associated with a priori knowledge or derived from the histogram of the image [12]. The advantage of thresholding is that it can guarantee a robust near real-time approach based on fast and automated processing [34]. To the best of our knowledge, there have not been other attempts to use changepoint detection in combination with threshold-based classification for the characterization of changes in urban areas.

The paper is organized into five sections: The Materials section presents the study area, the Sentinel data used for this study via the Terrascope platform and the ground truth used for validation. The Methods section is divided into three parts: the first part explains the feature extraction and the creation of temporal profiles, the second part investigates the change detection method chosen and the third part presents the methodologies used for the classification of the changes. The last three sections are the presentation of the results, the discussion and the conclusions.

2. Materials and Methods

2.1. Materials

2.1.1. Study Area

The study was performed in Wallonia, the southern part of Belgium that covers an area of about 17,000 km^2. The industrial development in this region took place mainly along the Haine–Sambre–Meuse–Vesdre river axis. In total, slightly over 2200 sites are distributed mainly along this particular path, for a total area of 3800 hectares (Figure 1). However, a certain number of sites are spread over the whole territory of Wallonia. As mentioned in the Introduction, the size of the RDSs themselves, depending on their original use, can vary greatly. Figure 2 shows a former industrial area presenting a large number of RDSs of different sizes.

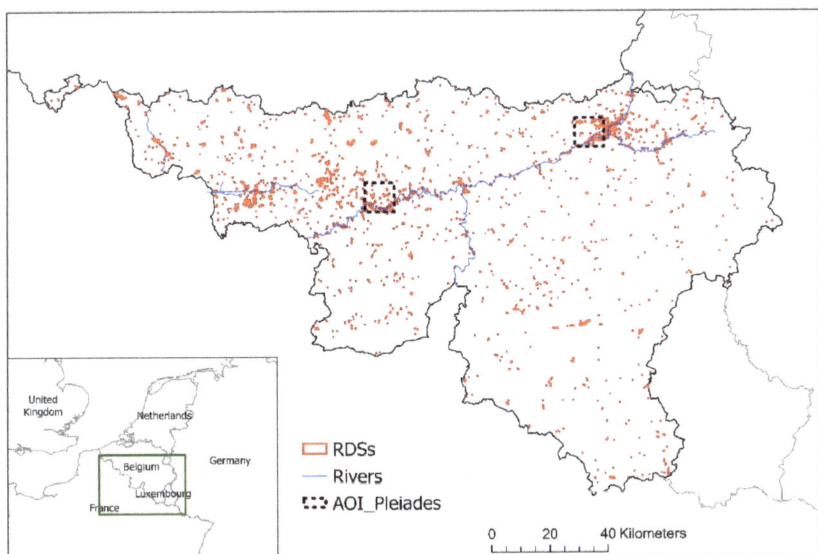

Figure 1. Study area (green mark), with the spatial distribution of the RDSs in Wallonia (red marks) and the Pléiades ground truth areas (black marks).

Figure 2. Close-up, illustrated with an orthophoto, of a former industrial area presenting several RDSs of different sizes.

2.1.2. Sentinel Data and Computing Environment

Sentinel-1 and Sentinel-2 data have been available since 2014 and 2015, respectively. Both missions consist of two satellites (A and B). Sentinel-1 mounts an SAR instrument that operates at a center frequency of 5.405 GHz and supports operation in dual polarization. For Belgium, the typical acquisition mode is Interferometric Wide (IW) in dual polarization

(VV+VH), which provides a resolution of around 5 × 20 m for Single Look Complex (SLC) products and around 20 × 20 m for Ground Range Detected (GRD) products [21]. Sentinel-2 carries an on-board Multi Spectral Instrument (MSI) measuring the reflected solar spectral radiances with 13 spectral bands ranging from visible to shortwave infrared (SWIR) bands [22]. The spatial resolution is 10 m, 20 m or 60 m depending on the spectral band. As regards the temporal resolution, Sentinel-1 and Sentinel-2 have a repeat cycle of 6 and 5 days, respectively, making them suitable for the creation of time series.

All the processing was carried out using the Terrascope platform [41], the Sentinel Collaborative Ground Segment for Belgium. Terrascope was chosen because it offers, in open access, up-to-date pre-processed Sentinel data, a computing environment, long-term maintenance and technical support. Concerning the Sentinel-2 data, the platform makes available atmospherically corrected images Level 2A Top-of-Canopy (TOC), downloaded from the ESA hubs. As regards Sentinel-1, along with the original SLC and GRD products, Terrascope also conveniently offers the corresponding calibrated and orthorectified images, which we ultimately used to avoid unnecessary pre-processing. Their spatial resolution is 20 × 20 m resampled at 10 m. The SARSAR service was run on a dedicated machine with a 6-core hyperthreading enabled CPU, 24 GB RAM, a boot volume of 2 TB and a data volume of 8 TB. Data storage was ensured by a PostgreSQL (11.11) server. The data processing was performed via a combination of Python (3.6) scripts, PostgreSQL stored procedures and PostGIS (3.1) functions. The whole processing chain was launched automatically and at predefined intervals thanks to CRON. Ultimately, the final users received notifications and reports by e-mail.

2.1.3. Ground Truth

For validation purposes, two ground truth datasets were created by visual analysis. The first ground truth is based on the orthophotos (25 cm resolution) taken in summer 2016 and 2018, and focuses on the RDSs, spread throughout the region, for which there are changes that can be observed from Sentinel data. This dataset was developed to account for major changes and for which we do not have information about the exact dates of change. The second ground truth is based on Pléiades images (50 cm resolution) acquired monthly between January 2019 and December 2020 on two specific areas (Figure 1) with a high concentration of RDSs. This provides complementary information compared to the orthophotos' ground truth. In fact, while the orthophotos' ground truth focuses on RDSs with significant changes, this dataset was created to take into account in a more balanced way the different types of change. Although, due to meteorological conditions, only 14 and 16 images, respectively, are available for each area, an estimation of the change dates was extracted taking into consideration that several dates can occur per site. In addition to the change dates for the whole period, information on the changes occurring between summer 2019 and summer 2020 was also extracted in order to provide a dataset that complements the one based on orthophotos.

In total, 141 and 161 sites are present in the orthophotos and Pléiades ground truth, respectively. For each of the 302 RDSs, changes were manually recorded for vegetation, buildings and soil. Overall, 152 of the sites presented at least one change and 150 remained unaltered. The breakdown of the changes into the three possible types is shown in Table 1, and Figure 3 provides two examples of changes.

Table 1. Number of changes per ground truth and breakdown into change types.

Ground Truth	Building	Vegetation	Soil	Total Changes	Total RDSs
Orthophotos	60	97	125	282	141
Pléiades	8	13	15	36	161
Total	68	110	140	318	302

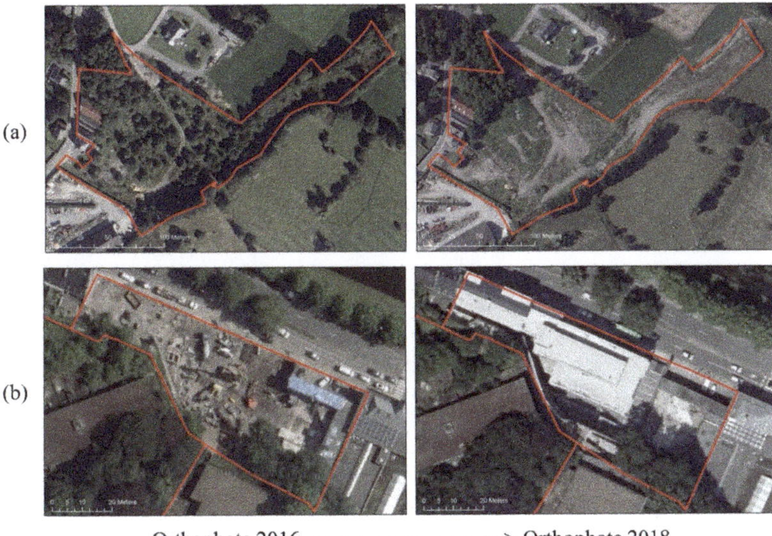

Figure 3. Close-ups, illustrated with orthophotos, of two RDSs showing (**a**) vegetation decrease and soil change ("Scierie Renard"); (**b**) building increase and soil change ("Immobilière Bouchoms").

2.2. Methods

Several methodologies for change detection and classification can be applied, and selecting the appropriate technique is related to the objective of the study [46]. Techniques such as single image differing or ratioing provide binary information (change/no-change), and if detailed information is required, such as the change direction, classification techniques are preferable. Another technique providing information on change type and direction is index differencing. These are mathematical transformations in multispectral mode and are produced separately; then, other change detection techniques (e.g., differencing or ratioing) can be applied [46]. Concerning the unit of analysis, on the one hand, in [46], it is explained that pixel-based change detection methods have been used traditionally, the main advantages of this unit of analysis being its suitability for large pixels; it does not generalize the data; and it is an effective methodology, especially when the relationship between pixel intensity and the land cover changes under investigation is strong [14]. On the other hand, the object-based approach allows the exploitation of the spatial context, reduces the noisy outputs of isolated changed pixels and allows direct object change detection (DOCD) by comparing spectral information [46]. One of the object-based units of analysis is the vector polygon, which is extracted from existing geodatabases; they group together pixels that are suitable for statistical analysis, the result of which may indicate changes within the corresponding polygons. On one hand, vector polygons provide a cartographically 'clean' basis for analysis [14], allow the exploitation of additional thematic information about the objects to obtain better results and enhance the interpretation of the image [47]. They also provide important information on the location of the objects to investigate for change detection. In [47], this type of object-based approach, combined with spectral indices, was used for the automatic change detection of buildings in an urban environment as it can handle the complexity of urban environments. On the other hand, vector polygons generalize the data, and the size and shape of objects cannot be compared [14]. As regards the current study, we opted for this latter methodology, where a set of features (multi-spectral indices and radar backscattering) are used to create what we could define as the temporal signatures of the RDSs. The methodology responds to the need to monitor the RDS polygons at a regional scale, and to have generalized information of changes detected for the three types of classes (vegetation, building and soil), thereby reducing the manual work [14].

The proposed overall methodology, whose main goal is to provide a shortlist of the sites that are likely to have changed and for which an on-field visit would be required, is shown in Figure 4. The first main block is the feature extraction, where the Sentinel-1 and Sentinel-2 images available in Terrascope and described in the previous section are processed to obtain the above-mentioned temporal profiles of the RDSs (each RDS has multiple temporal profiles—one per feature). The second main block is devoted to the characterization of the changes, which is carried out in two steps: (i) the change detection, which flags a site as changed (or not) and provides an estimate of the change date(s)—this is carried out once every two months; (ii) the change classification, which is divided into two separated processes. First, when a change date is detected, a rule-based classification is performed in order to provide additional information on the type of change: vegetation, building or soil. Second, the same methodology is applied once per year, considering summer average features in order to detect gradual changes.

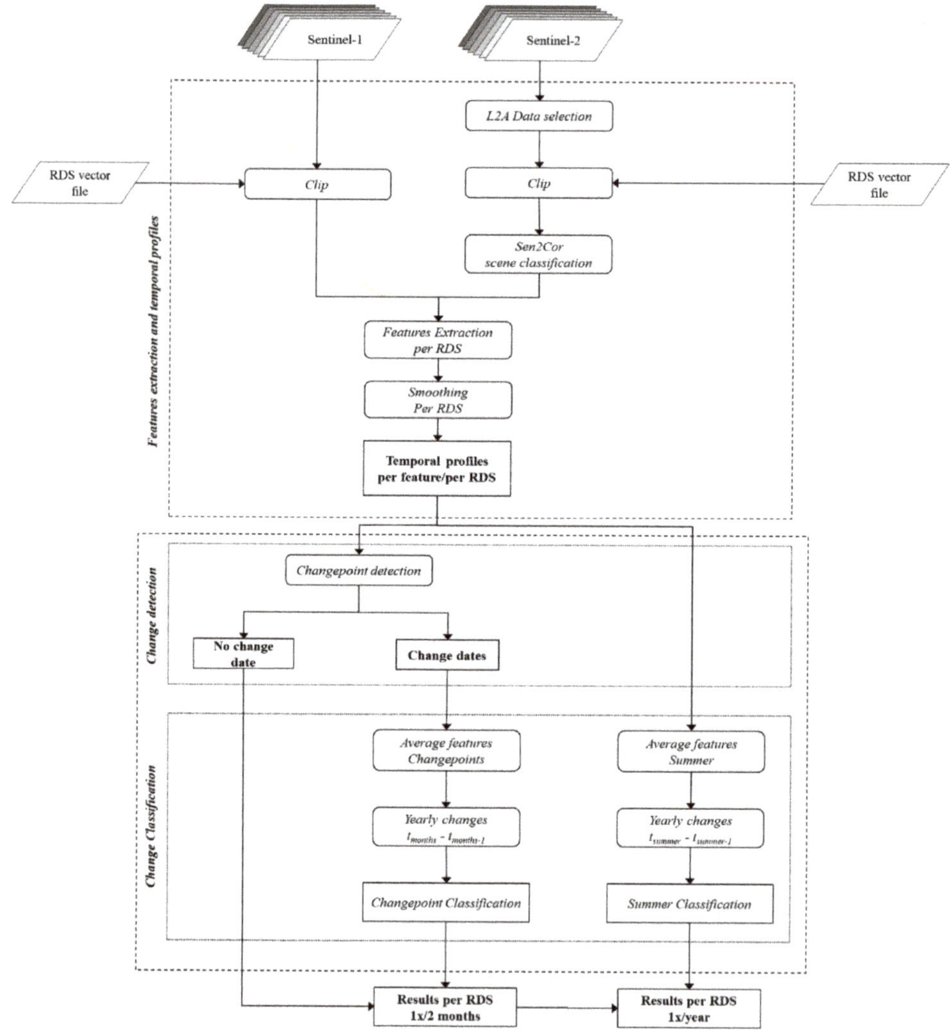

Figure 4. Workflow of the automatic change detection and classification of the RDSs.

The final output of the service is a csv file that is automatically delivered to the operator. For each RDS, this report includes: (1) information on whether a change has occurred or not, (2) the type of change and (3) the estimated date of the change (if available).

2.2.1. Features Extraction and Temporal Profiles

For each Sentinel-1 acquisition (more specifically, the VH band, which was found to be the most suitable for our scope) that contained the site of interest within the desired time frame, the average backscatter (sigma0) for that site was computed and used to populate the corresponding temporal profile. Since a site can be typically seen from 3 to 4 different viewing angles (considering both ascending and descending orbits), separate profiles were created for each satellite pass and then averaged to obtain a unique "sigma0_{VH}" feature.

Regarding Sentinel-2 data, all the L2A tiles over the area were analyzed. Only the tiles presenting less than 25% of clouds were selected, which greatly reduced the number of undetected cloud pixels. Then, each image was clipped based on the RDS vector polygons file. Image co-registration was ensured during this process. Then, the Scene Classification layer, a classification map generated via the Sen2Cor ESA processor that accompanies every L2A image and is directly available in Terrascope, was used to remove every single pixel classified as "No_Data", "Cloud_Shadows", "Cloud_Medium_Probability", "Cloud_High_Probability", "Thin_Cirrus" and "Snow". This allowed us to remove, site per site, the dates for which no data, shadows, clouds, or snow pixels were present.

In the feature extraction step, six widely used spectral indices were calculated that were to be used in the next processes: (1) the Built-Up Areas Index (BAI) [47], (2) the Brightness Index (BI) [48], (3) the Second Brightness Index (BI2) [48], (4) the Normalized Vegetation Index (NDVI) [49], (5) the second Normalized Difference Water Index (NDWI2) [50] and (6) the soil brightness index (SBI) [47]. The selection of the spectral indices was motivated by their widespread application in the literature and by considering that most built-up indices require SWIR bands, which are available only in a coarse resolution for Sentinel-2. The BI2 index has been tested for built-up detection after applying NDVI and NDWI2 to mask vegetation and water [51]. BAI has proven to be useful to detect asphalt and concrete surfaces [47], and SBI has been successfully investigated by [47] and [52]. For each index, each RDS and each available image since 2015, the average per RDS was calculated and used to generate the Sentinel-2 time series:

$$\text{BAI} = ((\text{B02} - \text{B08}))/((\text{B02} + \text{B08})) \tag{1}$$

$$\text{BI} = \sqrt{}(((\text{B04*B04}) + (\text{B03*B03}))/2) \tag{2}$$

$$\text{BI2} = \sqrt{}(((\text{B04*B04}) + (\text{B03*B03}) + (\text{B08*B08}))/3) \tag{3}$$

$$\text{NDVI} = ((\text{B08} - \text{B04}))/((\text{B08} + \text{B04})) \tag{4}$$

$$\text{NDWI2} = ((\text{B03} - \text{B08}))/((\text{B03} + \text{B08})) \tag{5}$$

$$\text{SBI} = \sqrt{}((\text{B04*B04}) + (\text{B08*B08})) \tag{6}$$

where B0n corresponds to the n-th Sentinel-2 band used for the calculation, here B02, B03, B04 and B08, all with a 10 m resolution.

To create the final temporal profiles (each RDS has multiple profiles, one per feature), a linear interpolation to fill in the gaps (1 data point per day) in the data and a smoothing using a Gaussian kernel with a standard deviation of 61 were performed.

2.2.2. Change Detection

The second processing block is the change detection, where some of the features extracted from the Sentinel images are jointly analyzed using the Pruned Exact Linear Time (PELT) [43]. The method is a well-known changepoint detection method that provides an exact segmentation of the time series with a linear time complexity.

Given a time series $s = (s_1, \ldots, s_k)$, the number n and time position $t_{1:n} = (t_1, \ldots, t_n)$ of the changepoints are obtained by solving the following penalized minimization problem:

$$Q_n(s_{1:k}, p) = \min_{n, t_{1:n}} \left\{ \sum_{i=1}^{n+1} \left[C\left(s_{(t_{i-1}+1):t_i} \right) \right] + p \right\} \tag{7}$$

where C is the segment-specific cost function

$$C(s_{a:b}) = \sum_{i=a+1}^{b} \| s_i - \bar{s}_{a:b} \|_2^2 \tag{8}$$

and $p = log(k)$ a penalty term to control overfitting.

In a preliminary study that we carried out on 22 test sites [44], we showed that the combined use of the Sentinel-1 sigma0$_{\mathrm{VH}}$ and Sentinel-2 NDVI returns more accurate change detection results than those of the single features. Figure 5 shows an example of the changepoints detected on an RDS where a building was demolished between summer 2017 and summer 2018, and some vegetation grew between summer 2018 and summer 2019. As can be seen, the combined use of Sentinel-1 and Sentinel-2 detection successfully returned the two dates. After the optimization phase of the change detection process, during which we performed several tests on an extended dataset using different combinations of features, the NDVI feature was replaced by NDWI2.

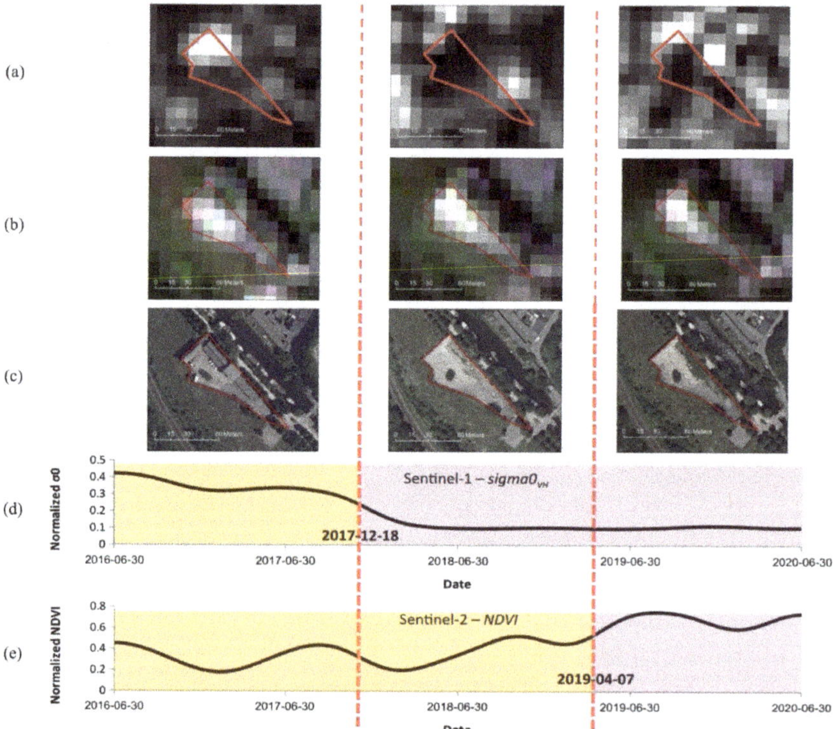

Figure 5. Changepoint analysis for the RDS "Service voirie d'Angleur" in Liège showing (**a**) Sentinel-1 image (left: July 2017; center: July 2018; right: July 2019); (**b**) Sentinel-2 images (left: July 2017; center: July 2018; right: July 2019); (**c**) orthophotos ground truth (left: summer 2017; center: summer 2018; right: summer 2019); (**d**) bi-dimensional time series sigma0$_{\mathrm{VH}}$ (Sentinel-1); (**e**) bi-dimensional time series NDVI (Sentinel-2).

The overall process returns either a list of changepoints dates (one or multiple) or no changepoints. When one or multiple changepoints are detected, these become the input of the next block—the change classification. When no changepoints are detected, this information is reported directly in the final report, "Results per RDS".

2.2.3. Change Classification

Determining the type of changes is essential in providing information about the changes to the local authorities. The Sentinel-1 sigma0$_{VH}$ and Sentinel-2 indices temporal profiles were analyzed to determine suitable threshold boundaries that would represent a change for each land cover type (vegetation, building or soil), and the data from the ground truth datasets were used to validate the method. Thresholding-specific indices have been proposed and successfully applied in many studies [47,51], e.g., thresholding NDVI has been used to qualify land-cover change [53] and detect forest cuts [25]. The use of Sentinel-1 data, which is radar sensitive to variations in height and shape, allowed us to complement the information provided by the Sentinel-2 indices and improve the characterization of the changes to buildings.

For each site, two separate processes were considered that allow, on the one hand, to provide information on the type of change for progressive changes and, on the other hand, to classify the changes associated with the detected changepoints.

The first one is solely based on Sentinel-2 data. It focuses on the summer months (t_{summer}), from May to August, as these are more appropriate for vegetation change. It also offers the best illumination conditions for the multi-spectral images considering the variation of the Sun–Zenith Angles due to the sensing time being the same throughout the year. This process, hereafter referred to as "summer classification", offers the opportunity to detect changes that occur gradually over a one-year period. The second process, the "changepoint classification", is based on both Sentinel-1 and Sentinel-2 features, and it is performed when one or multiple change dates are available from the previous block. It takes into account the average of the data available after the change date (t_{months}), namely, 2 months for Sentinel-2 data and 1 month for Sentinel-1 data. A calculation of the distance between the average features of the time period from the year of the change (t_{summer} and t_{months}) to the same time period the previous year (respectively, $t_{summer\text{-}1}$ and $t_{months\text{-}1}$) was performed. This distance was then compared to the thresholds of the different Sentinel-2 index and Sentinel-1 VH features in order to determine the chances of representing a type of change (Table 2). As described for the "summer classification", the one-year time step for the "changepoint classification" was chosen in order to limit the influence of the illumination for Sentinel-2 and the seasonality effect. In addition, while the "summer classification" considered the average features from May to August, the "changepoint classification" takes into account 2 months for the Sentinel-2 indices and 1 month for the Sentinel-1 VH feature. This discrepancy in the number of months used is based on the fact that valid Sentinel-2 data are typically fewer due to cloud cover. On top of the change classification, NDVI and sigma0$_{VH}$ helped us to determine the direction of the change (Table 2). Although the use of the VH band allowed this identification, the combination of the three indices, BI, BI2 and SBI, showed better results for the "summer classification", which is why these indices were selected. The detailed workflow for the evaluation of the type of changes is shown in Figure 4.

Table 2. Rule-based classifier for the determination of the types of changes.

Change Classification	t_{summer}–$t_{summer\text{-}1}$	t_{months}–$t_{months\text{-}1}$
Vegetation increase	NDVI $\geq$ 0.1	NDVI $\geq$ 0.1
Vegetation decrease	NDVI $\leq$ 0.1	NDVI ≥ -0.1
Building change	BI $\geq$ 150 or BI2 $\geq$ 150 or SBI $\geq$ 250	-
Building increase	-	VH $\geq$ 0.135
Building decrease	-	VH ≥ -0.135
Soil change	BAI $\geq$ 0.05	BAI $\geq$ 0.05

3. Results

3.1. Change Detection

The performance was assessed in terms of true positive rate (TPR) and false positive rate (FPR). The overall problem can be in fact seen as a binary classification where either a "change" (1) or a "no change" (0) has to be detected. In order to compare the results with the ground truth, the latter was coded so that any change in any of the three classes (building, vegetation and soil) was assigned the value 1; in the case of no change for all three classes, the ground truth was given the value 0. A confusion matrix was then generated so that the number of true positives (TPs), true negatives (TNs), false positives (FPs) and false negatives (FNs) could be used to compute the TPR and FPR. To provide a unique measure that takes into account both detection and miss rates, the F_1-score was also calculated. For the sake of completeness, the overall accuracy (OA) is also reported.

It is worth mentioning that, due to the specific way in which the ground truth is constructed, in order to generate the confusion matrix, we made the arbitrary assumption that only one change per site occurred in the considered period of time. This is a simplification that helped us to compare the results in a more straightforward way, but might not fully reflect the real situation, especially for the sites belonging to the orthophotos ground truth, as for a certain number of them it is more likely that multiple changes occurred at different times.

The change detection was performed using the sigma0$_{VH}$ and NDWI features, which amongst the other features ultimately provided the highest accuracy. The use of both Sentinel-1 and Sentinel-2 data, which provide complementary information (the VH band mostly about buildings and the NDWI index mostly about vegetation and soil), allowed a more effective identification and classification of changes. The results for the entire dataset are shown in the first row of Table 3. The number of sites for which we had an estimated change is 108, 91 of which were correctly classified. Among the unchanged sites, we missed 46 of them, resulting in an OA of 79%. In terms of correct and miss detection rates, we, therefore, obtained a TPR of 66% and an FPR of 10%, with an F_1-score of 0.74.

Table 3. Changepoint analysis: confusion matrix and performance metrics.

	TP	FP	FN	TN	TPR	FPR	F_1-Score	OA
Full dataset	91	17	46	148	66%	10%	0.74	79%
Pléiades	15	9	12	125	55%	7%	0.59	87%

In order to better understand the results of the following block, the change classification, it was helpful to separate the Pléiades detections from the full dataset. The results are provided in the second row of Table 1. For this dataset, the number of sites that were flagged as changed was 26, with nine FPs, whereas the correct detections of the unchanged sites were 125. As a result, the TPR and FPR decreased to 55% and 7%, respectively, and, consequently, the F_1-score dropped to 0.59. The OA, instead, increased to 87%, mainly due to the fact that the dataset was rather unbalanced.

3.2. Change Classification

3.2.1. Summer Classification

The "summer classification", as we discussed in the previous section, takes into account, for each of the 302 sites, the summer comparison between 2016 and 2018 for the orthophotos dataset and between 2019 and 2020 for the Pléiades dataset. Again, the performance was assessed by combining the two datasets and computing the TPR, FPR and the F_1-score for each class, along with the overall accuracy (see Table 4). The overall performance of the yearly classification based on summer values is satisfactory. The best results were obtained for the "vegetation" class, for which the OA was 90% and the TPR and FPR were 87% and 9%, respectively. The resulting F_1-score was 0.80. The performance for the "building" and "soil" classes were slightly lower, with an OA of 76% and 79%, respectively, yet still good, with an F_1-score above 0.7.

Table 4. "Summer classification" (full dataset): confusion matrix and performance metrics.

	TP	FP	FN	TN	TPR	FPR	F_1-Score	OA
Vegetation	59	21	9	213	87%	9%	0.80	90%
Building	87	49	23	143	79%	26%	0.71	76%
Soil	103	26	37	136	74%	16%	0.77	79%

To look deeper into the "vegetation" class, Table 5 also shows the results disaggregated by "increase", "decrease" and "no change" types, with the corresponding overall accuracy and omission/commission errors. As can be seen, for both the increase and decrease in vegetation, around 1 in 4 detections was a false alarm, whereas the percentage of missed changes were 20% and 12%, respectively. It is worth noting that there was no confusion between the two classes, as all the errors fell into the "no change" class. For this class, instead, the commission and omission errors were much lower, namely, 4% and 9%, respectively.

Table 5. "Summer classification" (full dataset): detailed confusion matrix for the "vegetation" class.

	Increase	Decrease	No Change	Total	Commission Errors
Increase	8	0	3	11	27%
Decrease	0	51	18	69	26%
no change	2	7	213	222	4%
Total	10	58	234	302	
Omission Errors	20%	12%	9%		OA = 90%

3.2.2. Changepoint Classification

The "changepoint classification" takes into consideration only the RDSs for which at least one changepoint date has been estimated within the change detection process. As multiple changes can occur in the same site during the considered time period, a yearly comparison was required for each estimated change date. This was only possible using the Pléiades dataset, as only for this ground truth are the exact change dates available. A performance assessment (Tables 6 and 7) was carried out for all the changepoint dates knowing that the overall accuracy of the changepoint dates themselves was shown in a previous section.

Table 6. "Changepoint classification" (Pléiades dataset): confusion matrix and performance metrics.

	TP	FP	FN	TN	TPR	FPR	F_1-Score	OA
Vegetation	6	1	3	16	67%	6%	0.75	84%
Building	7	1	3	15	70%	6%	0.78	85%
Soil	11	4	4	7	73%	36%	0.73	69%

Table 7. "Changepoint classification" (Pléiades dataset): detailed confusion matrix for the "vegetation" and "building" classes.

	Vegetation				
	Increase	**Decrease**	**No Change**	**Total**	**Commission Errors**
Increase	0	0	0	0	-
Decrease	0	6	1	7	14%
No change	0	3	16	19	16%
Total	0	9	17	26	
Omission Errors	-	33%	6%		OA = 85%
	Building				
	Increase	**Decrease**	**No Change**	**Total**	**Commission Errors**
Increase	2	0	0	2	0%
Decrease	0	5	1	6	17%
No change	2	1	15	18	17%
Total	4	6	16	26	
Omission Errors	50%	17%	6%		OA = 85%

Although some dates were during winter months, the results for the vegetation changes remained good, with an OA of 84% and a F_1-score of 0.75. With respect to the "summer classification", the main difference here was in the TPR, which was lower by 20 percentage points (87% for "summer classification" and 67% for Pléiades dataset). As regards the "building" class, there was the opposite trend for the Pléiades dataset, with both a higher OA and F_1-score than those obtained for the "summer classification". Although the TPR was slightly lower, the significant drop in the FPR improved the performance. Finally, for the "soil" class, all the metrics showed a drop in the performance, especially as far as the FPR is concerned.

To complete the analysis, the detailed confusion matrices for the classes "vegetation" and "building" are provided in Table 7. Once again, the results are disaggregated by "increase", "decrease" and "no change" types. For the "vegetation" class, no increase was reported within any site of the ground truth; therefore, no metric was calculated. Instead, out of nine "decrease" changes, six were correctly identified, resulting in a commission error of 14% and an omission error of 33%. If we look at the "no change" class, we had a similar false alarm rate, but a much lower miss rate. For the "building" class, half of the "increase" changes in buildings were missed (50% omission error). However, all the changes that were flagged as an increase were correct (0% commission error). Instead, the classification of a decrease was more accurate, with only one false alarm and one missed detection. Finally, the "no change" classification was the one providing the best performance, with a commission error of 17% and an omission error of 6%.

4. Discussion

The results described in the prior section provide answers to the several challenges that can be encountered when detecting changes on specific sites. Indeed, besides detecting the changes with their dates, there is a need to classify the type of changes and to detect gradual changes. Four main observations may be drawn from this research.

First, the proposed method provided satisfactory results for the change detection and the change classification for both ground truth datasets. As far as the change detection is concerned, thanks to the complementary information provided by the $sigma0_{VH}$ and NDWI features (the former mainly for buildings, and the latter mainly for vegetation/soil), we were able to achieve an overall accuracy for the full dataset of 79%. As far as the change classification is concerned, the OA ranged from 79% to 90%, depending on the type of change that was considered (vegetation, building and soil). The OA of 90% and the F_1-score of 0.80, obtained for the vegetation "summer classification", illustrate the well-known robustness of the selection of the NDVI as a vegetation indicator [25,49,53], especially in summer conditions. As previously shown in [47], the BAI was proven to be useful for soil detection. Regarding the classification of buildings, the results revealed the suitability of combining the BI, BI2 and SBI indices, as an OA of 76% and an F_1-score of 0.71 were obtained for the "summer classification". As mentioned in the Methods section, these indices were not used for the building classification rules of the "changepoint classification" and were replaced by the sigma0_{VH} feature. This is due to the fact that the probability of finding cloud-free images in other periods than the summer is lower and the radar backscatter helps improving building discrimination thanks to its sensitivity to variations in height and shape. For this reason, it will be useful to carry out additional tests to investigate whether the use of the sigma0_{VH} feature could be used also for the "summer classification". Moreover, further research could be conducted in regard to the number of Sentinel-2 images used for the "changepoint classification". Although data gaps were filled in through linear interpolation and the time series were smoothed using a Gaussian kernel, the cloud cover limits the number of usable images, especially during winter months. By only selecting the dates for which a certain number of S2 images are available, it is likely that the performance of the change classification would be improved.

Second, the "summer classification" is better suited for the detection of gradual changes. Figure 6 illustrates an ongoing vegetation growth leading to a soil decrease. This was not captured by the changepoint detection method but was classified as a vegetation increase and soil change thanks to the summer 2016–2018 comparison. The "summer classification" also provided better vegetation classification for change dates that occurred during winter, as seasonality strongly impacts the performance, as most vegetation is dormant during the winter. However, when comparing the "summer classification" and the "changepoint classification" results, it should be taken into account that the size of the two datasets is very different (302 vs. 26), and this had an impact on the results both in terms of representativeness and numerical accuracy.

Third, the use of vector polygons originating from the RDSs vector file to group the image pixels in the change analysis constitutes, at the same time, an advantage and a limitation. The fact that we averaged the information over the whole sites, on the one hand, helped reduce the noise (especially as far as Sentinel-1 is concerned) and filter out unnecessary details, but on the other hand, it may have led to the non-detection and/or non-classification of either small changes or bigger changes occurring on large sites, as the scales of the changes do not always match the scales of the vector polygons [14]. To partially overcome these issues, the polygon size could be reduced, for example, by segmenting each site either based on a fixed grid or external sources, such as WALlonie Occupation et Utilisation du Sol (WALOUS) [54,55]. However, this can lead to other problems, such as a significant increase in the computing power and and/or the creation of a large number of objects that would be too small compared to the Sentinel spatial resolution. Moreover, although external sources could, in principle, provide additional information on the type of change, this leads to the challenge of keeping these data up to date.

Fourth, the use of Sentinel data also has its limitations. First, as mentioned above, the spatial resolution reduces the number of RDSs for which the results can be reliable. For example, in total, 90.4% of the RDSs were larger than 400 square meters (roughly one Sentinel-1 pixel and four Sentinel-2 pixels). Moreover, although most of the sites are former industrial facilities with extensive infrastructure, changes may occur on only minor parts of the site, as illustrated in Figure 7. However, Sentinel images offer major advantages compared to orthophotos, which are open access but provided once a year, or Pléiades images, which can be obtained on demand and are costly. In fact, not only can they guarantee a much higher temporal coverage (especially if we consider the Sentinel-1 all-weather capabilities), but they are also completely free, which means that the operational costs of the tool are significantly reduced. Moreover, thanks to the Terrascope platform and its cloud computing environment, the method is automated and provides, every two months, results that are directly usable by regional authorities. Although the use of Sentinel data limits the number of RDSs that can be analyzed and the size of the changes detected, thanks to the results that we have shown, the regional authorities will be able to update the RDS inventory in a more efficient and less expensive way. Indeed, the SARSAR service enables the prioritization of the orthophotos analysis work and drastically limits field efforts. Table 8 shows a sample of bimestrial final change lists, and Figure 8 presents four RDSs, three for which a change date was detected and one with no change.

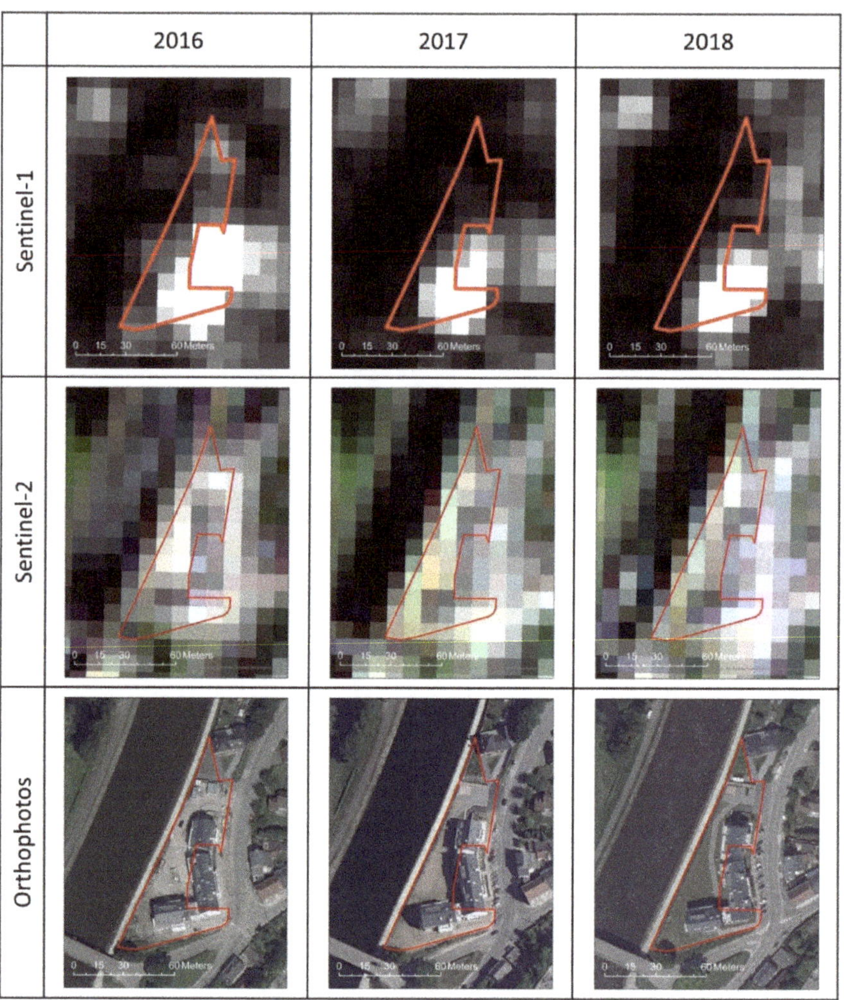

Figure 6. Close-ups of an RDS showing gradual vegetation increase ("Ets Biernaux"), between 2016 and 2018, illustrated at the top with Sentinel-1 images, in the middle with Sentinel-2 images and at the bottom with orthophotos.

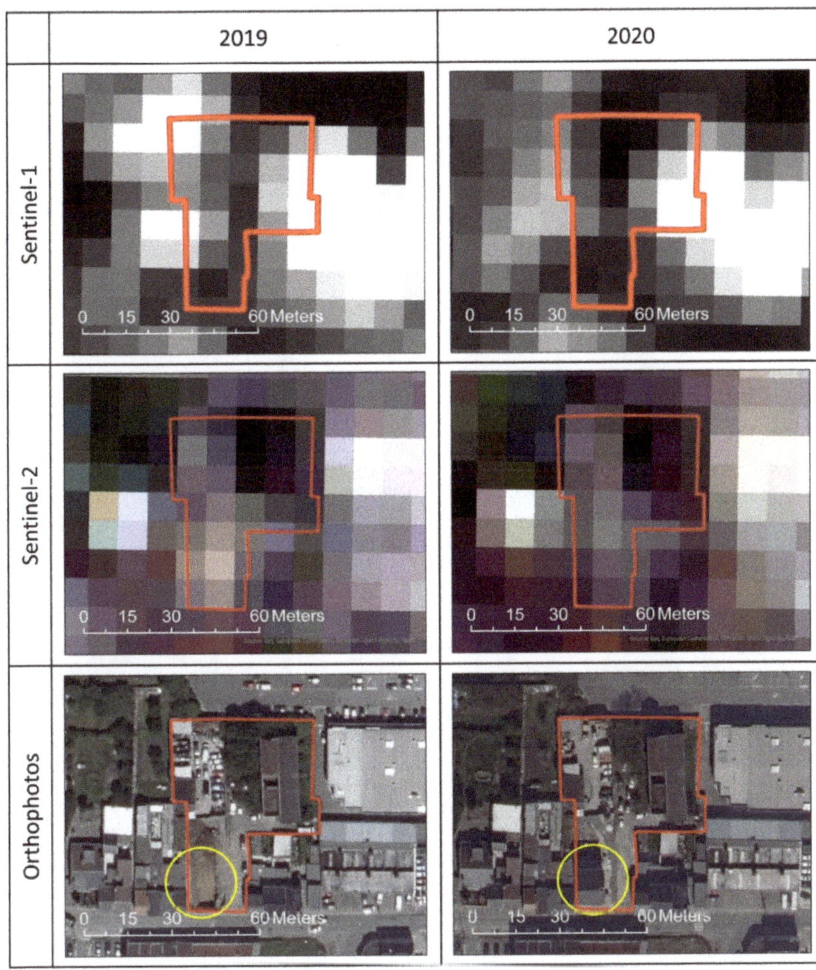

Figure 7. Close-ups of an RDS showing a building increase, between 2019 and 2020, too small for the Sentinel spatial resolution ("S.A.N.I. Carrelages"), illustrated at the top with Sentinel-1 images, in the middle with Sentinel-2 images and at the bottom with orthophotos.

Table 8. Example of bimestrial final change list for a sample of RDSs.

CODECARTO	RDS Name	Change Date	Estimated Change Date	Vegetation Change	Building Change	Soil Change
52011-ISA-0040-01	Cordial Bowling	Yes	20 April 2020	Yes, decrease	No	Yes
52011-ISA-0110-01	Carsid—Agglomération	Yes	12 March 2019	No	Yes, decrease	Yes
62063-ISA-0073-01	Patience et Beaujonc—site secondaire	Yes	31 March 2020	Yes, decrease	Yes, decrease	Yes
52011-ISA-0003-01	Technopôle de la Villette	No	NA	No	No	No

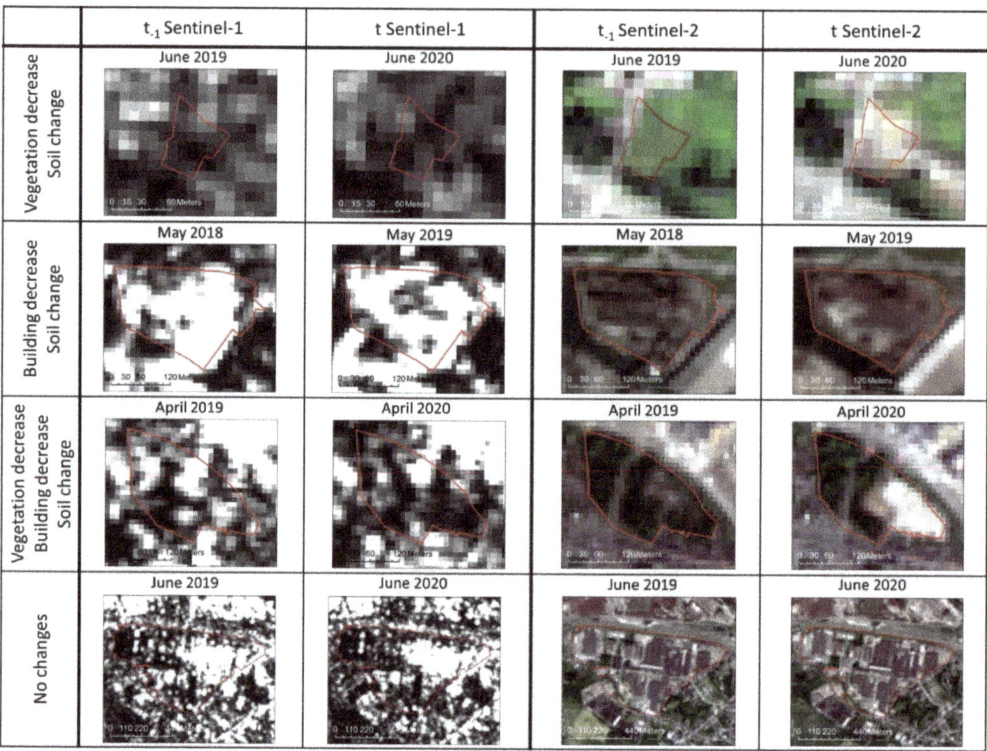

Figure 8. Examples of detected and classified changes, on Sentinel-1 images (left) and Sentinel-2 images (right). Details of the changes are explained in Table 8.

5. Conclusions

Managing former industrial lands is essential for urban planning and limiting the urbanization of new lands. In this article, we presented SARSAR, a new Earth observation service that has been developed to support the Walloon authorities' daily work by helping them update the RDS inventory in a more responsive, efficient and cost-effective manner.

The SARSAR service exploits Sentinel-1 and Sentinel-2 images, with their high spatial and temporal resolution and open data policy, and the cloud computing environment offered by Terrascope to generate and deliver a change report every two months directly to the Walloon authorities, who can integrate it into their management system. This saves time and effort compared to the current methods of updating the inventory (visual analysis of orthophotos and systematic field visits), enabling personnel to prioritize their work and focus on the RDSs showing evidence of significant changes. This service, which first performs a set of routines to extract and prepare the input data, is composed of two main processes: one for the flagging of the sites that are likely to have changed and one in charge of the classification of the changes.

The performance assessment provided satisfactory results, with an overall accuracy of around 80% for the change detection and in the range 70–90% for the change classification (depending on the class considered). The results highlight the relevance of using Sentinel-1 data, as well as a selection of Sentinel-2 indices, especially the NDVI for vegetation monitoring, and show the complementarity of the two processes in identifying both abrupt and gradual changes.

The results presented in this paper highlight opportunities not only for brownfield monitoring in other regions but also for multiple application domains and a larger user community, from land management and planning strategies, to agricultural and forestry areas monitoring, through disaster response mapping.

Author Contributions: Conceptualization, S.P., M.S., X.N. and E.H.; methodology, S.P., M.S., C.W., G.S., X.N. and E.H.; software, M.S., C.W. and G.S.; validation, S.P. and M.S.; formal analysis, S.P. and M.S.; investigation, S.P. and M.S.; resources; data curation, S.P., M.S., C.W. and G.S.; writing—original draft preparation, S.P.; writing—review and editing, S.P., M.S., C.W. and E.H.; visualization, S.P. and M.S.; supervision, X.N. and E.H.; project administration, S.P., M.S., X.N. and E.H.; funding acquisition, M.S., X.N. and E.H. All authors have read and agreed to the published version of the manuscript.

Funding: The research presented in this paper is funded by BELSPO (Belgian Science Policy Office) in the frame of the STEREO III programme—project SR/00/372.

Data Availability Statement: Not applicable.

Acknowledgments: The authors would like to thank the European Union's Earth Observation Programme Copernicus and VITO for the provision of the Sentinel images through Terrascope and for the virtual research environments. The authors would like to thank Airbus and BELSPO for the provision of the Pléiades images, and the Service Public de Wallonie for the orthophotos. The authors would also like to thank BELSPO and their STEREO program (Support To Exploitation and Research in Earth Observation). Finally, the authors would like to thank the members of the Remote sensing and geodata unit (ISSeP), in particular Odile Close and Benjamin Beaumont, for their support and paper review.

Conflicts of Interest: The authors declare no conflict of interest.

References

1. Ferber, U.; Tomerius, S. Brownfield redevelopment: Strategies and approaches in Europe and the United States. In *Private Finance and Economic Development: City and Regional Investment*; OECD Publishing: Paris, France, 2003. [CrossRef]
2. Turvani, M.; Tonin, S. Brownfields Remediation and reuse: An opportunity for urban sustainable development. In *Sustainable Development and Environmental Management: Experiences and Case Studies*; Clini, C., Musu, I., Gullino, M.L., Eds.; Springer: Dordrecht, The Netherlands, 2008; pp. 397–411. [CrossRef]
3. Rénover et Réaménager la Wallonie. Available online: http://lampspw.wallonie.be/dgo4/site_amenagement/site/directions/dao/sarpdf (accessed on 27 January 2022).
4. Sites à Réaménager. Available online: https://www.iweps.be/wp-content/uploads/2021/12/T008-SITES.REAM_.-122021_full1.pdf (accessed on 27 January 2022).
5. Service Public de Wallonie, Code du Développement Territorial. Available online: http://lampspw.wallonie.be/dgo4/tinymvc/apps/amenagement/views/documents/juridique/codt/CoDT_Fr.pdf (accessed on 27 January 2022).
6. Inventaire des Sites à Réaménager. Available online: http://lampspw.wallonie.be/dgo4/site_sar/index.php (accessed on 27 January 2022).
7. Schiller, G.; Blum, A.; Hecht, R.; Oertel, H.; Ferber, U.; Meinel, G. Urban infill development potential in Germany: Comparing survey and GIS data. *Build. Cities* **2021**, *2*, 36–54. [CrossRef]
8. Banzhaf, E.; Netzband, M. Detecting urban brownfields by means of high resolution satellite imagery. The International Archives of the Photogrammetry. *Remote Sens. Spat. Inf. Sci.* **2004**, *35*, 460–466.
9. Ferrara, V. Brownfield identification: Different approaches for analysing data detected by means of remote sensing. *WIT Trans. Ecol. Environ.* **2008**, *107*, 45–54. [CrossRef]
10. Xu, S.; Ehlers, M. Automatic detection of urban vacant land: An open-source approach for sustainable cities. *Comput. Environ. Urban Syst.* **2022**, *91*, 101729. [CrossRef]
11. Atturo, C.; Cianfrone, C.; Ferrara, V.; Fiumi, L.; Fontinovo, G.; Ottavi, C.M. Remote sensing detection techniques for brownfield identification and monitoring by GIS tools. *WIT Trans. Ecol. Environ.* **2006**, *94*, 241–250. [CrossRef]
12. Singh, A. Review Article Digital change detection techniques using remotely-sensed data. *Int. J. Remote Sens.* **1989**, *10*, 989–1003. [CrossRef]
13. Hecheltjen, A.; Thonfeld, F.; Menz, G. Recent advances in remote sensing change detection—A review. In *Land Use and Land Cover Mapping in Europe. Remote Sensing and Digital Image Processing*; Springer: Dordrecht, The Netherlands, 2014. [CrossRef]
14. Tewkesbury, A.P.; Comber, A.J.; Tate, N.J.; Lamb, A.; Fisher, P.F. A critical synthesis of remotely sensed optical image change detection techniques. *Remote Sens. Environ.* **2015**, *160*, 1–14. [CrossRef]
15. Wang, P.; Wang, L.; Leung, H.; Zhang, G. Super-Resolution Mapping Based on Spatial–Spectral Correlation for Spectral Imagery. *IEEE Trans. Geosci. Remote Sens.* **2021**, *59*, 2256–2268. [CrossRef]
16. Shi, W.; Zhang, M.; Zhang, R.; Chen, S.; Zhan, Z. Change Detection Based on Artificial Intelligence: State-of-the-Art and Challenges. *Remote Sens.* **2020**, *12*, 1688. [CrossRef]

17. Khelifi, L.; Mignotte, M. Deep Learning for Change Detection in Remote Sensing Images: Comprehensive Review and Meta-Analysis. *IEEE Access* **2020**, *8*, 126385–126400. [CrossRef]
18. Yan, J.; Wang, L.; Song, W.; Chen, Y.; Chen, X.; Deng, Z. A time- series classification approach based on change detection for rapid land cover mapping. *ISPRS J. Photogramm. Remote Sens.* **2019**, *158*, 249–262. [CrossRef]
19. Aminikhanghahi, S.; Cook, D.J. A survey of methods for time series change point detection. *Knowl. Inf. Syst.* **2017**, *51*, 339–367. [CrossRef] [PubMed]
20. Francini, S.; McRoberts, R.E.; Giannetti, F.; Mencucci, M.; Marchetti, M.; Scarascia Mugnozza, G.; Chirici, G. Near-real time forest change detection using PlanetScope imagery. *Eur. J. Remote Sens.* **2020**, *53*, 233–244. [CrossRef]
21. Torres, R.; Snoeij, P.; Geudtner, D.; Bibby, D.; Davidson, M.; Attema, E.; Potin, P.; Rommen, B.; Floury, N.; Brown, M.; et al. GMES Sentinel-1 mission. *Remote Sens. Environ.* **2012**, *120*, 9–24. [CrossRef]
22. Drusch, M.; del Bello, U.; Carlier, S.; Colin, O.; Fernandez, V.; Gascon, F.; Hoersch, B.; Isola, C.; Laberinti, P.; Martimort, P.; et al. Sentinel-2: ESA's Optical High-Resolution Mission for GMES Operational Services. *Remote Sens. Environ.* **2012**, *120*, 25–36. [CrossRef]
23. Belgiu, M.; Csillik, O. Sentinel-2 cropland mapping using pixel-based and object-based time-weighted dynamic time warping analysis. *Remote Sens. Environ.* **2018**, *204*, 509–523. [CrossRef]
24. Woodcock, C.E.; Loveland, T.R.; Herold, M.; Bauer, M.E. Transitioning from change detection to monitoring with remote sensing: A paradigm shift. *Remote Sens. Environ.* **2020**, *238*, 111558. [CrossRef]
25. Puletti, N.; Bascietto, M. Towards a Tool for Early Detection and Estimation of Forest Cuttings by Remotely Sensed Data. *Land* **2019**, *8*, 58. [CrossRef]
26. Radoux, J.; Chomé, G.; Jacques, D.C.; Waldner, F.; Bellemans, N.; Matton, N.; Lamarche, C.; D'Andrimont, R.; Defourny, P. Sentinel-2's Potential for Sub-Pixel Landscape Feature Detection. *Remote Sens.* **2016**, *8*, 488. [CrossRef]
27. Lefebvre, A.; Sannier, C.; Corpetti, T. Monitoring Urban Areas with Sentinel-2A Data: Application to the Update of the Copernicus High Resolution Layer Imperviousness Degree. *Remote Sens.* **2016**, *8*, 606. [CrossRef]
28. Close, O.; Beaumont, B.; Petit, S.; Fripiat, X.; Hallot, E. Use of Sentinel-2 and LUCAS Database for the Inventory of Land Use, Land Use Change, and Forestry in Wallonia, Belgium. *Land* **2018**, *7*, 154. [CrossRef]
29. Phiri, D.; Simwanda, M.; Salekin, S.; Nyirenda, V.R.; Murayama, Y.; Ranagalage, M. Sentinel-2 Data for Land Cover/Use Mapping: A Review. *Remote Sens.* **2020**, *12*, 2291. [CrossRef]
30. Pesaresi, M.; Corbane, C.; Julea, A.; Florczyk, A.J.; Syrris, V.; Soille, P. Assessment of the Added-Value of Sentinel-2 for Detecting Built-up Areas. *Remote Sens.* **2016**, *8*, 299. [CrossRef]
31. Reiche, J.; Hamunyela, E.; Verbesselt, J.; Hoekman, D.; Herold, M. Improving near-real time deforestation monitoring in tropical dry forests by combining dense Sentinel-1 time series with Landsat and ALOS-2 PALSAR-2. *Remote Sens. Environ.* **2018**, *204*, 147–161. [CrossRef]
32. Ban, Y.; Yousif, O.; Hu, H. Global urban monitoring and assessment through earth observation: Fusion of SAR and optical data for urban land cover mapping and change detection. In *Global Urban Monitoring and Assessment through Earth Observation*; CRC Press: Boca Raton, FL, USA, 2014.
33. Yousif, O.; Ban, Y. Fusion of SAR and optical data for unsupervised change detection: A case study in Beijing. In Proceedings of the Joint Urban Remote Sensing Event (JURSE), Dubai, United Arab Emirates, 6–8 March 2017.
34. Hirschmugl, M.; Deutscher, J.; Gutjahr, K.; Sobe, C.; Schardt, M. Combined use of SAR and optical time series data for near real-time forest disturbance mapping. In Proceedings of the 9th International Workshop on the Analysis of Multitemporal Remote Sensing Images (MultiTemp), Brugge, Belgium, 27–29 June 2017.
35. Satalino, G.; Mattia, F.; Balenzano, A.; Lovergine, F.P.; Rinaldi, M.; de Santis, A.P.; Ruggieri, S.; Nafría García, D.A.; Paredes Gómez, V.; Ceschia, E.; et al. Sentinel-1 & sentinel-2 data for soil tillage change detection. In Proceedings of the IGARSS 2018—2018 IEEE International Geoscience and Remote Sensing Symposium, Valencia, Spain, 22–27 July 2018.
36. Haas, J.; Ban, Y. Sentinel-1A SAR and sentinel-2A MSI data fusion for urban ecosystem service mapping. *Remote Sens. Appl. Soc. Environ.* **2019**, *8*, 41–53. [CrossRef]
37. Mercier, A.; Betbeder, J.; Rumiano, F.; Baudry, J.; Gond, V.; Blanc, L.; Bourgoin, C.; Cornu, G.; Ciudad, C.; Marchamalo, M.; et al. Evaluation of Sentinel-1 and 2 Time Series for Land Cover Classification of Forest-Agriculture Mosaics in Temperate and Tropical Landscapes. *Remote Sens.* **2019**, *11*, 979. [CrossRef]
38. Eckerstorfer, M.; Vickers, H.; Malnes, E.; Grahn, J. Near-Real Time Automatic Snow Avalanche Activity Monitoring System Using Sentinel-1 SAR Data in Norway. *Remote Sens.* **2019**, *11*, 2863. [CrossRef]
39. Pulvirenti, L.; Squicciarino, G.; Fiori, E.; Fiorucci, P.; Ferraris, L.; Negro, D.; Gollini, A.; Severino, M.; Puca, S. An Automatic Processing Chain for Near Real-Time Mapping of Burned Forest Areas Using Sentinel-2 Data. *Remote Sens.* **2020**, *12*, 674. [CrossRef]
40. Cerbelaud, A.; Roupioz, L.; Blanchet, G.; Breil, P.; Briottet, X. A repeatable change detection approach to map extreme storm-related damages caused by intense surface runoff based on optical and SAR remote sensing: Evidence from three case studies in the South of France. *ISPRS J. Photogramm. Remote Sens.* **2021**, *182*, 153–175. [CrossRef]
41. Terrascope. Available online: https://terrascope.be/en/ (accessed on 27 January 2022).

42. Everaerts, J.; Clarijs, D. Terrascope enables geohazard monitoring using the sentinel satellite constellation. In Proceedings of the 2021 IEEE International Geoscience and Remote Sensing Symposium IGARSS, Brussels, Belgium, 11–16 July 2021; pp. 1895–1898. [CrossRef]
43. Killick, R.; Fearnhead, P.; Eckley, I. Optimal detection of changepoints with a linear computational cost. *J. Am. Stat. Assoc.* **2012**, *107*, 1590–1598. [CrossRef]
44. Stasolla, M.; Petit, S.; Wyard, C.; Swinnen, G.; Neyt, X.; Hallot, E. Urban sites change detection by means of sentinel-1 and sentinel-2 time series. In Proceedings of the IEEE IGARSS'21, Brussels, Belgium, 11–16 July 2021; pp. 1065–1068.
45. Asokan, A.; Anitha, J. Change detection techniques for remote sensing applications: A survey. *Earth Sci. Inform.* **2019**, *12*, 143–160. [CrossRef]
46. Hussain, M.; Chen, D.; Cheng, A.; Wei, H.; Stanley, D. Change detection from remotely sensed images: From pixel-based to object-based approaches. *ISPRS J. Photogramm. Remote Sens.* **2013**, *80*, 91–106. [CrossRef]
47. Bouziani, M.; Goïta, K.; He, D.C. Automatic change detection of buildings in urban environment from very high spatial resolution images using existing geodatabase and prior knowledge. *ISPRS J. Photogramm. Remote Sens.* **2010**, *65*, 143–153. [CrossRef]
48. Escadafal, R. Remote sensing of arid soil surface color with Landsat thematic mapper. *Adv. Space Res.* **1989**, *9*, 159–163. [CrossRef]
49. Tucker, C.J. Red and Photographic Infrared Linear Combinations for Monitoring Vegetation. *Remote Sens. Environ.* **1979**, *8*, 127–150. [CrossRef]
50. McFeeters, S.K. The use of the Normalized Difference Water Index (NDWI) in the delineation of open water features. *Int. J. Remote Sens.* **1996**, *17*, 1425–1432. [CrossRef]
51. Gadal, S.; Ouerghemmi, W. Multi-Level Morphometric Characterization of Built-up Areas and Change Detection in Siberian Sub-Arctic Urban Area: Yakutsk. *ISPRS Int. J. Geo-Inf.* **2019**, *8*, 129. [CrossRef]
52. Réjichi, S.; Chaabane, F.; Tupin, F. Expert Knowledge-Based Method for Satellite Image Time Series Analysis and Interpretation. *IEEE J. Sel. Top. Appl. Earth Obs. Remote Sens.* **2015**, *8*, 2138–2150. [CrossRef]
53. Lunetta, R.S.; Knight, J.F.; Ediriwickrema, J.; Lyon, J.G.; Worthy, L.D. Land-cover change detection using multi-temporal MODIS NDVI data. *Remote Sens. Environ.* **2006**, *105*, 142–154. [CrossRef]
54. Bassine, C.; Radoux, J.; Beaumont, B.; Grippa, T.; Lennert, M.; Champagne, C.; de Vroey, M.; Martinet, A.; Bouchez, O.; Deffense, N.; et al. First 1-M Resolution Land Cover Map Labeling the Overlap in the 3rd Dimension: The 2018 Map for Wallonia. *Data* **2020**, *5*, 117. [CrossRef]
55. Beaumont, B.; Grippa, T.; Lennert, M. A user-driven process for INSPIRE-compliant land use database: Example from Wallonia, Belgium. *Ann. GIS* **2021**, *27*, 211–224. [CrossRef]

 land

Article

Land Cover Dynamics along the Urban–Rural Gradient of the Port-au-Prince Agglomeration (Republic of Haiti) from 1986 to 2021

Waselin Salomon [1,2,*], Yannick Useni Sikuzani [3], Kouagou Raoul Sambieni [4], Akoua Tamia Madeleine Kouakou [5], Yao Sadaiou Sabas Barima [5], Jean Marie Théodat [6,7] and Jan Bogaert [1]

[1] Unité Biodiversité et Paysage, Gembloux Agro-Bio Tech, Université de Liège, 2 Passage des Déportés, 5030 Gembloux, Belgium; j.bogaert@uliege.be
[2] Campus Henri Christophe de Limonade, Université d'Etat d'Haïti, 1130, Rte Nationale # 6 Limonade, Limonade HT 1130, Haiti
[3] Unité Ecologie, Restauration Ecologique et Paysage, Faculté des Sciences Agronomiques, Université de Lubumbashi, Lubumbashi 1825, Congo; sikuzaniu@unilu.ac.cd
[4] Ecole Régionale Postuniversitaire d'Aménagement et de Gestion Intégrée des Forêts et Territoires Tropicaux (ERAIFT), Kinshasa 7948, Congo; krsambieni@uliege.be
[5] Unité de Formation et de Recherche Environnement, Université Jean Lorougnon Guédé, Daloa 150, Côte d'Ivoire; kouakou.akoua@ujlg.edu.ci (A.T.M.K.); sabas.barima@ujlg.edu.ci (Y.S.S.B.)
[6] Faculté des Sciences, Université d'Etat d'Haïti (Haïti), URBATeR, URBALaB, Corner of Joseph and Mgr Guilloux Street, Port-au-Prince HT 6110, Haiti; theodat@univ-paris1.fr
[7] Laboratoire de Géographie PRODIG, Campus Condorcet Bâtiment Recherche Sud 5, Université Paris 1, 93 322 Paris, France
* Correspondence: w.salomon@uliege.be; Tel.: +32-465289793 or +509-33511330

Citation: Salomon, W.; Useni Sikuzani, Y.; Sambieni, K.R.; Kouakou, A.T.M.; Barima, Y.S.S.; Théodat, J.M.; Bogaert, J. Land Cover Dynamics along the Urban–Rural Gradient of the Port-au-Prince Agglomeration (Republic of Haiti) from 1986 to 2021. *Land* **2022**, *11*, 355. https://doi.org/10.3390/land11030355

Academic Editors: Sara Venafra, Carmine Serio and Guido Masiello

Received: 24 January 2022
Accepted: 19 February 2022
Published: 27 February 2022

Abstract: The landscape of the Port-au-Prince agglomeration in the Republic of Haiti has undergone profound changes linked to (peri-)urban expansion supported by rapid demographic growth. We quantify the land cover dynamics along the urban–rural gradient of the Port-au-Prince agglomeration using Landsat images from 1986, 1998, 1999, 2010, and 2021 coupled with geographic information systems and landscape ecology analysis tools. The results show that over 35 years the acreage of the urban zone increased seven-fold while that of the peri-urban area increased five-fold, to the detriment of the rural zone, which was reduced by 14%. The dynamics of the landscape composition along the urban–rural gradient are characterized by a rapid progression of built-up and bare land in urban and peri-urban zones and by fields in the rural zone, in contrast to the more accentuated regression of vegetation in the peri-urban zone. The landscape of the study area has undergone significant changes due to the high demand for housing resulting from rapid population growth, in the context of a lack of territorial development planning by public authorities. This impacts the sustainability of socio-economic and ecological processes in an area where populations are highly dependent on plant resources. Our results underline the necessity to orient territorial development planning in urban, peri-urban and rural zones through an integrated and participatory approach.

Keywords: remote sensing/GIS; spatial dynamics; landscape metrics; urban–rural gradient; urbanization

1. Introduction

Human impact on the natural landscape has been increasing since the advent of sedentarization coupled with the emergence of agriculture [1,2]. The creation and extension of cities resulting from rural exodus and natural demographic growth are among the phenomena that have amplified the human impact on natural environments in recent decades [3–6]. Indeed, in 1850 the proportion of the world's population living in urban areas was 6% [7] compared to 55% in 2020; that proportion is projected to reach 70% by 2050 [8].

In addition to the densification of existing built-up areas, many cities are experiencing reverse migration leading to low-density sprawl on land reserves at the urban–rural interface, an area known as the "peri-urban" zone [5,9–11]. In developing countries, the dynamics of peri-urban zones are characterized by spontaneous and/or anarchic urbanization [12], which constitutes a challenge for urban and landscape planners [6,12]. Thus, many countries in Latin America and the Caribbean have recorded a rapid spatial expansion of urban areas, for example, an urbanization rate of 76.2% in Trinidad has been noted [13]. Mexico City, in Mexico, experienced an annual spatial growth of 0.9% between 2000 and 2010 [14,15]. The Port-au-Prince agglomeration in the Republic of Haiti is no exception to this rule [16].

The uncontrolled peri-urbanization of the Port-au-Prince agglomeration (the capital of Haiti) is the result of changing lifestyles and ineffective land governance, all of which is prompted by galloping and uncontrolled urban population growth. Indeed, from 1982 to 2018, the area's population increased five-fold from approximately 720,000 to 4,000,000 inhabitants [17,18], and is forecasted to house more than five million inhabitants by 2030 [19]. The resulting spatial urban expansion leads to an intensified consumption of agricultural land and pressure on woody vegetation, especially for charcoal production and for the extraction of building materials, etc. [16,20]. Consequently, the green spaces in the Port-au-Prince agglomeration are rapidly disappearing. This, despite their performance as a valuable ecosystem service, which includes moderation of the urban heat island effect, cleansing of air and water, conservation of biodiversity, provision of recreational opportunities, and improvement of physical and psychological well-being for citizens.

However, due to a deficit of over 2.4 million quality housing units in the urban zone of Port-au-Prince [18], about 65% of its population has been relegated to precarious and informal neighborhoods in the peri-urban zone, where access to basic services remains insufficient [16,21]. It should be noted that this situation is also visible in several cities in Latin America and the Caribbean where economic restructuring induced by the process of peri-urbanization has led to significant disparities in development between different neighborhoods [15]. In addition, the growth of the Haitian capital "Port-au-Prince" is also to the detriment of the capital cities, departments, and districts at the country level. Indeed, the centralization of public expenditure and the concentration of the majority of the country's employment in Port-au-Prince favors a steadily increasing rural exodus. Consequently, the population seeks to ensure its housing in a difficult economic context and the absence of territorial development planning, with little concern for the sustainability of resources [16]. This situation is often exacerbated by natural disasters (earthquakes, cyclones, etc.) which lead to changes in the landscape followed by massive displacement of the population towards the capital Haiti. The population allows itself to create new unplanned urban spots wherever space is available [15,16]. This is notably the case regarding the informal district of Canaan, which was created after the 2010 earthquake to house the affected population [16,22].

If the trend continues at the current rate, in which each year more than 10,000 households spontaneously settle in peri-urban zones [16,18], the prosperity of the population could be compromised for many decades to come. It should be noted that most of the spontaneous growth of the peri-urban zones in Port-au-Prince reflects the overall poverty of Haitian society, where 80% of residents subsist on less than USD 1.50 per day [16]. In addition, urban governance in Port-au-Prince is challenged by the growing need for infrastructure provision and land management [21] in an urban core where land for building is becoming increasingly scarce and expensive [18].

Despite this alarming situation, research into quantifying the urban and peri-urban expansion of the Port-au-Prince agglomeration and assessments of the associated ecological consequences still remain limited, including in other Caribbean cities [20]. However, numerous studies establish the importance of understanding the local influence of urban expansion and the various associated anthropogenic activities on landscape dynamics [23] to assess the nature and basis of these changes from the perspective of rational natural resource management. Given that the urban–rural opposition is completed by accounting for an intermediate zone between both namely, the peri-urban zone [24], it was appropriate for the present study to separately assess the land cover dynamics in the urban, peri-urban, and rural zones of the Port-au-Prince agglomeration. For this reason, the urban–rural gradient approach [6,11,23] was employed.

Accordingly, we characterize the land cover dynamics along the urban–rural gradient of the Port-au-Prince agglomeration in the Republic of Haiti. We hypothesized that the rapid and uncontrolled spatial expansion of the built-up area in urban and peri-urban zones, coupled with the development of shifting agriculture in the rural zone, has led to a landscape dynamic. This dynamic has been marked by the fragmentation and spatial isolation of woody vegetation patches, the extent of which increases in the peri-urban zone of the Port-au-Prince agglomeration

2. Materials and Methods

2.1. Presentation of the Port-au-Prince Agglomeration

The study area represents a group of municipalities that constitute the Port-au-Prince district, namely Port-au-Prince, Delmas, Cité Soleil, Tabarre, Pétion-Ville, Carrefour, Kenscoff, and Gressier, and the municipalities attached to Port-au-Prince district (Croix des Bouquets and Léogane). The 10 municipalities examined by this study form the "Port-au-Prince agglomeration" and cover an acreage of 1755.63 km^2 in the western department of the Republic of Haiti, located between 18°20′–18°50′ north latitude and 72°0′–72°50′ west longitude (Table 1, Figure 1). The relief presents an altitudinal gradient that shifts from low-lying plains to a succession of mountains with peaks exceeding 2000 m [25,26]. According to Köppen's classification, the climate of the study area ranges from tropical savannah in the lowland areas (Aw) to tropical subhumid in the mountainous areas (Cwa), characterized by a total annual rainfall between 1047 mm and 2000 mm and mean annual temperatures between 20 and 26 °C [27]. The natural vegetation largely comprises mangrove forests, shrub savannahs, and stands of pine and hardwood [28]. The economic fabric in the urban zone of the Port-au-Prince agglomeration is dominated by the informal sector (small- and medium-sized enterprises), which accounts for more than two-thirds of GDP and almost 80% of employment [29]. In the surrounding rural zones, the main economic activities are agriculture, livestock, and wood exploitation [30]. The Port-au-Prince agglomeration concentrates the bulk of the country's economic potential, thus attracting large numbers of people from around the country in search of remunerative activities [8,21]. Due to the unprecedented pressure on space of this poorly educated population (the literacy rate in Haiti is 61%), the city limits were extended to the entire southern fringe of the Cul-de Sac Plain and the foothills of Morne l'Hôpital [16,22]. As a result, there are many threats to the environment in the Port-au-Prince agglomeration, including destruction of vegetation, gully erosion, flooding, and pollution [20,29]

Table 1. Population, area and geographic coordinates of the municipalities in the Port-au-Prince agglomeration [30].

Municipalities	Population	Area (Km^2)	Geographical Coordinates
Port-au-Prince	987,310	36.04	18°32′24″ N–72°20′24″ W
Delmas	395,260	27.74	18°33′00″ N–72°18′00″ W
Cité Soleil	265,072	21.81	18°35′00″ N–72°20′06″ W
Tabarre	130,283	24.47	18°35′00″ N–72°16′00″ W
Croix des Bouquets	249,628	634.62	18°35′00″ N–72°14′00″ W
Pétion-ville	376,834	165.49	18°31′00″ N–72°17′00″ W
Léogane	199,813	385.23	18°30′39″ N–72°38′02″ W
Gressier	36,453	92.31	18°27′00″ N–72°17′00″ W
Kenscoff	57,434	202.76	18°27′00″ N–72°17′00″ W
Carrefour	511,345	165.16	18°32′00″ N–72°24′00″ W
Total	3,209,432	1755.63	

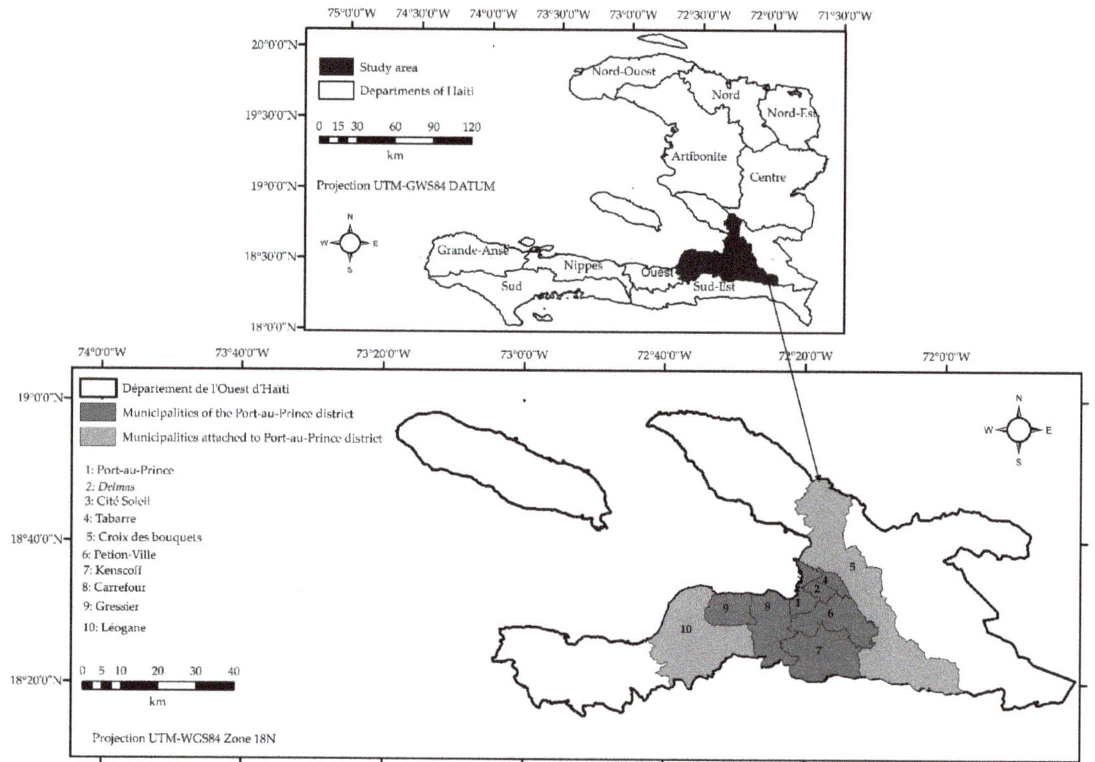

Figure 1. Geographical location of the municipalities of the Port-au-Prince district and attached municipalities in the western department of Haiti.

2.2. Methodology

2.2.1. Choice of Data and Materials Used

Landsat images downloaded from the site "https://earthexplorer.usgs.gov/" accessed on 27 March 2021 via the Multispectral Scanner System (acquired on 6 December 1986/13 December 1986), the Thematic Mapper (acquired on 14 December 1998/1 January 1999 and 22 January 2010/29 January 2010), and the Operational Land Imager (acquired on 4 January 2021/27 January 2021), with a spatial resolution of 30 m, were used to create the mosaics (two images per mosaic) from which the study area was extracted (Table 2). These images were chosen since they are free of charge and recommended for large-scale studies [31,32]. Moreover, they are particularly interesting for data-poor regions lacking recent and reliable spatial information [33]. For these reasons, the images meet the objectives of the study, despite their coarse resolution. All images were acquired during the winter dry season to minimize the effect of haze and clouds and thus facilitating the observation of larger spectral differences among landscape features [34,35]. Furthermore, the dates of acquisition of the Landsat images coincide with key periods marking the sociopolitical and economic life of the country and Port-au-Prince district in particular: (i) the fall of the Duvalier regime in 1986, the overthrow of President Aristide in 1991, and the subsequent embargo; (ii) the socioeconomic instability following the 2000 elections, the 2004–2008 hurricanes, and the 2010 earthquake; and (iii) the post-earthquake period (2010–2021). Additional data such as shapefiles illustrating the boundaries of the municipalities of the Port-au-Prince agglomeration from the Centre Nationale de l'Information Géographique et de Statistique (CNIGS) were used. ENVI 5.3 and ArcGiS 10.5.1 software was selected for the pre-processing and spatial analysis of the acquired satellite images.

Table 2. Satellite images characteristics.

Sensor	Dates	Path-Row	Spatial Resolution (m)
Landsat MSS	6 December 1986	008-047	30 m
	13 December 1986	009-047	30 m
Landsat TM	14 December 1998	009-047	30 m
	1 January 1999	008-047	30 m
	22 January 2010	008-047	30 m
	29 January 2010	009-047	30 m
Landsat OLI	4 January 2021	008-047	30 m
	27 January 2021	009-047	30 m

2.2.2. Landsat Image Processing and Classification

Pre-Processing

This work involved the development of a mosaic since the extent of the study area exceeded the scope of a remote sensing image [36]. However, a mosaic refers to the assembly of parts of images or contiguous images, preprocessed to be connectable geometrically and radiometrically [37,38]. Thus, the Landsat images used in this study were georeferenced in the UTM (Universal Transverse Mercator)/Zone 18 N, covering the study area, following the WGS 84 (World Geodesic System) reference ellipsoid. The 1986, 1988/1999, and 2010 images were geometrically corrected using 70 ground control points on the 2021 image, which was obtained as a reference. To ensure the efficiency of the change analysis, the geometric accuracy of the registration between the control points and the different Landsat images used was less than one pixel [39].

False Composite Color

A false composite color was created by combining the green, red, and near infrared channels, the last being understood as the most suitable for discriminating vegetation cover [40]. The composite color of the images provides the ability to select the training areas necessary to perform supervised classifications based on visual interpretation of the images supported by GPS data [41,42].

Determination of the Urban, Peri-Urban, and Rural Zones of the Port-au-Prince Agglomeration

To characterize the spatiotemporal dynamics of the different zones of the urban–rural gradient, the land cover was defined in urban, peri-urban, and rural zones according to the decision tree of the definitions of the zones present in the urban–rural gradient [11]. This decision tree, based on morphological characteristics, was preferred owing to its rapidity of execution, simplicity, and closeness to the ground reality, where there is a heterogeneous mix of land cover [6,43]. It should be noted that the urban zone is characterized by the dominance and continuity of the built-up area, which is otherwise dense. The peri-urban zone is characterized by the dominance of a discontinuous and less dense built-up area, while the dominance of vegetation indicates a rural zone [6,11].

The aforementioned decision tree was applied to map the different land cover (urban, peri-urban and rural) on each of the composite Landsat images by a supervised classification employing the maximum likelihood algorithm. This algorithm uses training sites to calculate the probability of each pixel belonging to one of the classes [44]. It should be noted that the urban zone is characterized by the dominance and continuity of the built-up area, which is otherwise dense. The peri-urban zone is characterized by the dominance of discontinuous and less dense built-up area, while the dominance of vegetation indicates a rural zone [6,11]. Thus, the training samples used for this classification were delineated through 219 fixed points acquired with a Garmin 66s GPS (accuracy 3 m) during November and December 2020. The classification accuracy was assessed using the Kappa coefficient and the overall accuracy, based on the confusion matrix generated with 387 validation points. The Kappa coefficient provides a more accurate estimate (which takes into account well-classified pixels) of the quality of the classification. The overall classification accuracy represents the average of the percentages of correctly classified pixels. The percentage of landscape, which indicates the relative abundance of each urban–rural gradient zone, was calculated.

Qualification of the Port-au-Prince Agglomeration's Municipalities in Urban, Peri-Urban, and Rural Zones

Subsequently, the morphological status of the municipalities along the urban–rural gradient of the Port-au-Prince agglomeration was defined according to the proportions of the different zones (urban, peri-urban, and rural) resulting from the supervised classification of the urban–rural gradient zones from the Landsat image of 2021. If the proportion of the built-up area dominates the landscape, a distinction is drawn between the urban and the peri-urban: if the urban dominates, the area is urban and if the peri-urban dominates, the area is peri-urban. If the co-dominance of urban and peri-urban is less than rural, the area is recognized as rural. Finally, if the co-dominance of urban and peri-urban is higher than rural, the area is considered peri-urban [43].

Classification and Assessment of Land Cover Changes along the Urban–Rural Gradient Zones

Based on knowledge of morphological status, the municipalities of the Port-au-Prince agglomeration were grouped into urban, peri-urban, and rural zones. In each group of municipalities, the land cover dynamics from 1986 to 2021 were assessed based on a second supervised classification. For this reason, the following land cover types were defined: built-up and bare soil (built-up area, bare ground, road), field (mono- or multi-crop agricultural areas, agroforestry systems), woody vegetation (wooded savannah, forest, mangrove) and grassy vegetation (grass, young fallow land, pastures). A total of 206 fixed points and plots obtained from these different land cover types were used in the definition of training samples for supervised classification, based on the maximum likelihood algorithm [45]. Finally, a confusion matrix generated from 497 ground points was employed to verify the classification accuracy, based on the Kappa coefficient and the overall accuracy—two appropriate indices for verifying the reliability of a supervised classification [46].

To assess the impact of peri-urbanization on land cover changes along the urban–rural gradient, the proportion of land cover types in each type of municipality (urban, peri-urban, and rural) was calculated based on the patch area. This index often indicates human impact on landscape morphology [47]. It may provide information on the fragmentation of a land cover type between two periods, particularly through its decrease (Equation (1)).

Rate of land cover change (Rc):

$$(\mathrm{Rc}) = \frac{(\mathrm{UA_{i+n}} - \mathrm{UA_i})}{\mathrm{UA_i}} \tag{1}$$

where $\mathrm{UA_i}$ is the extent occupied by a class in the initial year of a period, n is the interval between two evaluated years, and $\mathrm{UA_{i+n}}$ is the extent occupied by the same class in year i + n [48].

3. Results

3.1. Accuracy of Supervised Classifications

The overall accuracy values obtained were above 90% (Table 3), and the Kappa coefficient indicated values between 92 and 99%, thus suggesting a better distinction between land cover types.

Table 3. Overall accuracy and Kappa coefficient values from supervised classifications of Landsat image mosaics of the Port-au-Prince agglomeration from 1986, 1998–1999, 2010 and 2021 based on the maximum likelihood algorithm.

	Classification 1		Classification 2	
Image Mosaics Classified	**Overall Accuracy (%)**	**Kappa (%)**	**Overall Accuracy(%)**	**Kappa (%)**
1986	94.08	97.04	95.36	98.78
1998–1999	98.11	94.08	98.44	97.66
2010	94.52	96.38	96.46	97.36
2021	95.09	92.52	94.35	92.43

Classification 1 refers to the segmentation and qualification of the urban–rural gradient zones, and classification 2 to the land cover types classifications within the urban, peri-urban and rural zones.

3.2. Mapping and Quantification of the Spatial Changes in the Urban, Peri-Urban, and Rural Zones in the Port-au-Prince Agglomeration

A total of four land cover maps were produced following the supervised classification of Landsat images, illustrating the dynamics of the urban, peri-urban, and rural zones of the Port-au-Prince agglomeration in 1986, 1998–1999, 2010, and 2021 (Figure 2). The visual analysis of the spatial dynamics shows that the urban and peri-urban zones are in constant spatial progression between 1986 and 2021 in the north and east of the study area on a rural matrix that has registered a regressive dynamic (Figure 2).

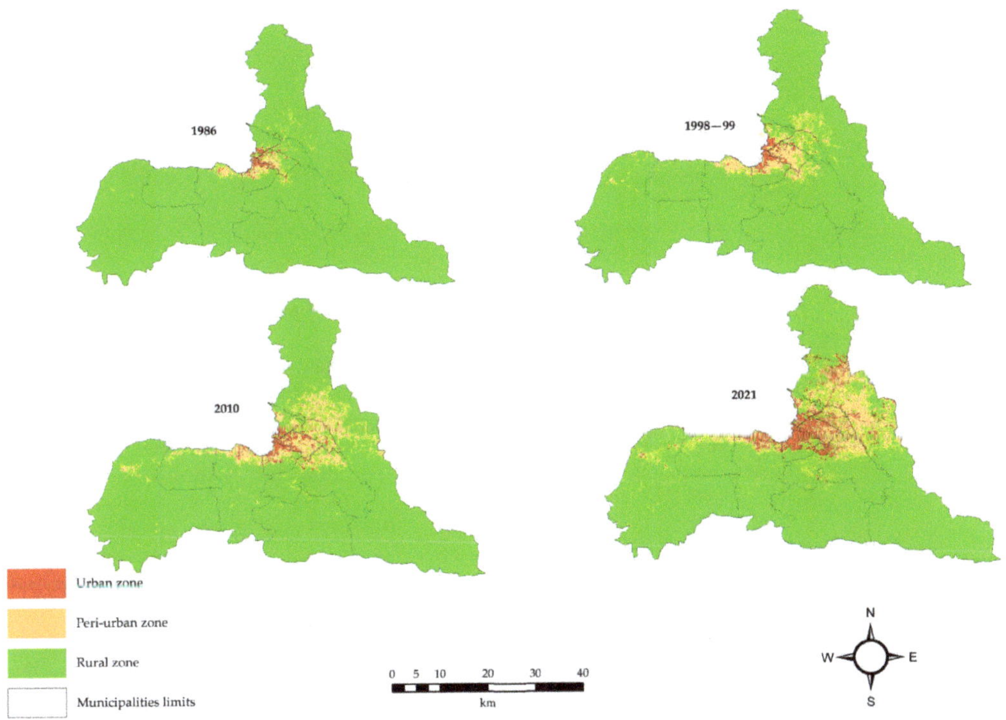

Figure 2. Land cover maps of the Port-au-Prince agglomeration obtained from supervised classification of Landsat images from 1986, 1998–1999, 2010 and 2021 based on the maximum likelihood algorithm. The black lines correspond to the boundaries of the municipalities.

The urban zone experienced a net increase of 612.33% in the landscape between 1986 and 2021, with its area increasing from 8.19 km^2 to 58.34 km^2. However, it should be noted that the most dramatic spatial increase in the urban zone occurred between 2010 and 2021, with a net increase of 229.42%. The peri-urban zone also increased in acreage from 45.57 km^2 in 1986 to 242.93 km^2 in 2021, with a rate of change of 433.09%. In contrast to the urban and peri-urban zones, the rural zone experienced a regression in acreage from 1697.47 km^2 in 1986 to 1449.82 km^2 in 2021, a net loss of 14.59% compared to 1986 (Table 4).

Table 4. Net area increase between 1986–1998/1999, 1998/1999–2010, 2010–2021 and 1986–2021 of the different zones (urban, peri-urban, rural) corresponding to the agglomeration of Port-au-Prince.

	Area 1986–1998/1999 (km^2)	Area 1998/1999 (km^2)	Net Increase/Decrease(%)
Urban zone	8.19	12.37	51.04
Peri-urban zone	45.57	102.52	124.97
Rural zone	1701.87	1640.74	−3.59
	Area 1998/1999 (km^2)	**Area 2010 (km^2)**	**Net Increase/Decrease (%)**
Urban zone	12.37	17.71	43.17
Peri-urban zone	102.52	165.1	61.04
Rural zone	1640.74	1572.82	−4.14
	Area 2010 (km^2)	**Area 2021 (km^2)**	**Net Increase/Decrease (%)**
Urban zone	17.71	58.34	229.42
Peri-urban zone	165.1	242.93	47.14
Rural zone	1572.82	1454.36	−7.53
	Area 1986 (km^2)	**Area 2021 (km^2)**	**Net Increase/Decrease (%)**
Urban zone	8.19	58.34	612.33
Peri-urban zone	45.57	242.93	433.09
Rural zone	1701.87	1454.36	−14.54

3.3. *Mapping and Quantification of Land Use Dynamics along the Urban–Rural Gradient of the Port-au-Prince Agglomeration*

Table 5 displays the morphological urbanization status of the 10 municipalities within the Port-au-Prince agglomeration and the land cover change that occurred within each morphological type of municipality. First, four municipalities exhibit a dominance of built-up area, notably Port-au-Prince and Delmas, which bear an urban zone status, as opposed to Cité Soleil and Tabarre, which have a peri-urban status. In the municipalities with an urban zone status, the "built-up and bare soil" class increased to become the landscape matrix (dominant land cover type) in 2021, while the proportion of fields (the dominant land cover type in 1986), woody vegetation, and grassy vegetation decreased (Figures 3 and 4). This seems to suggest the replacement of vegetation under the influence of expansion and built-up densification. Regarding the municipalities with peri-urban morphological status, the evolution of land cover shows a transition marked by the replacement of fields, which constituted the landscape matrix in 1986, 1998–1999, and 2010, by the built-up area and bare soil that became the dominant land cover type of the peri-urban zone in 2021. During the same period, the proportion of woody and grassy vegetation decreased in the peri-urban zone between 1986 and 2021 (Figures 3 and 4).

In contrast, the municipalities of Croix des Bouquets, Pétion-Ville, Léogane, Gressier, Kenscoff and Carrefour are characteristic of rural zones (Table 5). Within these municipalities, a degradation of woody vegetation (the dominant land cover type in 1986) and grassy vegetation was noted, marked by their replacement with fields, which increased in proportion to become the new landscape matrix in 1998–1999, 2010, and 2021. In these municipalities with a rural morphological status, the area of "built-up and bare soil" increased three-fold in the landscape over the entire study period, with a more marked evolution between 1998–1999 and 2021 (Figures 3 and 4).

Table 5. Morphological status of the municipalities along the urban–rural gradient of the Port-au-Prince agglomeration according to [43] typology. These results are derived from the supervised classification of the Landsat image mosaics of 2021 based on the maximum likelihood algorithm.

Municipalities	Urban Area in km^2 (%)	Peri-Urban Areain km^2 (%)	Rural Area in km^2 (%)	Zone Status
Port-au-Prince	12.7 (35.2)	7.94 (22.0)	15.4 (39.9)	Urban
Delmas	14.2 (51.9)	11.7 (42.9)	2.37 (8.5)	Urban
Cité Soleil	5.8 (26.6)	11.1 (51.3)	4.8 (22.)	peri-urban
Tabarre	2.8 (11.4)	17.1 (69.9)	4.6 (18.8)	peri-urban
Croix des Bouquets	13.9 (2.2)	115.9 (18.3)	504.8 (79.5)	Rural
Pétion-ville	5.5 (3.3)	36.9 (22.4)	123.5 (74.6)	Rural
Léogane	0.9 (0.2)	17.8 (4.6)	366.5 (95.1)	Rural
Gressier	0.4 (0.5)	12.9 (13.2)	78.3 (86.3)	Rural
Kenscoff	0.2 (0.1)	2.0 (1.0)	200.0 (98.6)	Rural
Carrefour	6.0 (3.7)	16.3 (10.0)	142.2 (86.4)	Rural

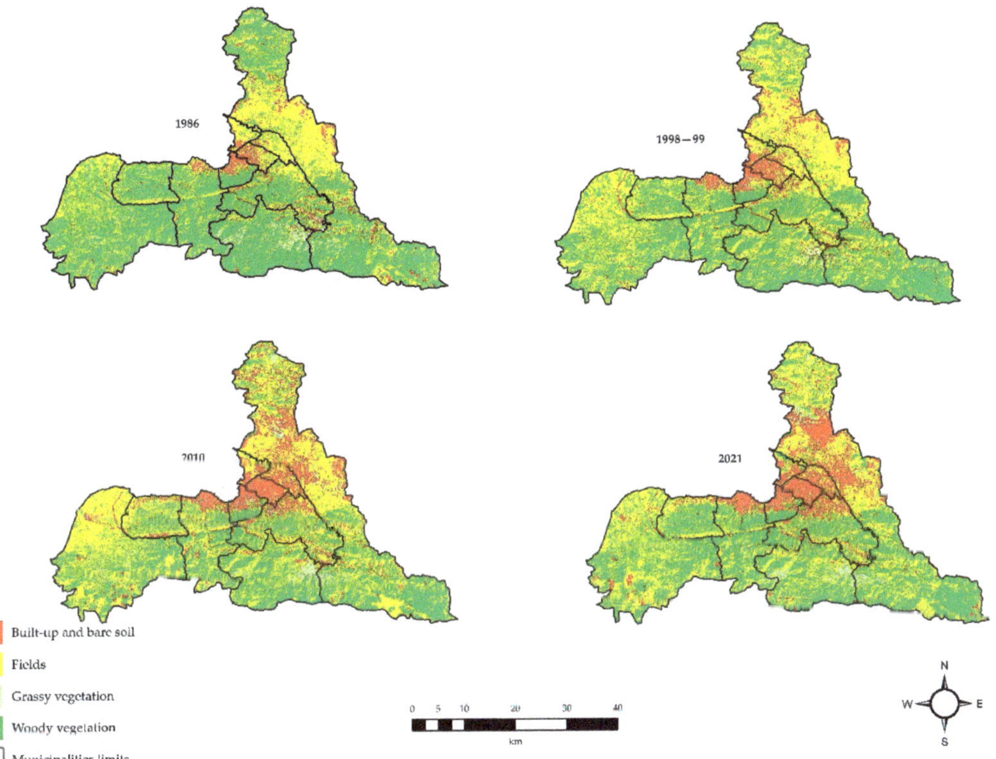

Figure 3. Land cover maps of the Port-au-Prince agglomeration from supervised classification of Landsat image mosaics from 1986, 1998–1999, 2010 and 2021 based on the maximum likelihood algorithm. The black lines correspond to the boundaries of the municipalities.

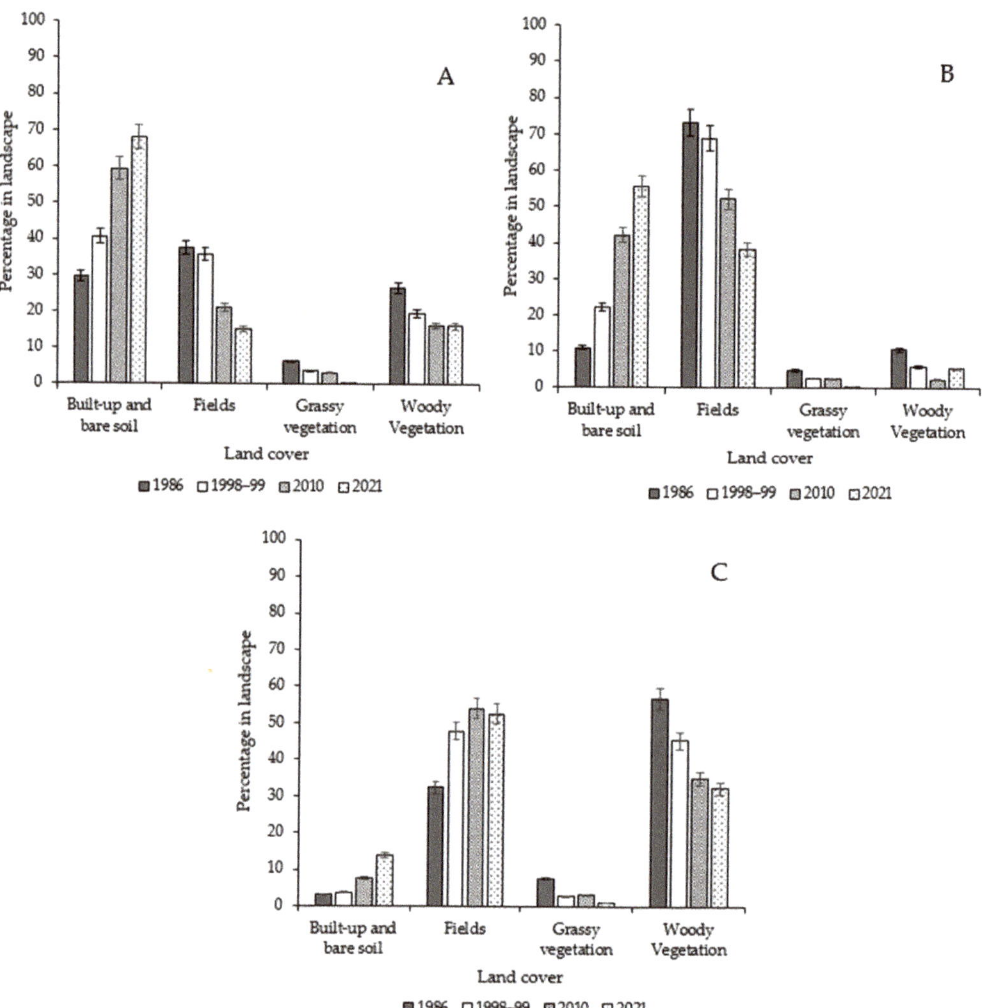

Figure 4. Evolution of the percentage of landscape of different land cover types in municipalities with urban (**A**), peri-urban (**B**) and rural (**C**) zones morphological status. These results were obtained on the basis of supervised classification of Landsat image mosaics from 1986, 1998–1999, 2010 and 2021 based on the maximum likelihood algorithm. The vertical bars represent the standard deviation.

4. Discussion

4.1. Dynamics of the Urban-Rural Gradient Zones of the Port-au-Prince Agglomeration

Since 1986, the various sociopolitical crises that have occurred in Haiti have led to a massive influx of rural populations into the Port-au-Prince agglomeration. In addition, the rapid population growth of the Port-au-Prince agglomeration is largely dependent on unplanned and informal urbanization to meet its housing needs [16]. As a result, the Port-au-Prince agglomeration has experienced rapid spatial urban expansion, particularly towards the north-east, and densification of preexisting built-up areas. The significant spatial expansion of the urban zone in the Port-au-Prince agglomeration seems to indicate a spatial densification of the built-up area in the urban core. These findings should be viewed within the context of an increase in built-up density closer in proximity to the otherwise more densely populated urban core [16]. This exacerbates the vulnerability of this disadvantaged population due to the mixing of highly densified marginal and risky urban and peri-urban spaces [22,28].

Furthermore, to meet the additional need for housing, the Port-au-Prince agglomeration tends to connect with peripheral municipalities [16], thus justifying the regressive dynamics of the rural zone to the benefit of the peri-urban zone. The rapid spatial urban expansion of Port-au-Prince city towards peripheral areas leads to the discontinuity of urban patches, further suggesting that the geographical space represents a limited resource [6].

The current pattern of urban expansion seems to be influenced by a more favorable topography (the Cul-de-Sac Plain). It has been recognized that topography could influence the expansion of urban areas [31,49]. However, in recent years, urbanization continues to progress, particularly in the south of the study area, in the foothills of the Massif de la Selle, especially on Morne l'Hôpital, despite its status as a reserved area [28]. Indeed, within a context of buildable land becoming scarce and relatively expensive and where the cost of living does not allow for the rental of flats in the urban zone, poor populations settle in risky areas, which lack urban planning infrastructure, and construct houses with salvaged materials [18]. These observations are similar to those of [50] in Cap-Haitian (the second largest city in Haiti), which shows the settlement of the poor population in risky areas such as mangrove forests and mountainsides. Moreover, urban growth is linked to the occurrence of natural disasters in the country (hurricanes and earthquakes), which have led the population to relocate to spaces reserved for agricultural use, mostly unsuitable for building, etc. [51]. Thus, it was revealed that the decade of 2010–2021 was characterized by a stronger urban expansion than other periods studied. Indeed, the urban dynamic during this period seems to have been determined by the 2010 earthquake, which pushed residents without housing and those coming from rural zones to occupy vacant spaces without basic infrastructure [18]. Indeed, new townships, including Canaan with nearly 250,000 inhabitants, emerged after the 2010 earthquake in the municipality of Croix des Bouquets, which bears a rural morphological status [16,18].

4.2. Landscape Dynamics of the Urban Core towards the Rural Areas Adjacent to the Port-au-Prince Agglomeration

The rapid evolution of the peri-urbanization process in the Port-au-Prince agglomeration between 1986 and 2021 is manifested by the anarchic expansion of built-up land to the detriment of the fields. Indeed, in recent decades, agricultural areas have been increasingly transformed into housing and roads [51]. This trend is similar to the findings of [50] on the city of Cap-Haitian (Haiti) and [52] in the French Antilles (Guadeloupe and Martinique), according to which agricultural land in peri-urban areas is constantly being invaded by anarchic buildings. However, agricultural activity is essential to boost the economy of the city and the peri-urban area, to regreen it, and to protect against food insecurity [53].

Moreover, the process of peri-urbanization contributes to the regression of woody vegetation, which is becoming scarcer in both lowland and mountain areas due to their accessibility [28]. This situation risks creating an imbalance between rainwater infiltration and groundwater exploitation in the Cul de Sac Plain, given that the quantity of water drawn from the aquifer is estimated between 63 and 86% of the annual recharge for a growing population [54]. It should be noted that, with an increasingly low poverty line, the population of the Port-au-Prince agglomeration is exploiting and destroying vegetation in favor of subsistent farming activities.

The landscape dynamics of the municipalities located in the rural zone of the Port-au-Prince agglomeration are marked by a decrease in woody vegetation in favor of fields. Indeed, the socioeconomic situation of the rural population, characterized by increasing poverty, has pushed a large proportion of the population into agriculture, particularly slash-and-burn agriculture. Despite the low average productivity of the agricultural sector and the low economic surplus generated, it remains the refuge sector *par excellence* for the population in the rural zone [55]. In addition, due to the increasing demand for charcoal by the urban and peri-urban population [56], pressure on vegetation in the rural zone is intensifying, especially since charcoal accounts for more than 70% of the country's energy needs. The degradation of vegetation in the Port-au-Prince agglomeration leads to a reduction in its resilience and could thus lead to an increase in flooding in the (peri-)urban zone and an increase in the risk of landslides and rockfalls [28]. In addition, this anthropization of the Port-au-Prince landscape could also lead to runoff and silting of the drainage networks during each rainfall event in the urban sectors located downstream of the mountain, thus obstructing the city's drainage infrastructure, which causes recurrent damage in the lowest areas [20,28].

4.3. *Proposals and Perspectives*

4.3.1. For the Government and Planners

The current spatial challenge of peri-urbanization in Port-au-Prince consists of adapting or readapting human settlements in such a way as to respond sustainably to the socio-spatial needs of city dwellers and thus to reduce environmental degradation as much as possible. It thus requires efficient planning and settlement policies coupled with a better understanding of the spatial and temporal evolution of the (peri-)urban areas of the Port-au-Prince agglomeration provided by this study. Our results deliver a basis for promoting better planning and efficient spatial organization of the (peri-)urban areas of the Port-au-Prince agglomeration aiming at sustainable development. Moreover, it would be important to anticipate peri-urbanization in currently rural areas that are destined to become potential peri-urban areas within the framework of a territorial development plan, in order to ensure the food security of the population. Indeed, agricultural land continues to be invaded by housing, according to our results. Conversely, for the preexisting urbanized spaces, there is an urgent need to reverse the current socio-spatial imbalance from the perspective of establishing dynamic balances of the mid-place, especially concerning vegetation [57]. Finally, it would be necessary to address the land issue, corruption and also the establishment and enforcement of legal frameworks appropriate to urbanization and the implementation of peri-urban agriculture in the design of a development plan.

It is necessary to delay the growth rate of the Port-au-Prince agglomeration and to reduce the demographic and economic gap between it and other chief towns of the departments and districts of Haiti. This implies the elaboration and application of a true spatially-balanced growth strategy and to work towards decentralization, economic and political deconcentration through the development of different departmental cities.

4.3.2. For Scientific Research Institutions

This study has rendered it possible to characterize (peri-)urban growth in the Port-au-Prince agglomeration and to evaluate its consequences along the urbanization gradient. However, there remain many aspects to be investigated in order to identify a sustainable solution that will enable reconciliation of the conservation of biodiversity and the satisfaction of the spatial needs of an ever-growing population. In this sense, it is up to scientific research institutions to contribute, among other things, to the evaluation of the impact of the degree of urbanization on the ecosystem services mainly provided by green spaces in the Port-au-Prince agglomeration; to develop indicators for monitoring the health of (semi-)natural ecosystems; to integrate the notion of ecosystem services in the planning of territorial development; and to provide scientific assistance to the conservation and development of green spaces.

4.3.3. For the Public

An integrated and sustainable management of the landscape is therefore a very important issue. To achieve this, the populations will have to become involved in the conservation of (semi-)natural ecosystems in the urban and peri-urban landscape of the Port-au-Prince agglomeration, as vegetation directly influences the urban soil and climate while providing beneficial ecosystem services to city dwellers [9,42]. It is important to diversify energy sources and to adopt new techniques and practices to reduce the collection of wood for energy production, as wood resources tend to decrease along the urban–rural gradient of Port-au-Prince. The scarcity of wood resources bears socioeconomic consequences: the lack of wood energy limits the amount of food cooked and therefore has consequences for nutrition and health, loss of jobs, and income for charcoal producers. It should be noted that the rapid development of the charcoal network is a popular reaction to the lack of alternative energy sources, particularly electricity, in Haitian cities [58]. Participatory land use mapping is urgently needed and the population should be made aware of the preservation of agricultural and (semi-)natural areas in view of the various socio-ecological benefits they provide. Urban fragmentation through building densification should be controlled in urban areas, as it could pose a threat to the preservation of vegetation in the plots.

5. Conclusions

This study sought to highlight the spatial dynamics of land use that prevails along the urban–rural gradient of the Port-au-Prince agglomeration. Our results confirm a change in the spatial pattern along the urban–rural gradient, characterized over 35 years by a rapid progression of built-up and bare soil in urban and peri-urban zones, and of fields in the rural zone. The expansion of these anthropogenic land cover types leads to a regression in the patch area of woody and grassy vegetation among the landscape. This represents an indication of the anthropogenic impact on landscape dynamics along the urban–rural gradient of the Port-au-Prince agglomeration, the extent of which has intensified over the years in the peri-urban zone. This study provides basic information that should lead to an improved understanding of the spatial urban and peri-urban growth of the Port-au-Prince agglomeration and its impact on the different land cover types along the urban–rural gradient. This information remains crucial for the implementation of territorial development planning measures through an integrated and participatory approach.

Author Contributions: Conceptualization, W.S., Y.U.S., K.R.S. and J.B.; Methodology, W.S., Y.U.S., K.R.S., A.T.M.K., Y.S.S.B. and J.B.; Software, W.S.; Validation, W.S., Y.U.S., K.R.S., A.T.M.K., Y.S.S.B., J.M.T. and J.B.; Formal Analysis, W.S., Y.U.S., K.R.S., A.T.M.K., Y.S.S.B. and J.B.; Investigation, W.S.; Resources, W.S., Y.U.S. and J.B.; Data Curation, W.S.; Writing—Original Draft Preparation, W.S.; Writing—Review & Editing W.S., Y.U.S., K.R.S., A.T.M.K., Y.S.S.B., J.M.T. and J.B.; Visualization, W.S., Y.U.S., K.R.S., A.T.M.K., Y.S.S.B. and J.B.; Supervision, W.S., Y.U.S., K.R.S., A.T.M.K., J.M.T. and J.B.; Project Administration, W.S., Y.U.S., J.M.T. and J.B.; Funding Acquisition, J.B. All authors have read and agreed to the published version of the manuscript.

Funding: This research was supported by the Académie de Recherche et d'Enseignement Supérieur-Commission de la Coopération au Développement (ARES-CCD. Exceptional scholarship program) and the Agence Universitaire de la Francophonie (AUF, Antenor Firmin Scholarship) and Fédération Wallonie Bruxelles (FWB, Exceptional Covid-19 measures).

Institutional Review Board Statement: Not applicable.

Informed Consent Statement: Not applicable.

Data Availability Statement: Not applicable.

Conflicts of Interest: The authors declare no conflict of interest.

References

1. Mazoyer, M.; Roudart, L. *Histoire des Agricultures du Monde: Du Néolithique à la Crise Contemporaine*; Histoire; Éditions du Seuil: Paris, France, 2002.
2. Bogaert, J.; Vranken, I.; André, M. Anthropogenic Effects in Landscapes: Historical Context and Spatial Pattern. In *Biocultural Landscapes*; Hong, S.-K., Bogaert, J., Min, Q., Eds.; Springer: Dordrecht, The Netherlands, 2014; pp. 89–112. [CrossRef]
3. Nguimalet, C.R. Population et Croissance Spatiale: Diagnostic et Implications Pour Une Gestion Urbaine de Bangui (République Centrafricaine). In Proceedings of the PRIPODE Workshop on Urban Population, Development and Environment Dynamics in Developing Countries, Cicred, Aphrc, Pern, Ciesin, Nairobi, Kenya, 11–13 June 2007. [CrossRef]

4. Weeks, J.R. Defining Urban Areas. In *Remote Sensing of Urban and Suburban Areas*; Remote Sensing and Digital Image Processing; Rashed, T., Jürgens, C., Eds.; Springer: Dordrecht, The Netherlands, 2010; Volume 10, pp. 33–45. [CrossRef]
5. Besussi, E.; Chin, N.; Batty, M.; Longley, P. The Structure and Form of Urban Settlements. In *Remote Sensing of Urban and Suburban Areas*; Remote Sensing and Digital Image, Processing; Rashed, T., Jürgens, C., Eds.; Springer: Dordrecht, The Netherlands, 2010; Volume 10, pp. 13–31. [CrossRef]
6. Bogaert, J.; Biloso, A.; Vranken, I.; André, M. Peri-urban dynamics: Landscape ecology perspectives. In *Territoires Périurbains: Développement, Enjeux et Perspectives Dans les Pays du Sud*; Bogaert, J., Halleux, J.M., Eds.; Les presses agronomiques de Gembloux: Gembloux, Belgique, 2015; pp. 63–73. Available online: http://hdl.handle.net/2268/188554 (accessed on 18 January 2022).
7. The World Bank. *World Development Report. 1984: Recovery or Relapse in the World Economy? Population Change and Development, Population Data Supplement, World Development Indicators*; Oxford University Press: Oxford, UK, 1984. [CrossRef]
8. United Nations, Department of Economic and Social Affairs, Population Division. World Urbanization Prospects: The 2014 Revision, Highlights (ST/ESA/SER.A/352). 2014. Available online: https://population.un.org/wup/publications/files/wup2014-report.pdf (accessed on 18 January 2022).
9. Alberti, M. *Advances in Urban Ecology: Integrating Humans and Ecological Processes in Urban Ecosystems*; Springer: New York, NY, USA, 2008.
10. Chapman, R. Urbanism in the Age of Climate Change. *J. Urban Des.* **2014**, *19*, 149. [CrossRef]
11. André, M.; Mahy, G.; Lejeune, P.; Bogaert, J. Vers une synthèse de la conception et d'une définition des zones dans le gradient urbain-rural. *Biotechnol. Agron. Soc. Environ.* **2014**, *18*, 61–74.
12. Tchékoté, H.; Ngouanet, C. Périurbanisation anarchique et problématique de l'aménagement du territoire dans le périurbain de Yaoundé. In *Territoires Périurbains: Développement, Enjeux et Perspectives Dans les Pays du Sud*; Bogaert, J., Halleux, J.M., Eds.; Les Presses Agronomiques de Gembloux: Gembloux, Belgique, 2015; pp. 259–270.
13. Moullet, D.; Saffache, P.; Transler, A.-L. L'urbanisation Caribeenne: Effets et Contrastes. *Etudes Caribeennes* **2007**, *7*. [CrossRef]
14. Souchaud, S.; Prévôt-Schapira, M.-F. Introduction: Transitions métropolitaines en Amérique latine: Densification, verticalisation, étalement. *Problèmes D'amérique Lat.* **2013**, *90*, 5. [CrossRef]
15. Darbouze, J.; Simonneau, C.; Hanin, Y. Précisions sémantiques, état des lieux dans les pays du Sud et introduction à l'aire métropolitaine de Port-au-Prince. In *Rapport du Programme de Recherche Dans le Champ de L'Urbain FED/2015/360-478, Perspectives de Développement de L'Aire Métropolitaine de Port-au-Prince, Horizon 2030*; Rapport; UQAM: Montréal, QC, Canada, 2018; pp. 41–57.
16. Théodat, J.-M. Port-au-Prince en sept lieues. *Outre-Terre* **2013**, *1*, 123. [CrossRef]
17. Persée, Le Recensement Haitien de 1982. *Population* **1983**, *38*, 1055. [CrossRef]
18. Lizarralde, G.; Petter, A.-M.; Julien, O.J.; Bouchereau, K. L'habitat Dans la Zone Métropolitaine de Port-au-Prince: Principales Représentations, Défis, Opportunités et Prospectives. 2018. Available online: https://www.researchgate.net/publication/326876764 (accessed on 30 December 2021).
19. Bodson, P.; Benoît, J.; Duval, C.J.; Thérasmé, K. la Population de l'aire métropolitaine de port-au-prince 2009–2030. In *Rapport du Programme de Recherche Dans le Champ de L'Urbain FED/2015/360-478, Perspectives de Développement de L'Aire Métropolitaine de Port-au-Prince, Horizon 2030*; Rapport; UQAM: Montréal, QC, Canada, 2018; pp. 89–118.
20. Dehoorne, O.; Cao, H.; Ilies, D. Étudier La Ville Caribéenne. *Etudes Caribeennes* **2018**, 39–40. [CrossRef]
21. Belvert, A. Etude de L'urbanisation du Secteur Sud du Littoral de Port-au-Prince: Cas des Quartiers de Cite Michel et de Ruelle Assade. Faculté des Sciences Économiques, Sociales, Politiques et de Communication, Université Catholique de Louvain. Prom: Emmanuelle PICCOLI; Claire SIMONNEAU. 2019. Available online: http://hdl.handle.net/2078.1/thesis:22484 (accessed on 18 January 2022).
22. Herrera, J.; Lamaute-Brisson, N.; Milbin, D.; Roubaud, F.; Saint-Macary, C.; Torelli, C.; Zanuso, C. L'Evolution des Conditions de vie en Haïti Entre 2007 et 2012. La Réplique Sociale du Séisme. IHSI, DIAL, Paris, Port-au-Prince. 2014. Available online: https://horizon.documentation.ird.fr/exl-doc/pleins_textes/divers14-08/010062827.pdf (accessed on 11 November 2021).
23. Mcdonald, R.I.; Kareiva, P.; Forman, R.T.T. The Implications of Current and Future Urbanization for Global Protected Areas and Biodiversity Conservation. *Biol. Conserv.* **2008**, *141*, 1695–1703. [CrossRef]
24. Useni, S.Y.; Sambiéni, K.R.; Maréchal, J.; Ilunga wa Ilunga, E.; Malaisse, F.; Bogaert, J.; Munyemba, K.F. Changes in the Spatial Pattern and Ecological Functionalities of Green Spaces in Lubumbashi (the Democratic Republic of Congo) in Relation with the Degree of Urbanization. *Trop. Conserv. Sci.* **2018**, *11*, 194008291877132. [CrossRef]
25. Moise, P. Etude de Marché de l'aire Métropolitaine de Port-au-Prince: Première étape à la Mise en Place d'une Ferme de Production Maraîchère à Dume (Croix-des-Bouquets), Travail de Fin d'étude en Master de Spécialisation en Production Intégrée et Préservation des Ressources Naturelles en Milieu Urbain et Péri-Urbain, ULiège Gembloux Agro-Bio Tech (Belgique). 2017. Available online: http://hdl.handle.net/2268.2/3087 (accessed on 22 July 2021).
26. MDE. Sixième Rapport National sur la Biodiversité d'Haïti. gef/CBD/6NR/UNDP. Ministère de L'Environnement. Direction de La Biodiversité. 2019. Available online: https://www.cbd.int/doc/nr/nr-06/ht-nr-06-fr.pdf (accessed on 6 May 2021).
27. Climate-Data.org. Available online: https://fr.climate-data.org/amerique-du-nord/haiti/departement-de-l-ouest/port-au-prince-3571/ (accessed on 23 January 2022).
28. Fifi, U. Impacts des Eaux Pluviales Urbaines Sur les Eaux Souterraines dans les Pays en Développement Mécanismes de Transfert des Métaux Lourds à Travers un sol Modèle de Port-au-Prince, Haïti. Ph.D. Thesis, L'Institut National des Sciences Appliquées de Lyon, Lyon, France, 2010.

29. MDE, DDC, HSI. Plan de gestion 2017–2022 du Parc National Naturel de l'Unité 2 de la Forêt des Pins, Rapport d'étude élaboré dans le cadre du Projet Valorisation de la Biodiversité à l'Unité II du Parc National Naturel de la Forêt des Pins, 2017.
30. IHSI. Population Totale, de 18 Ans et plus. Ménages et Densités Estimés en 2015. Port-au-Prince. 2015. Available online: https://www.humanitarianresponse.info/sites/www.humanitarianresponse.info/files/documents/files/estimat_poptotal_18ans_menag2015.pdf (accessed on 20 July 2021).
31. Bamba, I.; Yedmel, M.S.; Bogaert, J. Effets des routes et des villes sur la forêt dense dans la province orientale de la République Démocratique du Congo. *Eur. J. Sci. Res.* **2010**, *43*, 417–429.
32. Mama, A.; Sinsin, B.; De Canniere, C.; Bogaert, J. Anthropisation et dynamique des paysages en zone soudanienne au nord du Benin. *Tropicultura* **2013**, *31*, 78–88.
33. Pham, H.M.; Yamaguchi, Y.; Bui, T.Q.A. Case Study on the Relation between City Planning and Urban Growth Using Remote Sensing and Spatial Metrics. *Landsc. Urban Plan.* **2011**, *100*, 223–230. [CrossRef]
34. Clerici, M.; Mantiero, D.; Lucchini, G.; Longhese, M.P. The Saccharomyces cerevisiae Sae2 protein negatively regulates DNA damage checkpoint signalling. *EMBO Rep.* **2006**, *7*, 212–218. [CrossRef]
35. Oszwald, J.; Gond, V.; Dolédec, S.; Lavelle, P. Identification d'indicateurs de Changement d'occupation Du Sol Pour Le Suivi Des Mosaïques Paysagères. *Bois For. Trop.* **2011**, *307*, 7. [CrossRef]
36. Ouattara, G.; Koffi, G.B.; Kouakou, Y.K.A. Contribution des images satellitales Landsat 7 ETM+ a la cartographie lithostructurale du Centre-Est de la Côte d'Ivoire (Afrique de l'Ouest). *IJIAS* **2012**, *1*, 61–75.
37. Leruth, F. Les mosaïques d'images. *Bull. Soc. Géogr. Liège* **2000**, *38*, 95–106.
38. Girard, M.-C.; Girard, C.M. *Traitement des Données de Télédétection: Environnement et Ressources Naturelles*, 2nd ed.; Technique et Ingénierie: Paris, France, 2017.
39. Mas, J.F. Une Revue Des Méthodes et Des Techniques de Télédétection Du Changement. *Can. J. Remote Sens.* **2000**, *26*, 349–362. [CrossRef]
40. Vital, J.A. Land Use/Cover Change Using Remote Sensing and Geographic Information Systems: Pic Macaya National Park, Haiti. Master's Thesis, Michigan Technological University, Houghton, MI, USA, 2008. [CrossRef]
41. Koné, M.; Aman, A.; Adou Ayo, C.Y.; Coulibaly, L.; N'Guessan, K.E. Suivi diachronique par télédétection spatiale de la couverture ligneuse en milieu de savane soudanienne en Côte d'Ivoire. *Télédétection* **2007**, *7*, 433–446.
42. Useni, S.Y.; Cabala, K.S.; Malaisse, F.; Nkuku, K.C.; Amisi, M.Y.; Bogaert, J.; Munyemba, K.F. Vingt-cinq ans de monitoring de la dynamique spatiale des espaces verts en réponse à l'urbanisation dans les communes de la ville de Lubumbashi (Haut-Katanga, R.D. Congo). *Tropicultura* **2017**, *4*, 300–311. [CrossRef]
43. Sambieni, K.R.; Messina Ndzomo, J.-P.; Biloso Moyenne, A.; Halleux, J.-M.; Occhiuto, R.; Bogaert, J. Les statuts morphologiques d'urbanisation des communes de Kinshasa en République Démocratique du Congo. *Tropicultura* **2018**, *3*, 520–530. [CrossRef]
44. Bonn, F.J.; Rochon, G. *Précis de Télédétection*; Universités francophones; Presses de l'Université du Québec: Sillery, QC, Canada, 1992.
45. Inoussa, M.M.; Mahamane, A.; Mbow, C.; Saadou, M.; Yvonne, B. Dynamique spatio-temporelle des forêts claires dans le Parc national du W du Niger (Afrique de l'Ouest). *Sci. Chang. Planét. Sécher.* **2011**, *22*, 108–116. [CrossRef]
46. Skupinski, G.; BinhTran, D.; Weber, C. Les Images Satellites Spot Multi-Dates et La Métrique Spatiale Dans l'étude Du Changement Urbain et Suburbain—Le Cas de La Basse Vallée de La Bruche (Bas-Rhin, France). *Cybergeo* **2009**. [CrossRef]
47. Krummel, J.R.; Gardner, R.H.; Sugihara, G.; O'Neill, R.V.; Coleman, P.R. Landscape Patterns in a Disturbed Environment. *Oikos* **1987**, *48*, 321. [CrossRef]
48. Barima, Y.S.S.; Barbier, N.; Bamba, I.; Traoré, D.; Lejoly, J.; Bogaert, J. Dynamique Paysagère En Milieu de Transition Forêt-Savane Ivoirienne. *Bois For. Trop.* **2009**, *299*, 15. [CrossRef]
49. Useni, S.Y.; Andre, M.; Mahy, G.; Cabala, K.S.; Malaisse, F.; Munyemba, K.F.; Bogaert, J. Interpretation paysagere du processus d'urbanisation à Lubumbashi (RD Congo): Dynamique de la structure spatiale et suivi des indicateurs ecologiques entre 2002 et 2008. In *Anthropisation des Paysages Katangais*; Bogaert, J., Colinet, G., Mahy, G., Eds.; Presses Agronomique de Gembloux: Gembloux, Belgique, 2018; pp. 281–296.
50. Salomon, W.; Useni, S.Y.; Kouakou, A.T.M.; Barima, Y.S.S.; Kaleba, S.C.; Barthelemy, J.-P.; Bogaert, J. Caractérisation de la dynamique de l'occupation du sol en zone urbaine et périurbaine de la ville du Cap-Haïtien (Haïti) de 1986 à 2017. *Tropicultura* **2020**, *38*, 1438. [CrossRef]
51. Milian, J.; Tamru, B. Port-Au-Prince, Ville Du Risque? Un Mythe Au Prisme d'une Urbanisation Vulnérable. *Etudescaribeennes* **2018**, 39–40. [CrossRef]
52. Audebert, C. Les Antilles françaises à la croisée des chemins: De nouveaux enjeux de développement pour des sociétés en crise. *Les Cah. D'Outre-Mer. Rev. De Géographie De Bordx.* **2011**, *64*, 523–549. [CrossRef]
53. Tambwe, N.A. Urban agriculture, land sustainability. The case of Lubumbashi. In *Territoires Périurbains: Développement, Enjeux et Perspectives dans les Pays du Sud*; Bogaert, J., Halleux, J.M., Eds.; Les Presses Agronomiques de Gembloux: Gembloux, Belgique, 2015; pp. 153–162.
54. Chérubin, D.C.; Jébrak, Y.; Isabelle, T. Infrastructures et Grands Équipements, Vulnérabilité et Constats. In *Rapport du Programme de Recherche Dans le Champ de l'urbain FED/2015/360-478, Perspectives de Développement de L'Aire Métropolitaine de Port-au-Prince, Horizon 2030*; rapport; UQAM: Montréal, QC, Canada, 2018; pp. 227–228.

55. Montas, R. Estimation du poids relatif de l'économie informelle dans l'économie haïtienne en 2016. In *Rapport du Programme de Recherche Dans le Champ de L'Urbain FED/2015/360-478, Perspectives de Développement de l'Aire Métropolitaine de Port-au-Prince, Horizon 2030*; rapport; UQAM: Montréal, QC, Canada, 2018; pp. 141–144.
56. Angelier, J.P. Analyse de la Substitution Entre Combustibles Dans le Secteur Résidentiel en Haïti. 2005. Available online: https://halshs.archives-ouvertes.fr/halshs-00120739/fr/ (accessed on 18 January 2022).
57. Sambieni, K.R. Dynamique du Paysage de la Ville province de Kinshasa sous la Pression de la Périurbanisation: L'Infrastructure Verte Comme Moteur d'Aménagement, Thèse de Doctorat en Cotutelle, École Régionale Post-Universitaire d'Aménagement et de Gestion Integrés des Forêts et Territoires tropicaux (ÉRAIFT) et Université de Liège (ULIEGE). 2020. Available online: http://hdl.handle.net/2268/234317 (accessed on 23 January 2022).
58. Salomon, W.; Sikuzani, Y.U.; Kouakou, A.T.M.; Barima, S.S.; Théodat, J.-M.; Bogaert, J. Monitoring of Anthropogenic Effects on Forest Ecosystems within the Municipality of Vallières in the Republic of Haiti from 1984 to 2019. *Trees For. People* **2021**, *6*, 100135. [CrossRef]

 land

Article

Normalizing the Local Incidence Angle in Sentinel-1 Imagery to Improve Leaf Area Index, Vegetation Height, and Crop Coefficient Estimations

Gregoriy Kaplan [1], Lior Fine [1,2], Victor Lukyanov [1], V. S. Manivasagam [1,3], Josef Tanny [1,4] and Offer Rozenstein [1,*]

1 Institute of Soil, Water and Environmental Sciences, Agricultural Research Organization—Volcani Institute, HaMaccabim Road 68, P.O. Box 15159, Rishon LeZion 7528809, Israel; grigorii@volcani.agri.gov.il (G.K.); liorf@volcani.agri.gov.il (L.F.); viclukyanov@gmail.com (V.L.); vs_manivasagam@cb.amrita.edu (V.S.M.); tanai@volcani.agri.gov.il (J.T.)
2 Department of Soil and Water Sciences, Faculty of Agriculture, Food and Environment, The Hebrew University of Jerusalem, P.O. Box 12, Rehovot 76100, Israel
3 Amrita School of Agricultural Sciences, Amrita Vishwa Vidyapeetham, J. P. Nagar, Arasampalayam, Myleripalayam, Coimbatore 642 109, Tamil Nadu, India
4 HIT—Holon Institute of Technology, Holon 58102, Israel
* Correspondence: offerr@volcani.agri.gov.il

Abstract: Public domain synthetic-aperture radar (SAR) imagery, particularly from Sentinel-1, has widened the scope of day and night vegetation monitoring, even when cloud cover limits optical Earth observation. Yet, it is challenging to combine SAR images acquired at different incidence angles and from ascending and descending orbits because of the backscatter dependence on the incidence angle. This study demonstrates two transformations that facilitate collective use of Sentinel-1 imagery, regardless of the acquisition geometry, for agricultural monitoring of several crops in Israel (wheat, processing tomatoes, and cotton). First, the radar backscattering coefficient (σ^0) was multiplied by the local incidence angle (θ) of every pixel. This transformation improved the empirical prediction of the crop coefficient (K_c), leaf area index (LAI), and crop height in all three crops. The second method, which is based on the radar brightness coefficient (β^0), proved useful for estimating K_c, LAI, and crop height in processing tomatoes and cotton. Following the suggested transformations, R^2 increased by 0.0172 to 0.668, and RMSE improved by 5 to 52%. Additionally, the models based on the suggested transformations were found to be superior to the models based on the dual-polarization radar vegetation index (RVI). Consequently, vegetation monitoring using SAR imagery acquired at different viewing geometrics became more effective.

Keywords: Sentinel-1; SAR; RVI; incidence angle; crop coefficient; leaf area index

Citation: Kaplan, G.; Fine, L.; Lukyanov, V.; Manivasagam, V.S.; Tanny, J.; Rozenstein, O. Normalizing the Local Incidence Angle in Sentinel-1 Imagery to Improve Leaf Area Index, Vegetation Height, and Crop Coefficient Estimations. *Land* **2021**, *10*, 680. https://doi.org/10.3390/land10070680

Academic Editors: Carmine Serio, Guido Masiello and Sara Venafra

Received: 29 May 2021
Accepted: 22 June 2021
Published: 28 June 2021

1. Introduction

Spaceborne monitoring of agricultural landscapes is predominantly performed using optical sensors and synthetic-aperture radar (SAR). The use of passive optical remote sensing in the visual, near-infrared, shortwave infrared, and thermal spectral regions for the estimation of agricultural variables is well established [1–10]. However, optical sensors are limited by cloud cover. To overcome this problem, previous studies have suggested combining observations acquired at different times by several optical sensors [11–15], but even this approach does not always produce enough cloud-free observations to monitor cloudy regions effectively. Moreover, leaf area index (LAI) estimation from optical imagery suffers from a saturation effect when the LAI is greater than 3 [16–20]. Overcoming this limitation is desirable since LAI is commonly used as a measure of crop growth, nitrogen, and fertilization status estimation [21]. The LAI is also a good proxy for vegetation vigor [22,23], and a good yield predictor [24–27].

This study proposes complementing optical remote sensing with SAR to overcome these obstacles in monitoring vegetation properties and to facilitate better agricultural practices [28]. SAR penetration of the canopy can mitigate saturation in LAI estimation [29–31]. Moreover, since SAR can penetrate clouds, it produces high-quality imagery even in adverse weather conditions [32]. In addition to the LAI, remote sensing can be used to estimate other variables such as the crop coefficient (K_c) and height. K_c-based estimation of crop water consumption is one of the most commonly used irrigation management methods [33,34]. Crop height is a good predictor of the aboveground biomass [35] and is commonly used by growers as a proxy for crop development. Therefore, deriving reliable SAR-based LAI, K_c, and height estimation models can facilitate better agricultural monitoring, especially in cloudy regions.

Several studies have employed spaceborne SAR for agricultural purposes [36–40] and demonstrated that quad-polarization SAR (e.g., RADARSAT-2, TerraSAR-X/TanDEM-X) could be used for crop monitoring. However, quad-polarization SAR images currently come at a high cost that limits their use in routine monitoring of crops and in research. Since 2014, the Sentinel-1 mission, consisting of two polar-orbiting satellites, provides a dual-polarization alternative at no cost to the user. These satellites have a revisit time of six days at 30° latitude at the same viewing geometry and a 10 × 10 m pixel size, thus having significant potential for agricultural applications.

One of the most critical challenges in creating time series of SAR imagery is the dependence of radar backscatter on the incidence angle [41]. The incidence angle is defined by the incident radar beam and the vertical (normal) to the surface. More specifically, the local incidence angle (θ) takes into account the local relief. The backscatter is weaker in images acquired at shallow incidence angles compared to images acquired at steeper incidence angles; therefore, the same object has different and uncomparable backscatter values in images acquired with different incidence angles. Given the dependence of the backscatter's intensity on the incidence angle, previous studies have underlined the need to correct this effect [42,43]. Until now, many studies using C-band SAR imagery from Sentinel-1, RADARSAT-2, and RISAT-1 for agricultural monitoring only used a subset of the available imagery acquired from either ascending or descending orbits with a limited range of incidence angles. Accordingly, these studies discarded imagery acquired at incidence angles that fell outside certain margins (Table 1). This practice might exclude more than half of the available images from the time series. Moreover, empirical models developed based on these limited datasets are likely applicable only for the same range of incidence angles. Therefore, the practice of excluding images from the time series reduces the applicability of SAR-based models.

Table 1. Summary of the incidence angle range considered in past studies.

Study	Incidence Angle (°)	Satellite	Crop	Application
Van Tricht et al. (2018) [44]	32–42	Sentinel-1	Many crops	Crop classification
Inoue et al. (2014) [45]	25–35	RADARSAT-2	Paddy rice	Various biophysical variables
Veloso et al. (2017) [46]	38–41	Sentinel-1	Wheat, rapeseed, maize, soybean, sunflower	Temporal behavior
Bousbih et al. (2017) [47]	39–40	Sentinel-1	Cereals	Crop height and LAI
Nasirzadehdizaji et al. (2019) [48]	39–40	Sentinel-1	Maize, sunflower, wheat	Crop height and canopy coverage
Navarro et al. (2016) [49]	38.87–39.26	Sentinel-1	Maize, soybean, bean, pasture	Crop water requirements
Inglada et al. (2016) [50]	38.89–39.05	Sentinel-1	Wheat, rapeseed, barley, corn, sunflower	Crop classification
Hosseini et al. (2018) [51]	20.63–28.16	RADARSAT-2	Corn	Biomass

Table 1. *Cont.*

Study	Incidence Angle (°)	Satellite	Crop	Application
Phan et al. (2021) [52]	42–44	Sentinel-1	Rice	Various biophysical variables
Molijn et al. (2019) [53]	36.0–36.6	Sentinel-1	Sugarcane	Productivity mapping
Demarez et al. (2019) [54]	30	Sentinel-1	Maize	Crop mapping
Srivastava et al. (2019) [55]	31	RISAT-1	Wheat	Crop height
Srivastava et al. (2018) [56]	32	RISAT-1	Paddy	LAI
Benabdelouahab et al. (2018) [57]	23.3	ERS-1	Wheat	Irrigation supply detection
Han et al. (2019) [58]	42.5	Sentinel-1	Wheat	Crop water content
Yadav et al. (2019) [59]	40	Sentinel-1	Wheat	LAI
Chauhan et al. (2018) [60]	38	RISAT-1	Wheat	Various biophysical variables
Harfenmeister et al. (2019) [61]	Constant. Undisclosed.	Sentinel-1	Wheat, barley	Various biophysical variables
Song and Wang (2019) [62]	Constant. Undisclosed.	Sentinel-1	Wheat	Crop classification and phenology monitoring
Nihar et al. (2019) [63]	Constant. Undisclosed.	Sentinel-1	Cotton, maize	Crop classification
Vreugdenhil et al. (2018) [64]	Constant. Undisclosed.	Sentinel-1	Corn, cereals, oilseed rape	Various biophysical variables

Several different incidence angle normalization procedures were carried out in previous studies. For example, [44] normalized their selected subset of imagery (incidence angles between 32° and 42°) to 37° using a simplified correction method based on Lambert's law of optics. However, this method is insufficiently effective because it is relatively reliable only at the center of the image [41,65]. Two new effective methods for incidence angle normalization were proposed by [65], but environmental conditions limited the applicability of these methods, and they have been used mostly for ocean monitoring. Other methods for incidence angle normalization, such as simplified normalization [43], radiative transfer-based models, and statistical methods, can be applied only under specific ground conditions [66]. Therefore, despite past attempts to deal with the heterogeneity of the incidence angle in the SAR time series, the challenge of incidence angle normalization remains.

Therefore, the main goal of this study was to propose methods to reduce the backscatter dependence on the local incidence angle to permit the use of all available Sentinel-1 images in a single dataset without defining a range of allowed incidence angles and omitting images that extend beyond it. The second goal of this study was to use the proposed methods to accurately estimate vegetation properties (Kc, LAI, and crop height) based on incidence angle-normalized Sentinel-1 imagery.

2. Materials and Methods

2.1. Test Sites and Field Measurements

The field measurements used in this study were carried out during two seasons of winter wheat, three seasons of processing tomatoes, and two seasons of cotton in different locations in Israel (Table 2, Figure 1). LAI was measured by a SunScan Canopy Analysis System—SS1 developed by Delta-T Company (Cambridge, United Kingdom) during two wheat seasons and two processing tomato seasons. The SunScan is an accurate, nondestructive LAI measurement system successfully employed in many previous studies [25,67]. Each LAI value is an average of at least 30 consecutive field measurements taken at 20-cm intervals along a transect perpendicular to the row direction. Vegetation height was measured at a precision of 1 cm using a tape measure and represented by an average of at least 30 plants per measurement date. LAI and vegetation height in wheat and processing tomatoes were measured throughout the growing seasons; therefore, they represent the full range of these variables. Cotton height was measured during the middle and late stages of one growing season. The backscattering coefficient (σ^0) and the radar brightness coefficient (β^0) were used in linear scale. Sentinel-1 backscatter values were averaged for a polygon that represented the eddy covariance measurement footprint calculated based on a two-dimensional footprint model [68]. All the empirical regression models in this study utilized the average values of either σ^0 or β^0, and the local incidence angle (θ) within the areas of interest and same-date field measurements. In cases of gaps in the time series, linearly interpolated values of field measurements from adjacent dates (crop height and LAI) were used. The number of SAR images used for the derivation of the various models was not uniform because each model was based on the period for which field measurements were available, resulting in different numbers of corresponding satellite images. For example, LAI could not be measured using the SunScan system when the plants were very small, while vegetation height was easily measured at any time using a ruler. Accordingly, the LAI models were based on shorter time spans and fewer images than were plant height models. In-field paths and their surrounding area were masked out from analysis polygons of the processing tomato experiments which took place in 2019 to remove bare soil areas and avoid border effects. These excluded areas consisted of approximately 20% of the overall areas of interest. Therefore, the Gadash 2019 area of interest consisted of four vegetated regions separated by paths, and the Gadot 2019 area of interest consisted of two regions.

Table 2. Summary of seven field experiments conducted at six locations in Israel.

Experiment Area	Crop	Period *	# Crop Height Measurements	# LAI Measurements	Area Size (# Sentinel-1 Pixels)	Nearest Meteorological Station ET_0 Data	Distance and Bearing to the Meteorological Station
Saad	Wheat	1-Jan-2018 9-Apr-2018	8	6	260	Dorot	9.5 km NE
Yavne	Wheat	18-Dec-2018 10-Apr-2019	7	7	550	-	-
Tel Nof	Cotton	6-Jun-2016 17-Sep-2016	7	-	1300	Revadim	5 km S
Negba	Cotton	25-Jul-2017 11-Sep-2017	-	-	460	Negba	2.5 km SW
Gadash	Processing tomatoes	9-May-2018 30-Jul-2018	8	-	250	-	-
Gadash	Processing tomatoes	3-May-2019 24-Jul-2019	7	6	500	Gadash	250 m SE
Gadot	Processing tomatoes	25-Apr-2019 14-Aug-2019	11	11	300	Gadot	1.5 km SW

Note: * Period indicates the starting and ending dates.

2.2. Agro-Meteorological Measurements

Agro-meteorological measurements of the reference evapotranspiration (ET_0) and actual evapotranspiration (ET_c) were performed to derive the crop coefficient (K_c) as $K_c = ET_c/ET_0$. K_c is an important variable used to determine the irrigation dose [69]. Daily ET_c was derived from water vapor flux measurements by eddy covariance systems [6,70]. The daily ET_0 was calculated according to the Penman–Monteith method [33] based on meteorological measurements of air temperature, relative humidity, wind speed, and solar irradiance at the meteorological station closest to the field or at the flux tower itself (Table 2). Meteorological station data are publicly available at http://www.meteo.co.il/ (accessed on 24 June 2021) and http://www.mop-zafon.net/ (accessed on 24 June 2021). The K_c data used for developing the processing tomato models were smoothed in Python with the SciPy library using the cubic and the second-order splines. Smoothing spline is a non-parametric regression technique, which was previously used in various remote sensing applications [15,71–73]. ET_c was measured throughout the growing seasons of wheat and processing tomatoes and from the middle of the cotton growing seasons. ET_c data collected during the Gadash processing tomato experiment in 2018 and the Yavne 2019 wheat experiment were not used for the K_c model development because more than half the data were lost due to technical difficulties that emerged during the experiments.

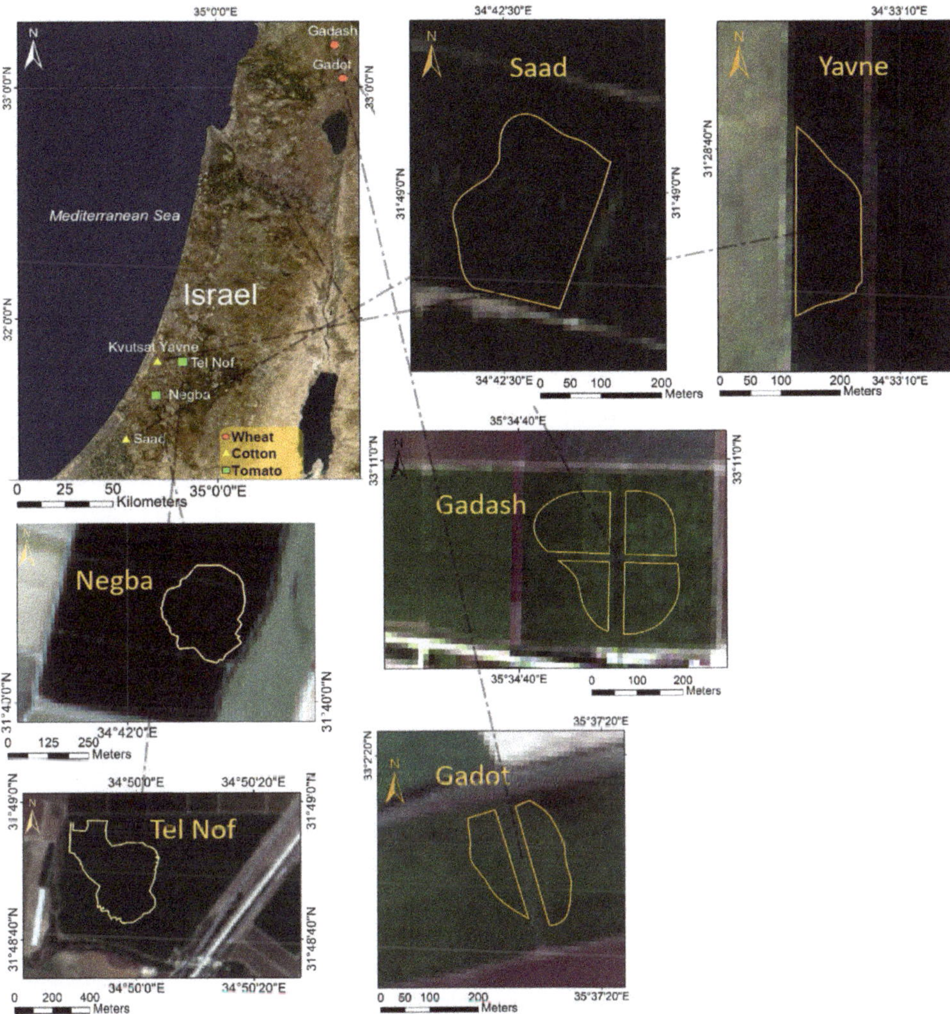

Figure 1. Locations of analysis polygons in the experiments conducted in Israel in 2016–2019.

2.3. Satellite Imagery

Sentinel-1 is part of the European Copernicus program for Earth observation. The payload of the two Sentinel-1 satellites includes a dual-polarization (VV and VH) C-band SAR instrument that is an active phased array antenna working at 5.405 GHz frequency (corresponding wavelength 5.55 cm). The resolution of the Level-1 Ground Range Detected (GRD) Interferometric Wide (IW) mode that was used in this study is 20 × 22 m, with a pixel size of 10 × 10 m, swath width of 250 km, and a revisit time of six days for images with the same geometry. Sentinel-1A and Sentinel-1B were launched on 3 April 2014 and on 25 April 2016, respectively. The Sentinel-1 incidence angle in the IW mode ranges approximately between 29° and 46°. Figure 2 shows the graphical representation of the local incidence angle. In this study, some sites were close to the edge of the images, resulting in an incidence angle range from 30.8° to 45.8°, and in local incidence angle values from 30.3° to 47.7° (Figure 3). Therefore, this study was based on a wide range of incidence angles. The areas within SAR imagery used in the present study are not affected by adverse geometrical effects, such as radar shadow, foreshortening, and layover. The SAR imagery used in this study was downloaded from the ESA Copernicus site (https://scihub.copernicus.eu/dhus/#/home, accessed on 23 June 2021). Overall, 38 SAR images were used to derive models for wheat (Table S1), 19 for cotton (Table S2), and 94 for processing tomatoes (Table S3), Figure 3).

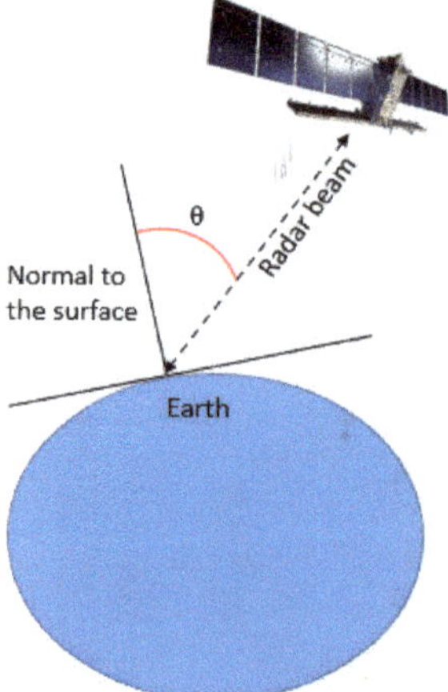

Figure 2. Local incidence angle (θ). The local incidence angle (θ) is defined as the angle between the incidence radar beam and a line that is normal to the surface, considering local relief, typically derived from a DEM.

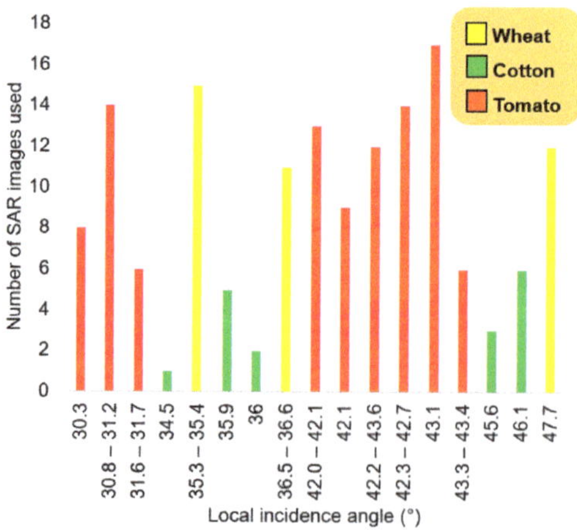

Figure 3. The number of SAR images acquired at different local incidence angles used in the modeling of different crops.

2.4. Image Processing

All images were pre-processed in the Sentinel Application Platform (SNAP versions 6.0 and 7.0, European Space Agency). The sequential pre-processing of the Sentinel-1 imagery was as follows: subsetting a region around the target area, applying the latest orbit file to correct for the satellite path, thermal noise removal, calibration to β^0 and σ^0 in a natural scale, range Doppler terrain correction using the Shuttle Radar Topography Mission (SRTM, [74]) digital elevation model (DEM) with 30 m resolution. These pre-processing steps were performed in the same sequence as in Song et al. [62], with the addition of the thermal noise removal step that was also performed by Van Tricht et al. [44]. The speckle filtering operation was avoided; therefore, the spatial resolution was retained [75–77].

2.5. Dual-Polarized RVI Algorithm

One of the conventional approaches for agricultural monitoring from quad-polarization SAR data is the calculation of the radar vegetation index (RVI) [40]. An adaptation for Sentinel-1 data assumes that $\sigma^0_{VV} \approx \sigma^0_{HH}$ [78–80], such that

$$\mathrm{RVI} = \frac{4*\sigma^0_{VH}}{\sigma^0_{VH} + \sigma^0_{VV}} \tag{1}$$

The radar backscatter coefficient, σ^0, also known as the radar cross-section (RCS) per unit area, is the conventional measure of the intensity of the signal reflected by the surface. It is a normalized dimensionless number that varies significantly with the incidence angle, wavelength, and polarization, as well as with properties of the scattering surface itself [81].

Each RVI model in the present study was based only on one dataset with the maximum number of images acquired at an ascending orbit with the same incidence angle. These are considered to be the most favorable conditions for determining the RVI without incidence angle normalization. An RVI-based K_c model for cotton was not produced because of the lack of imagery acquired at the same incidence angle over the specific fields where the agro-meteorological measurements were performed. The models based on the suggested methods for local incidence angle normalization methods described in Sections 2.6 and 2.7 were compared to the models based on the RVI.

2.6. σ^0-Based Local Incidence Angle Normalization

The radar backscatter intensity depends on the incidence angle, with σ^0 decreasing proportionally to the incidence angle increase in the intermediate range of incidence angles typical for Sentinel-1 and the majority of spaceborne SAR missions [43,82–84]. Based on this understanding, σ^0 was normalized by multiplying it with the local incidence angle (θ) in the decimal degree scale:

$$\sigma^0_{\mathrm{Norm}} = \sigma^0 * \theta \tag{2}$$

The normalization of σ^0 is achieved by multiplying lower σ^0 values obtained under shallower local incidence angles by higher θ values than the higher σ^0 values acquired under steeper local incidence angles. In this study, different VV and VH polarization combinations of normalized σ^0 values were used to model K_c, LAI, and crop height. The models described below (Equations (3)–(5)) were produced using polarization combinations that showed the best R^2 and RMSE values. The following polarization combination was used to model K_c and LAI in wheat, and LAI in processing tomatoes:

$$V = \sigma^0_{\mathrm{Norm,\ VH}} + \sigma^0_{\mathrm{Norm,\ VV}} \tag{3}$$

where, and afterward, V is a vegetation variable being estimated.

The following polarization combination was used to model wheat and cotton height:

$$V = \sigma^0_{\mathrm{Norm,\ VH}} - \sigma^0_{\mathrm{Norm,\ VV}} \tag{4}$$

Processing tomato and cotton K_c estimation models were based on

$$V = \sigma^0_{\text{Norm, VV}} \tag{5}$$

The descending winter wheat imagery showed a very low correlation with wheat variables and, therefore, was not used for the development of wheat models.

2.7. β^0-Based Local Incidence Angle Normalization Method for Tomato and Cotton Height, LAI, and K_c Estimation

A radar beam transmitted at a shallow angle travels longer distances through the vegetation canopy than a beam transmitted under a steep angle; thus, the attenuation of the former is typically higher than that of the latter. Apart from the beam two-way travel distance through the vegetation, the radar backscatter is affected by soil roughness, dielectric properties, and a combination of different types of scatterers that exist in each pixel [85,86]. The wheat fields in this study are flat, and the vegetation growth is uniform. Hence, scattering from the soil surface is mostly specular in the early part of the season, and the volume scattering component increases as the vegetation develops [87,88]. Unlike the wheat fields, the structure of processing tomatoes and cotton fields is more complex, with mounds and furrows. The distance between planted mound centers in all three processing tomato fields is two meters, and it is one meter in cotton. The difference between the elevation of the mounds is up to 15 cm in processing tomatoes and 12 cm in cotton. Consequently, the standard deviation of the surface height is up to 7.5 cm and 6 cm in processing tomatoes and cotton, respectively. According to the Peake and Oliver roughness criterion [89], the surface is considered rough if

$$h_{\text{rms}} > \frac{\lambda}{4 * \cos\ \delta} \tag{6}$$

where h_{rms} is the standard deviation of the surface height variation; λ is the wavelength; and δ is the incidence angle. The incidence angle is slightly different from the local incident angle for slopeless surfaces, but this difference does not affect the calculation of the roughness criterion. Accordingly, in C-band SAR with an incidence angle range of 30°–45°, the roughness threshold is $h_{\text{rms}} > 1.5$ cm for an incidence angle of 30°, and $h_{\text{rms}} > 1.8$ cm for an incidence angle of 45°. Therefore, the processing tomato and cotton fields are rough, decreasing backscatter dependence on the incidence angles [85], and modifying the rate of backscatter change as the incidence angle increases [90]. Unlike the smooth wheat fields, every pixel in processing tomato and cotton fields contains multiple types of scattering: specular (plant-free furrows), double bounce (corners between furrows and mounds), and volume scattering in the canopy. Moreover, at some incidence angles, Bragg scattering caused by the row frequency might occur [91,92].

Owing to the complex surface structure in cotton and processing tomato fields, another transformation method specific to these fields was derived empirically in addition to the σ^0 normalization method. This new method is based on the polynomial regression between plant variables multiplied by the newly derived attenuation coefficient $\sin(\text{Radians}(90-\theta)^3)$ and radar brightness (β^0):

$$V * \ \sin(\text{Radians}(90-\theta)^3) = A * \left(\beta^0\right)^2 + B * \left(\beta^0\right) + C \tag{7}$$

where V is the plant variable (such as height, LAI, or K_c); θ is the local incidence angle in degrees; A, B, and C are the specific model coefficients; and β^0 is the radar brightness coefficient [41] at either VV (processing tomatoes) or the sum of VV and VH polarizations (cotton). β^0 is a dimensionless coefficient that corresponds to the reflectivity per unit area in the slant range. β^0 is used because the radiometric correction of σ^0 is based on a sea-level ellipsoid [41,93] and, therefore, less suitable for monitoring of rough surfaces and areas with a rugged topography [94]. Previous studies found β^0 to be the best unencumbered estimate

SAR measurement [41,95,96]. Using radar brightness is an established practice for research in space [97,98] and common practice in the analysis of RADARSAT-1 imagery [85,99], but not of Sentinel-1 data.

The attenuation coefficient $\sin(\text{Radians}(90-\theta)^3)$ is used to account for the dependence of beam attenuation on the local incidence angle. The SAR beam interaction with objects on the ground can be described as a triangle in which the radar beam is the hypotenuse, and the local incidence angle θ is the angle between the hypotenuse and vertical cathetus normal to the surface (Figure 2). β^0 values are reconstructed to a normalized value by applying a sine function to a cubed value of the $(90-\theta)$ value in radians. The attenuation coefficient is linearly and inversely proportional to the local incidence angle, as shown in Figure 4. Therefore, by applying the suggested normalization, higher β^0 values obtained under steeper (closer to vertical) local incidence angles are divided by higher coefficient values compared to β^0 values acquired under shallower local incidence angles. The main difference between the σ^0 and β^0 methods is that the former applies a steeper increase to the radar backscatter (σ^0) as the local incidence angle increases than the latter (β^0). This difference in the behavior of the methods was created to take into account that as the incidence angle increases, the radar backscatter decreases more slowly for rough surfaces than for smooth surfaces [90,100].

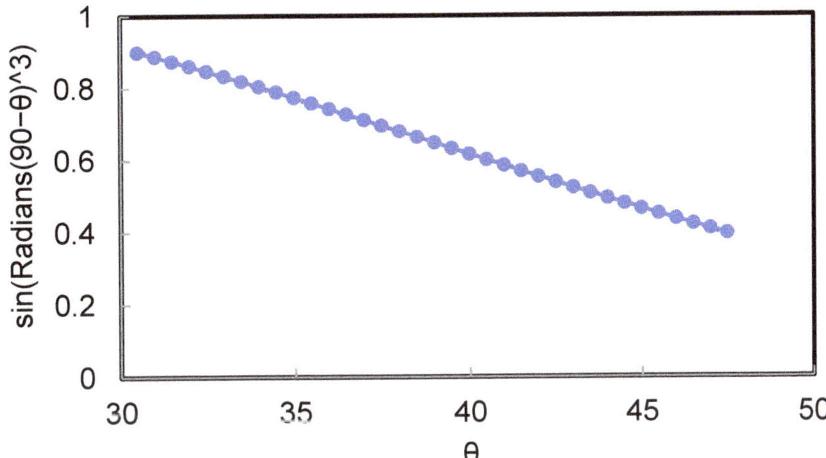

Figure 4. The attenuation coefficient ($\sin(\text{Radians}(90-\theta)^3)$), as a function of θ (in degrees), and the local incidence angle. This attenuation coefficient was used to correct for the dependence of radar brightness β^0 values on the local incidence angle.

Therefore, the normalized β^0 value can be written as

$$\beta^0_{\text{Norm}} = \frac{\beta^0}{\sin(\text{Radians}(90-\theta)^3)} \tag{8}$$

where β^0_{Norm} is the normalized radar brightness in VV or VH polarization, β^0 is the radar brightness in VV or VH polarization, and $\sin(\text{Radians}(90-\theta)^3)$ is the attenuation coefficient.

The processing tomato LAI model utilizes the sum of normalized β^0 in both polarizations:

$$\text{LAI} = \frac{\beta^0_{\text{VH}}}{\sin(\text{Radians}(90-\theta)^3)} + \frac{\beta^0_{\text{VV}}}{\sin(\text{Radians}(90-\theta)^3)} \tag{9}$$

2.8. Calibration and Validation of Empirical Vegetation Variable Estimation Models

The field-measured vegetation variables were used in regression models against the uncorrected radar backscatter parameters. Further, the proposed local incidence angle normalization methods were applied to the SAR images, and new empirical regression models were derived. Finally, the models based on the data prior to normalization and post-normalization and on the dual-polarized RVI were compared to assess the performance of the normalization process (Figure 5).

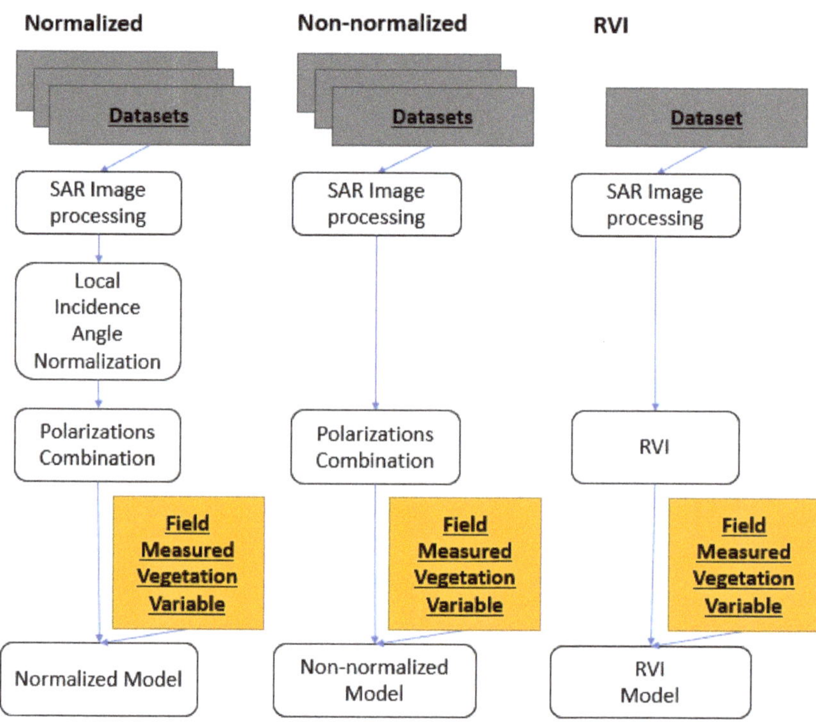

Figure 5. Derivation of the dual-polarized normalized (with σ^0 or β^0 normalization), non-normalized, and RVI models.

In every case, the same type of regression model (either linear or polynomial) and the same polarization combination were used in the comparison. Coefficients of determination in models based on non-normalized data were symbolized as R_0^2, and their root-mean-square error was symbolized as $RMSE_0$. The differences in the R^2 and RMSE following the normalization procedures were calculated. In addition, the Steiger variation [101] of the two-tailed Fisher Z-score tests [102] was performed to determine whether the difference in the models' R^2 is significant ($\alpha \leq 0.05$). The significance of the RMSE difference was calculated using the two-tailed Wilcoxon signed-rank test [103] to determine whether the difference in the models' RMSE was significant ($\alpha \leq 0.05$). According to the goals set in this study and due to a finite amount of available SAR imagery and ground truth data, all the available data were used to calibrate the empirical models to achieve the models' maximum reliability and estimation accuracy [104]. In order to additionally validate the models' estimation performance, the RMSE values of normalized models applied separately to each dataset (experiment) were also calculated and are presented in Tables S4–S8.

3. Results

3.1. Wheat, Processing Tomato, and Cotton Height, LAI, and K_c Models Based on the σ^0 Normalization Method

The effect of the proposed normalization (Equation (2)) on the SAR backscatter from two incidence angles is illustrated in Figure 6. Following the normalization process, the difference is greatly reduced, and a considerable improvement in the R^2 and RMSE of all the σ^0-based models is observed. In wheat, processing tomatoes, and cotton, the height, LAI, and K_c models' R^2 improved in the range of 0.0172–0.367, and the RMSE improved in the range of 5–52%. Table 3, Tables S4–S6, and Figure 7 show the performance of σ^0-based height, LAI, and K_c models in wheat, processing tomatoes, and cotton.

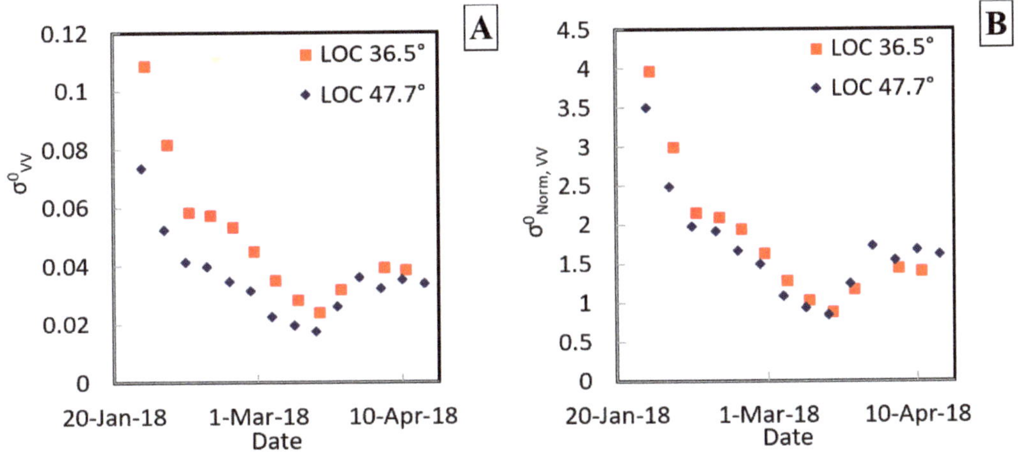

Figure 6. A time series of VV polarization values recorded by Sentinel-1 on its ascending overpasses with local incidence angles of 36.5° and 47.7° during the wheat experiment in Saad: (**A**) prior to applying the σ^0 normalization; (**B**) post-normalization.

Table 3. R^2 and RMSE improvements following the local incidence angle normalization procedure. Significance is marked by *. The percentage values in the brackets show improvement in RMSE after normalization (i.e., the reduction in prediction error).

Model	# Images	R^2	RMSE	R^2 Improvement	RMSE Improvement (%)
Wheat height	38	0.8566	6 cm	0.0738 *	2 cm, (25%)
Wheat LAI	34	0.7194	0.6	0.1639 *	0.2, (25%)
Wheat K_c	11	0.6722	0.073	0.1601	0.016, (18%)
Tomato K_c σ^0-based	59	0.8549	0.0871	0.0172	0.005, (5%) *
Tomato LAI σ^0-based	50	0.7881	1.0	0.1001	1.1, (52%) *
Tomato height σ^0-based	94	0.4201	11 cm	0.0446	1 cm, (8%)
Tomato K_c β^0-based	59	0.871	0.0821	0.1143 *	0.0307, (27%)
Tomato LAI β^0-based	50	0.8341	0.9	0.352 *	0.7, (44%)
Tomato height β^0-based	94	0.8107	9 cm	0.3442 *	2 cm, (18%)
Cotton height σ^0-based	11	0.8721	5 cm	0.367 *	5 cm, (50%)
Cotton K_c σ^0-based	12	0.3742	0.0511	0.3543 *	0.0128, (21%) *
Cotton height β^0-based	11	0.9467	8 cm	0.668 *	5 cm, (38%)
Cotton K_c β^0-based	12	0.707	0.1293	0.6353	0.0379, (23%) *

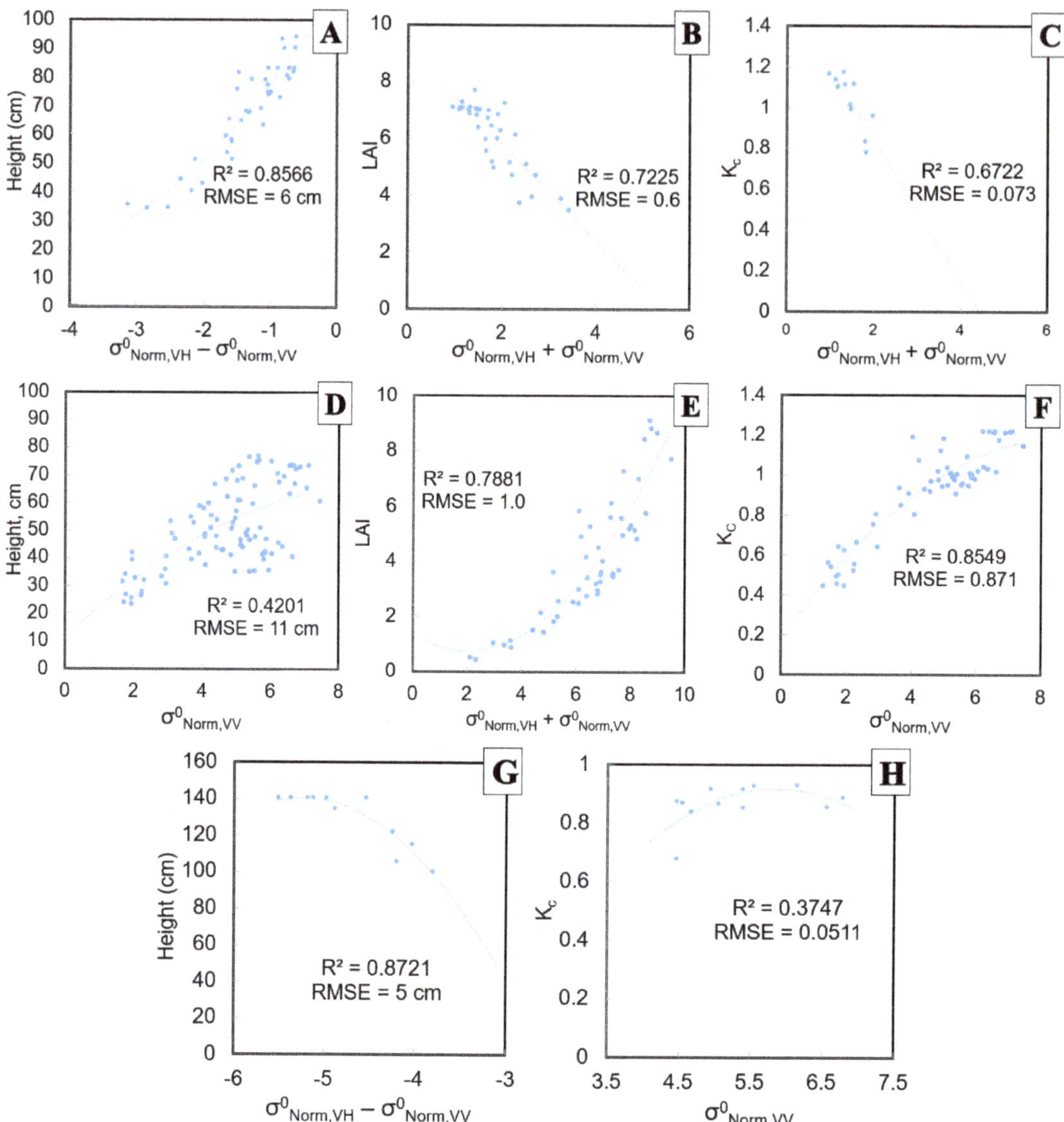

Figure 7. Models based on field measurements and the σ^0 normalization method: (**A**) wheat height; (**B**) wheat LAI; (**C**) wheat K_c; (**D**) processing tomato height; (**E**) processing tomato LAI; (**F**) processing tomato K_c; (**G**) cotton height; (**H**) cotton K_c.

3.2. Processing Tomato and Cotton Height, LAI, and K_c Models Based on the β^0 Normalization Method

The effect of the β^0-based normalization (Equation (8)) that reduces the difference in β^0 images acquired at different angles is shown in Figure 8.

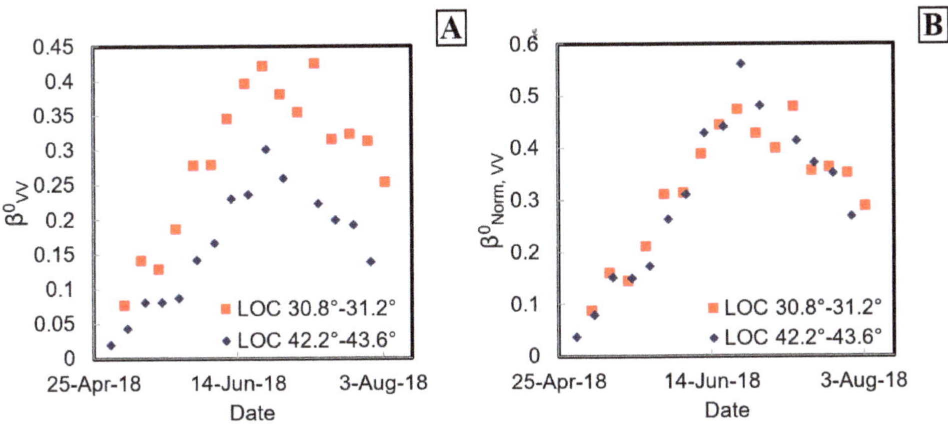

Figure 8. A time series of VV polarization values recorded by Sentinel-1 on its descending overpasses with local incidence angles in the ranges of 30.8°–31.2° and 42.2°–43.6° during the processing tomato experiment in Gadash, 2018: (**A**) prior to applying the β^0 normalization; (**B**) post-normalization.

The β^0-based normalization method permitted achieving the improvement in the R^2 and RMSE of all the β^0-based models. For the processing tomato and cotton height, LAI, and K_c models, the R^2 improved in the range of 0.1143–0.668, and the RMSE improved in the range of 18–44%. Table 3, Tables S7 and S8, and Figure 9 show the performance of processing tomato and cotton β^0-based height, LAI, and K_c models. Table 3 shows the performance of all the σ^0-based and β^0-based normalized models developed in this study and their R^2 and RMSE improvements over the non-normalized models.

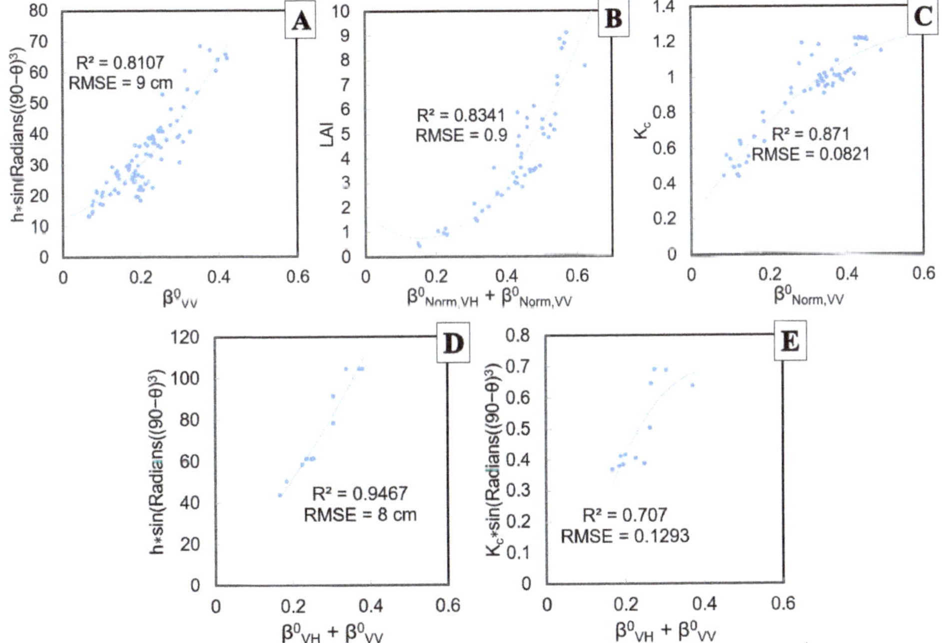

Figure 9. Models based on field measurements and the β^0 normalization method: (**A**) processing tomato height; (**B**) processing tomato LAI; (**C**) processing tomato K_c; (**D**) cotton height; (**E**) cotton K_c.

3.3. Performance of the Dual-Polarized RVI

The dual-polarized RVI-based wheat, processing tomato, and cotton models are shown in Table 4. The wheat and cotton RVI models were compared against the σ^0-based models, while the processing tomato RVI models were compared to the β^0-based processing tomato models.

Table 4. RVI models for wheat, processing tomato, and cotton height, LAI, and K_c. The differences in R^2 and RMSE indicate the difference in performance compared to models based on either the σ^0 (wheat and cotton models) or β^0 (processing tomatoes) local incidence angle normalization methods in Table 3. Negative values represent lower R^2 and higher RMSE of the RVI models.

	Height	LAI	K_c
Wheat			
Overpass	Asc	Asc	Asc
# SAR images used	26	25	6
Local incidence angle (°)	35.3–36.6	35.3–36.6	47.7
R^2	0.4248	0.1389	0.2912
R^2 difference	−0.2626	−0.5805	−0.381
RMSE	13 cm	1.6	0.102
RMSE difference	−4 cm	−1.0	−0.029
(%)	(−44)	(−167)	(−40)
Processing tomatoes			
Overpass	Asc	Asc	Asc
# SAR images used	25	31	27
Local incidence angle (°)	42.0–43.1	42.0–43.1	42.0–43.1
R^2	0.1584	0.3425	0.5635
R^2 difference	−0.6523	−0.4916	−0.3075
RMSE	14 cm	1.9	0.2488
RMSE difference	−5 cm	−1.0	−0.1667
(%)	(−56)	(−111)	(−203)
Cotton			
Overpass	Asc		
# SAR images used	5		
Local incidence angle (°)	35.9		
R^2	0.3297		
R^2 difference	−0.5424		
RMSE	12 cm		
RMSE difference	−7 cm		
(%)	(−140)		

4. Discussion

In contrast with previous studies that mostly used images acquired under fixed or within narrow ranges of incidence angles, the correction derived in this study facilitates the use of imagery acquired under all typical geometrical conditions. By applying simple transformations to Sentinel-1 imagery acquired under a wide range of incidence angles, the dependency of σ^0 and β^0 on the local incidence angle decreased, and the empirical modeling of several crop properties was improved. This achievement is remarkable because monitoring crops using the full temporal resolution of SAR imagery is much more useful than using imagery acquired at a narrow range of angles. Moreover, vegetation variable estimation models calibrated in one region using the proposed methods can be applied to other areas.

The improvement in the R^2 and RMSE of the models following the local incidence angle normalization procedure was found to be significant in many of the models: wheat height and LAI models; β^0-based processing tomato LAI, height, and K_c models; σ^0-based cotton K_c and height models; and β^0-based cotton height model. Since the statistical significance of the difference between correlations is dependent on the number of images used for model development, some of the models yielded a difference that was not significant (wheat K_c model, processing tomato σ^0-based LAI, height, and K_c models, and cotton K_c β^0-based model). However, the trend of improvement following the proposed normalization procedures is clear, and the practical usefulness of the proposed methods can be better represented by the RMSE improvements because RMSE represents the vegetation variable estimation accuracy. The RMSE improvement was found to be significant in the following models: processing tomato σ^0-based LAI and K_c models; and cotton K_c σ^0 and β^0-based models. The RMSE improvement was not significant in the following models: wheat, processing tomato σ^0-based LAI, and processing tomato β^0-based models, and cotton height σ^0- and β^0-based models. The R^2 and RMSE of all the models calibrated in the present study improved following incidence angle normalization. The range of RMSE improvements varied from model to model (Table 3), from 5 to 52%. Moreover, the performance of the newly developed β^0-based local incidence angle normalization method shows potential for overcoming the limitations of σ^0-based modeling for agricultural purposes. The use of β^0 to improve vegetation variable estimation is particularly useful in fields with a rough soil surface geometry. Using β^0 is not common for Sentinel-1 imagery, and the users' community could benefit from adopting this approach.

The models presented here for wheat and processing tomatoes were calibrated based on measurements taken throughout the entire duration of growing seasons and can, therefore, be applied at any time during crop development without restrictions. Nevertheless, the RMSE of LAI and height estimations was slightly higher at the peak of the season compared to the rest of the season. The relatively high accuracy of the models calibrated in this study and their independence from the incidence angle following the new normalization methods are advantageous compared to previous studies [47,55,60,61,64], in which the images used were limited to a narrow range of incidence angles. In addition, in contrast to [48] that presented models that can only be reliably applied to certain vegetation heights, the wheat and processing tomato models presented here are applicable to any height within the range measured in our experiments: 34–95 cm (wheat) and 24–77 cm (processing tomatoes). A comparison between several studies that used C-band SAR to estimate vegetation height and LAI is shown in Tables 5 and 6.

The models for LAI estimation show a better performance than previous studies. Previous estimation based on imagery acquired under a narrow range of incidence angles and dual-polarization [47] only achieved R^2 = 0.25. Moreover, the models in this study performed similarly to quad-polarization RADARSAT-2-based models for corn and soybean LAI estimation that utilized imagery acquired under a narrow range of incidence angles [39] and achieved R^2 = 0.66 and RMSE = 0.75 and R^2 = 0.64 and RMSE = 0.63, respectively. Another study [60] presented a wheat LAI estimation model, which has better prediction performance than the models obtained in the present study (RMSE = 0.4), but as in other previous models, it was based on images acquired under only one incidence angle. Unlike the LAI estimation based on optical imagery, the wheat and processing tomato LAI models developed in this study were not saturated even at the peak of vegetation development (wheat LAI_{max} = 7.7, processing tomato LAI_{max} = 9.1). Therefore, the LAI models in this paper might be applied throughout the whole season duration, which is useful because the LAI is a proxy for many vegetation variables [23], including crop productivity [105].

Table 5. Comparison of vegetation height estimation models based on Sentinel-1 and RISAT-1 C-band SAR.

Model	Satellite	Crop	Incidence Angle (°)	R^2	Accuracy (RMSE)
Wheat (this study)	Sentinel-1	Wheat	34.6–45.8	0.8566	6 cm
Processing tomatoes σ^0-based (this study)	Sentinel-1	Tomato	30.8–43.1	0.4201	11 cm
Processing tomatoes β^0-based (this study)	Sentinel-1	Tomato	30.8–43.1	0.8107	9 cm
Bousbih et al. (2017) [47]	Sentinel-1	Cereals	39–40	0.54	Not given
Nasirzadehdizaji et al. (2019) [48]	Sentinel-1	Wheat	39–40	0.67 (<53 cm) 0.07 (≥53 cm)	Not given
Srivastava (2019) [55]	RISAT-1	Wheat	31	0.37	18 cm
Vreugdenhil et al. (2018) [64]	Sentinel-1	Cereals	Constant	0.68	Not given
Harfenmeister et al. (2019) [61]	Sentinel-1	Wheat	Constant	0.41	Not given

Table 6. Comparison of vegetation LAI estimation models based on Sentinel-1 and RISAT-1 C-band SAR.

Model	Satellite	Crop	Incidence Angle (°)	R^2	Accuracy (RMSE)
Wheat (this study)	Sentinel-1	Wheat	34.6–45.8	0.7225	0.6
Processing tomatoes σ^0-based (this study)	Sentinel-1	Tomato	30.8–43.0	0.7881	1.0
Processing tomatoes β^0-based (this study)	Sentinel-1	Tomato	30.8–43.0	0.8341	0.9
Chauhan et al. (2018) [60]	RISAT-1	Wheat	38	0.76	0.4
Bousbih et al. (2017) [47]	Sentinel-1	Cereals	39–40	0.25	Not given
Vreugdenhil et al. (2018) [64]	Sentinel-1	Cereals	Constant	0.30	Not given
Harfenmeister et al. (2019) [61]	Sentinel-1	Wheat	Constant	0.48	Not given

The use of SAR for agricultural purposes has also been significantly enhanced by this study. While several previous studies used SAR to estimate the wheat LAI and crop height, processing tomatoes were not studied enough. Moreover, estimating K_c of wheat, processing tomatoes, and cotton by SAR, to the best of our knowledge, was not previously conducted. Previously, the crop water requirement estimation of maize, soybean, pasture, and bean using SAR imagery acquired under a narrow range of incidence angles was conducted [49]. Another study showed a non-crop-specific region-wise correlation between only one Sentinel-1 image and the crop water stress index derived through the LANDSAT-8 image [106]. Finally, [107] used smoothed time series of Sentinel-1 backscatter values in different polarization combinations to estimate K_c in vineyards. Therefore, the wheat and processing tomato K_c estimation models derived in this study pave the way to accurate K_c estimation using all available SAR imagery. This study stands out by overcoming the limits imposed by the range of incidence angles typical for SAR imagery. As a result, the newly developed normalized wheat and processing tomato K_c estimation models can be used with confidence during the entire duration of a growing season.

Although the cotton models calibrated in this study showed good performance, they are based on the data recorded from the middle to late stages of growing seasons. Therefore, future studies should improve upon this by including the early stages of the growing seasons. In addition, we did not calibrate an LAI model for cotton in this paper, but this should be feasible given good field measurements. Therefore, additional field experiments should be carried out to calibrate models for crop variables throughout the growing season. Even though the cotton models developed in this study might have only limited use, all four cotton models showed a sizeable improvement in the R^2 and RMSE over the non-normalized models. This result confirms the effectiveness of the novel angle normalization approach suggested in the present study.

The performance of models based on the new transformation was favorable compared to models based on the dual-polarized RVI. Although the RVI-based models in the present study were calibrated under the most favorable conditions possible, using only ascending overpass imagery acquired under only one incidence angle, the new models based on local incidence angle normalization methods outperformed them: the RMSE of RVI-based models was 40–203% higher. It should be noted that the assumption $\sigma^0_{VV} \approx \sigma^0_{HH}$ underlying the dual-polarized RVI is in contradiction to previous findings that show a typical difference of 5 dB between σ^0_{VV} and σ^0_{HH} in the intermediate range of incidence angles in the C band [91,108,109]. Therefore, we conclude that the dual-polarized RVI is not recommended where the assumption of the equality of backscatter in the two polarizations cannot be made.

Unlike previous studies that used only fields with rows perpendicular to the SAR beam [110], in this study, all the fields were used in model calibration. While this row geometry is less noticeable in wheat fields, particularly in the middle and later stages of the season, it should be noted that cotton and processing tomatoes are planted in rows of earth mounds with furrows between them. In addition, the spatial orientation of the rows in the fields in this study was not uniform between the locations. For example, in the processing tomato fields in Gadash, the rows were oriented from west to east, while in Gadot, the orientation was from west-southwest to east-northeast. This difference in the spatial orientation of rows affects the backscatter because the target's radar cross-section depends on its angle relative to the satellite [111], and even minimal changes in the target aspect significantly affect the RCS [112,113]. Nevertheless, the processing tomato models were not sensitive to the crop row orientation because they showed a similar RMSE (Tables S5 and S7) when they were applied to different fields. Therefore, the proposed models seem to be insensitive to the row orientation and could likely be used in other fields with different row orientations relative to the satellite orbit. However, this should be further tested in future studies.

Despite the overall reliable performance of the newly developed models, it should be pointed out that winter images in descending orbits have much weaker correlations with the vegetation height, LAI, and K_c compared to images from ascending orbits. Consequently, SAR images acquired in descending orbits could not be used for the development of the wheat model. In the summer crops tested in this study, this phenomenon did not occur, rendering the imagery acquired from descending orbits usable for the modeling of crop variables.

A likely explanation for the weaker performance of wheat models based on imagery from descending orbits might be related to the higher relative humidity in the early morning (descending images were acquired around 03:40 GMT) compared to the relative humidity in the evening (ascending images were acquired around 15:40 GMT). This observation is confirmed by our meteorological measurements in Saad and Kvutsat Yavne, which showed a regular diurnal pattern of a decrease in relative humidity following sunrise: from up to 100% in early morning hours to 40–60% in the afternoons. At night and in the early morning, the relative humidity is very high, and the formation of fog and dew, along with increased topsoil moisture, causes increased scattering and attenuation of the SAR beam [114,115]. Additionally, the SAR beam can be affected by common atmospheric inhomogeneities in the morning hours over Israel that create radar echoes [116] and increase the atmospheric reflectance and attenuation of the transmitted energy [117]. In previous studies, datasets affected by these effects were binned. For example, [61] omitted a dataset that was affected by dew. The issue of the relatively lower performance of descending orbit-based models is an interesting direction that can be studied by analyzing data from other regions and coupling them with the complementary ground and atmospheric measurements.

Although the newly proposed local incidence angle normalization methods were tested on the typical incidence angle range of Sentinel-1 and most other spaceborne SAR missions, they are not expected to be effective for very steep incidence angles near to the "nadir hole" region [118] or for very shallow angles because of the non-linear dependence of the radar backscatter on the incidence angle in these ranges [90,91].

The proposed σ^0 local incidence angle normalization method can be used not only for agricultural purposes but also for other SAR applications. Additional studies need to be carried out to determine if this method is ideal for general use. The β^0 local incidence angle normalization method might be used for the vegetation variable estimation of crops other than processing tomatoes and cotton grown on rough soil surfaces. Future studies should pursue this.

5. Conclusions

The proposed σ^0 and β^0 local incidence angle normalization methods facilitate the use of all the images acquired by the Sentinel-1 constellation under the full range of typical incidence angles. This is supported by an improvement in the correlations between the SAR measurements and crop variables such as LAI, crop height, and K_c following these normalization procedures in three crops: cotton, tomatoes, and wheat. Models based on the suggested normalization of the incidence angle show considerable R^2 and RMSE improvements over the models that were not based on these transformations. This increase in performance is the most notable for the wheat height and LAI models, processing tomato σ^0-based LAI and β^0-based height models, and the cotton models. Most K_c, LAI, and height models worked well with imagery acquired from ascending and descending orbits, but winter imagery performed better with ascending orbits. This approach to estimate vegetation variables is useful for routine vegetation and agricultural monitoring, having a higher temporal resolution and accuracy than the previous approaches. Despite these results, we wish to stress that the most important achievement is not only the improvement in the models' performance but also the enablement of the conjoint use of images acquired under different incidence angles and even different orbits.

Supplementary Materials: The following are available online at https://www.mdpi.com/2073-445X/10/7/680/s1; Table S1: Sentinel-1 image inventory used in the development of models for wheat. Each line represents a dataset with specific geometrical parameters; Table S2: Sentinel-1 image inventory used in the development of models for cotton. Each line represents a dataset with specific geometrical parameters; Table S3: Sentinel-1 image inventory used in the development of models for processing tomatoes. Each line represents a dataset with specific geometrical parameters; Table S4: Wheat height, LAI, and K_c models; Table S5: Processing tomato height, LAI, and K_c models based on the σ^0 normalization method; Table S6: Cotton height and K_c models based on the σ^0 normalization method; Table S7: Processing tomato height, LAI, and K_c models based on the β^0 normalization methods; Table S8: Cotton height and K_c models based on the β^0 normalization method.

Author Contributions: Conceptualization, O.R., G.K.; methodology, G.K., software, G.K., L.F., V.L.; validation, G.K.; formal analysis, G.K.; investigation, G.K.; fieldwork, G.K., V.L., L.F., V.S.M.; writing—original draft preparation, G.K., O.R.; writing—review and editing, O.R., G.K., V.S.M., J.T.; visualization, G.K., V.S.M., O.R.; supervision, O.R.; project administration, O.R., J.T.; funding acquisition, O.R., J.T. All authors have read and agreed to the published version of the manuscript.

Funding: The cotton field measurements were funded by the Chief Scientist of the Ministry of Agriculture, Israel, under grant number 304-0505, and the wheat and tomato field measurements were funded by the Ministry of Science and Technology, Israel, under grant numbers 3-14559, 3-15605. Gregoriy Kaplan was supported by an absorption grant for new immigrant scientists provided by the Israeli Ministry of Immigrant Absorption. Manivasagam V.S. was supported by the ARO Postdoctoral Fellowship Program from the Agriculture Research Organization, Volcani Institute, Israel.

Institutional Review Board Statement: Not applicable.

Informed Consent Statement: Not applicable.

Data Availability Statement: Sentinel-1 data were obtained from the ESA Copernicus Open Access Hub website (https://scihub.copernicus.eu/dhus/#/home, accessed on 26 June 2021).

Acknowledgments: We thank Leonid Dinevich from Tel-Aviv University for his helpful advice. We thank Nitai Haymann for his contribution to software processing and fieldwork. We thank all the growers. We also thank the reviewers.

Conflicts of Interest: The authors declare no conflict of interest.

References

1. Broge, N.H.; Thomsen, A.G.; Andersen, P.B. Comparison of Selected Vegetation Indices as Indicators of Crop Status. Available online: http://www.earsel.org/symposia/2002-symposium-Prague/pdf/083.pdf (accessed on 24 June 2021).
2. González-Dugo, M.P.; Mateos, L. Spectral vegetation indices for benchmarking water productivity of irrigated cotton and sugarbeet crops. *Agric. Water Manag.* **2008**, *95*, 48–58. [CrossRef]
3. Johnson, L.F.; Trout, T.J. Satellite NDVI Assisted Monitoring of Vegetable Crop Evapotranspiration in California's San Joaquin Valley. *Remote Sens.* **2012**, *4*, 439–455. [CrossRef]
4. Mróz, M.; Sobieraj, A. Comparison of several vegetation indices calculated on the basis of a seasonal SPOT XS time series, and their suitability for land cover and agricultural crop identification. *Technol. Sci.* **2004**, *7*, 39–66. [CrossRef]
5. Purevdorj, T.S.; Tateishi, R.; Ishiyama, T.; Honda, Y. Relationships between percent vegetation cover and vegetation indices. *Int. J. Remote Sens.* **1998**, *19*, 3519–3535. [CrossRef]
6. Rozenstein, O.; Haymann, N.; Kaplan, G.; Tanny, J. Estimating cotton water consumption using a time series of Sentinel-2 imagery. *Agric. Water Manag.* **2018**, *207*, 44–52. [CrossRef]
7. Santos, C.L.M.d.O.; Lamparelli, R.A.C.; Figueiredo, G.K.D.A.; Dupuy, S.; Boury, J.; Luciano, A.C.d.S.; Torres, R.d.S.; le Maire, G. Classification of crops, pastures, and tree plantations along the season with multi-sensor image time series in a subtropical agricultural region. *Remote Sens.* **2019**, *11*, 334. [CrossRef]
8. Viña, A.; Gitelson, A.A.; Nguy-Robertson, A.L.; Peng, Y. Comparison of different vegetation indices for the remote assessment of green leaf area index of crops. *Remote Sens. Environ.* **2011**, *115*, 3468–3478. [CrossRef]
9. Nguy-Robertson, A.L.; Peng, Y.; Gitelson, A.A.; Arkebauer, T.J. Agricultural and Forest Meteorology Estimating green LAI in four crops Potential of determining optimal spectral bands for a universal algorithm. *Agric. For. Meteorol. J.* **2014**, *192–193*, 140–148. [CrossRef]
10. Rozenstein, O.; Qin, Z.; Derimian, Y.; Karnieli, A. Derivation of land surface temperature for Landsat-8 TIRS using a split window algorithm. *Sensors* **2014**, *14*, 5768–5780. [CrossRef]
11. Manivasagam, V.S.; Kaplan, G.; Rozenstein, O. Developing Transformation Functions for VENμS and Sentinel-2 Surface Reflectance over Israel. *Remote Sens.* **2019**, *11*, 1710. [CrossRef]
12. Flood, N. Continuity of reflectance data between Landsat-7 ETM+ and Landsat-8 OLI, for both top-of-atmosphere and surface reflectance: A study in the Australian landscape. *Remote Sens.* **2014**, *6*, 7952–7970. [CrossRef]
13. Bannari, A. Synergy Between Sentinel-MSI and Landsat-OLI to Support High Temporal Frequency for Soil Salinity Monitoring in an Arid Landscape. *Res. Dev. Saline Agric.* **2019**. [CrossRef]
14. Padró, J.C.; Pons, X.; Aragonés, D.; Díaz-Delgado, R.; García, D.; Bustamante, J.; Pesquer, L.; Domingo-Marimon, C.; González-Guerrero, Ò.; Cristóbal, J.; et al. Radiometric correction of simultaneously acquired Landsat-7/Landsat-8 and Sentinel-2A imagery using Pseudoinvariant Areas (PIA): Contributing to the Landsat time series legacy. *Remote Sens.* **2017**, *9*, 1319. [CrossRef]
15. Kaplan, G.; Fine, L.; Lukyanov, V.; Manivasagam, V.S.; Malachy, N.; Tanny, J.; Rozenstein, O. Estimating Processing Tomato Water Consumption, Leaf Area Index, and Height Using Sentinel-2 and VENμS Imagery. *Remote Sens.* **2021**, *13*, 1046. [CrossRef]
16. Asrar, G.; Fuchs, M.; Kanemasu, E.T.; Hatfield, J.L. Estimating Absorbed Photosynthetic Radiation and Leaf Area Index from Spectral Reflectance in Wheat. *Agron. J.* **1984**, *76*, 300. [CrossRef]
17. Reichenau, T.G.; Korres, W.; Montzka, C.; Fiener, P.; Wilken, F.; Stadler, A.; Waldhoff, G.; Schneider, K. Spatial heterogeneity of Leaf Area Index (LAI) and its temporal course on arable land: Combining field measurements, remote sensing and simulation in a Comprehensive Data Analysis Approach (CDAA). *PLoS ONE* **2016**, *11*, 1–24. [CrossRef] [PubMed]
18. Sun, Y.; Qin, Q.; Ren, H.; Zhang, T.; Chen, S. Red-Edge Band Vegetation Indices for Leaf Area Index Estimation from Sentinel-2/MSI Imagery. *IEEE Trans. Geosci. Remote Sens.* **2020**, *58*, 826–840. [CrossRef]
19. Wang, C.; Feng, M.C.; Yang, W.D.; Ding, G.W.; Sun, H.; Liang, Z.Y.; Xie, Y.K.; Qiao, X.X. Impact of spectral saturation on leaf area index and aboveground biomass estimation of winter wheat. *Spectrosc. Lett.* **2016**, *49*, 241–248. [CrossRef]
20. Kaplan, G.; Rozenstein, O. Spaceborne Estimation of Leaf Area Index in Cotton, Tomato, and Wheat Using Sentinel-2. *Land* **2021**, *10*, 505. [CrossRef]
21. Simic Milas, A.; Romanko, M.; Reil, P.; Abeysinghe, T.; Marambe, A. The importance of leaf area index in mapping chlorophyll content of corn under different agricultural treatments using UAV images. *Int. J. Remote Sens.* **2018**, *39*, 5415–5431. [CrossRef]
22. Ewert, F. Modelling plant responses to elevated CO2: How important is leaf area index? *Ann. Bot.* **2004**, *93*, 619–627. [CrossRef]
23. Herrmann, I.; Pimstein, A.; Karnieli, A.; Cohen, Y.; Alchanatis, V.; Bonfil, D.J. LAI assessment of wheat and potato crops by VENμS and Sentinel-2 bands. *Remote Sens. Environ.* **2011**, *115*, 2141–2151. [CrossRef]
24. Heuvelink, E.; Bakker, M.J.; Elings, A.; Kaarsemaker, R.; Marcelis, L.F.M. Effect of leaf area on tomato yield. *Acta Hortic.* **2005**, *691*, 43–50. [CrossRef]

25. Sadeh, Y.; Zhu, X.; Dunkerley, D.; Walker, J.P.; Zhang, Y.; Rozenstein, O.; Manivasagam, V.S.; Chenu, K. Fusion of Sentinel-2 and PlanetScope time-series data into daily 3 m surface reflectance and wheat LAI monitoring. *Int. J. Appl. Earth Obs. Geoinf.* **2021**, *96*, 102260. [CrossRef]
26. Manivasagam, V.S.; Rozenstein, O. Practices for upscaling crop simulation models from field scale to large regions. *Comput. Electron. Agric.* **2020**, *175*, 105554. [CrossRef]
27. Manivasagam, V.S.; Sadeh, Y.; Kaplan, G.; Bonfil, D.J.; Rozenstein, O. Studying the Feasibility of Assimilating Sentinel-2 and PlanetScope Imagery into the SAFY Crop Model to Predict Within-Field Wheat Yield. *Remote Sens.* **2021**, *13*, 2395. [CrossRef]
28. Qi, J.; Wang, C.; Inoue, Y.; Zhang, R.; Gao, W. Synergy of optical and radar remote sensing in agricultural applications. *Ecosyst. Dyn. Agric. Remote Sens. Model. Site-Specific Agric.* **2004**, *5153*, 153. [CrossRef]
29. Luo, P.; Liao, J.; Shen, G. Combining Spectral and Texture Features for Estimating Leaf Area Index and Biomass of Maize Using Sentinel-1/2, and Landsat-8 Data. *IEEE Access* **2020**, *8*, 53614–53626. [CrossRef]
30. Jin, X.; Yang, G.; Xu, X.; Yang, H.; Feng, H.; Li, Z.; Shen, J.; Zhao, C.; Lan, Y. Combined multi-temporal optical and radar parameters for estimating LAI and biomass in winter wheat using HJ and RADARSAR-2 data. *Remote Sens.* **2015**, *7*, 13251–13272. [CrossRef]
31. Wang, J.; Xiao, X.; Bajgain, R.; Starks, P.; Steiner, J.; Doughty, R.B.; Chang, Q. Estimating leaf area index and aboveground biomass of grazing pastures using Sentinel-1, Sentinel-2 and Landsat images. *ISPRS J. Photogramm. Remote Sens.* **2019**, *154*, 189–201. [CrossRef]
32. Reamer, R.E.; Stockton, W.O.; Stromfors, R.D. New military uses for Synthetic Aperture Radar (SAR). *Airborne Reconnaiss. XVI* **1993**, 113–119. [CrossRef]
33. Allen, R.G.; Pereira, L.S.; Dirk, R.; Smith, M. *Crop Evapotranspiration-Guidelines For Computing Crop Water Requirements-FAO Irrigation And Drainage Paper 56*; FAO: Rome, Italy, 1998; Volume 300, p. D05109. Available online: http://www.fao.org/3/X0490E/X0490E00.htm (accessed on 24 June 2021).
34. Pereira, L.S.; Paredes, P.; Hunsaker, D.J.; López-Urrea, R.; Mohammadi Shad, Z. Standard single and basal crop coefficients for field crops. Updates and advances to the FAO56 crop water requirements method. *Agric. Water Manag.* **2021**, *243*, 1–35. [CrossRef]
35. Trevisan, R.G.; Junior, N.d.S.V.; Portz, G.; Eitelwein, M.T.; Molin, J.P. Use of crop height and optical sensor readings to predict mid-season cotton biomass. In Proceedings of the Precision agriculture '15, Volcani Center, Israel, 12–16 July 2015; p. 8.
36. Asilo, S.; Nelson, A.; Bie, K.D.; Skidmore, A.; Laborte, A.; Maunahan, A.; Quilang, E.J.P. Relating X-band SAR Backscattering to Leaf Area Index of Rice in Different Phenological Phases. *Remote Sens.* **2019**, *11*, 1462. [CrossRef]
37. Canisius, F.; Shang, J.; Liu, J.; Huang, X.; Ma, B.; Jiao, X.; Geng, X.; Kovacs, J.M.; Walters, D. Tracking crop phenological development using multi-temporal polarimetric Radarsat-2 data. *Remote Sens. Environ.* **2018**, *210*, 508–518. [CrossRef]
38. McNairn, H.; Champagne, C.; Shang, J.; Holmstrom, D.; Reichert, G. Integration of optical and Synthetic Aperture Radar (SAR) imagery for delivering operational annual crop inventories. *ISPRS J. Photogramm. Remote Sens.* **2009**, *64*, 434–449. [CrossRef]
39. Hosseini, M.; McNairn, H.; Merzouki, A.; Pacheco, A. Estimation of Leaf Area Index (LAI) in corn and soybeans using multi-polarization C- and L-band radar data. *Remote Sens. Environ.* **2015**, *170*, 77–89. [CrossRef]
40. Kim, Y.; Jackson, T.; Bindlish, R.; Lee, H.; Hong, S. Radar vegetation index for estimating the vegetation water content of rice and soybean. *IEEE Geosci. Remote Sens. Lett.* **2012**, *9*, 564–568. [CrossRef]
41. Small, D. Flattening Gamma: Radiometric Terrain Correction for SAR Imagery. *IEEE Trans. Geosci. Remote Sens.* **2011**, *49*, 3081–3093. [CrossRef]
42. Makynen, M.; Karvonen, J. Incidence angle dependence of first-year sea ice backscattering coefficient in Sentinel-1 SAR imagery over the Kara Sea. *IEEE Trans. Geosci. Remote Sens.* **2017**, *55*, 6170–6181. [CrossRef]
43. Widhalm, B.; Bartsch, A.; Goler, R. Simplified normalization of C-band synthetic aperture radar data for terrestrial applications in high latitude environments. *Remote Sens.* **2018**, *10*, 1–18. [CrossRef]
44. Van Tricht, K.; Gobin, A.; Gilliams, S.; Piccard, I. Synergistic Use of Radar Sentinel-1 and Optical Sentinel-2 Imagery for Crop Mapping: A Case Study for Belgium. *Remote Sens.* **2018**, *10*, 1642. [CrossRef]
45. Inoue, Y.; Sakaiya, E.; Wang, C. Capability of C-band backscattering coefficients from high-resolution satellite SAR sensors to assess biophysical variables in paddy rice. *Remote Sens. Environ.* **2014**, *140*, 257–266. [CrossRef]
46. Veloso, A.; Mermoz, S.; Bouvet, A.; Le Toan, T.; Planells, M.; Dejoux, J.F.; Ceschia, E. Understanding the temporal behavior of crops using Sentinel-1 and Sentinel-2-like data for agricultural applications. *Remote Sens. Environ.* **2017**, *199*, 415–426. [CrossRef]
47. Bousbih, S.; Zribi, M.; Lili-Chabaane, Z.; Baghdadi, N.; El Hajj, M.; Gao, Q.; Mougenot, B. Potential of Sentinel-1 Radar Data for the Assessment of Soil and Cereal Cover Parameters. *Sensors* **2017**, *17*, 2617. [CrossRef] [PubMed]
48. Nasirzadehdizaji, R.; Balik Sanli, F.; Abdikan, S.; Cakir, Z.; Sekertekin, A.; Ustuner, M. Sensitivity Analysis of Multi-Temporal Sentinel-1 SAR Parameters to Crop Height and Canopy Coverage. *Appl. Sci.* **2019**, *9*, 655. [CrossRef]
49. Navarro, A.; Rolim, J.; Miguel, I.; Catalão, J.; Silva, J.; Painho, M.; Vekerdy, Z. Crop Monitoring Based on SPOT-5 Take-5 and Sentinel-1A Data for the Estimation of Crop Water Requirements. *Remote Sens.* **2016**, *8*, 525. [CrossRef]
50. Inglada, J.; Vincent, A.; Arias, M.; Marais-Sicre, C. Improved early crop type identification by joint use of high temporal resolution SAR and optical image time series. *Remote Sens.* **2016**, *8*, 362. [CrossRef]
51. Hosseini, M.; McNairn, H.; Mitchell, S.; Davidson, A.; Robertson, L.D. Combination of optical and SAR sensors for monitoring biomass over corn fields. *Int. Geosci. Remote Sens. Symp.* **2018**, *2018*, 5952–5955. [CrossRef]

52. Phan, H.; Le Toan, T.; Bouvet, A. Understanding Dense Time Series of Sentinel-1 Backscatter from Rice Fields: Case Study in a Province of the Mekong Delta, Vietnam. *Remote Sens.* **2021**, *13*, 921. [CrossRef]
53. Molijn, R.; Iannini, L.; Vieira Rocha, J.; Hanssen, R. Sugarcane Productivity Mapping through C-Band and L-Band SAR and Optical Satellite Imagery. *Remote Sens.* **2019**, *11*, 1109. [CrossRef]
54. Demarez, V.; Helen, F.; Marais-Sicre, C.; Baup, F. In-Season Mapping of Irrigated Crops Using Landsat 8 and Sentinel-1 Time Series. *Remote Sens.* **2019**, *11*, 118. [CrossRef]
55. Srivastava, H.S.; Sivasankar, T. Potential of C-Band Hybrid Polarimetric RISAT-1 SAR Data To Estimate Wheat Crop Height. In Proceedings of the ISPRS-GEOGLAM-ISRS International Workshop on 'Earth Observations for Agricultural Monitoring', New Delhi, India, 18–20 February 2019.
56. Srivastava, H.S.; Sivasankar, T.; Patel, P. The sensitivity of C-band hybrid polarimetric RISAT-1 SAR data to leaf area index of paddy crop. *ISPRS Ann. Photogramm. Remote Sens. Spat. Inf. Sci.* **2018**, *4*, 215–222. [CrossRef]
57. Benabdelouahab, T.; Dominique, D.; Hayat, L.; Hadria, R.; Tychon, B.; Boudhar, A.; Balaghi, R.; Lebrini, Y.; Maaroufi, H.; Barbier, C. Using SAR Data to Detect Wheat Irrigation Supply in an Irrigated Semi-arid Area. *J. Agric. Sci.* **2018**, *11*, 1916–9760. [CrossRef]
58. Han, D.; Liu, S.; Du, Y.; Xie, X.; Fan, L.; Lei, L.; Li, Z.; Yang, H.; Yang, G. Crop Water Content of Winter Wheat Revealed with Sentinel-1 and Sentinel-2 Imagery. *Sensors* **2019**, *19*, 4013. [CrossRef]
59. Yadav, V.P.; Prasad, R.; Bala, R. Leaf area index estimation of wheat crop using modified water cloud model from the time-series SAR and optical satellite data. *Geocarto Int.* **2019**, *36*, 791–802. [CrossRef]
60. Chauhan, S.; Srivastava, H.S.; Patel, P. Wheat crop biophysical parameters retrieval using hybrid-polarized RISAT-1 SAR data. *Remote Sens. Environ.* **2018**, *216*, 28–43. [CrossRef]
61. Harfenmeister, K.; Spengler, D.; Weltzien, C. Analyzing temporal and spatial characteristics of crop parameters using Sentinel-1 backscatter data. *Remote Sens.* **2019**, *11*, 1569. [CrossRef]
62. Song, Y.; Wang, J. Mapping Winter Wheat Planting Area and Monitoring Its Phenology Using Sentinel-1 Backscatter Time Series. *Remote Sens.* **2019**, *11*, 449. [CrossRef]
63. Ashmitha, M.N.; Ahamed, M.J.; Pazhanivelan, S.; Kumaraperumal, R.; Ganesha Raj, K. Estimation of cotton and maize crop area in Perambalur district of Tamil Nadu using multi-date Sentinel-1A SAR data. *Int. Arch. Photogramm. Remote Sens. Spat. Inf. Sci. ISPRS Arch.* **2019**, *42*, 67–71. [CrossRef]
64. Vreugdenhil, M.; Wagner, W.; Bauer-Marschallinger, B.; Pfeil, I.; Teubner, I.; Rüdiger, C.; Strauss, P. Sensitivity of Sentinel-1 backscatter to vegetation dynamics: An Austrian case study. *Remote Sens.* **2018**, *10*, 1396. [CrossRef]
65. Topouzelis, K.; Singha, S. Incidence angle Normalization of Wide Swath SAR Data for Oceanographic Applications. *Open Geosci.* **2016**, *8*, 450–464. [CrossRef]
66. Mladenova, I.E.; Jackson, T.J.; Bindlish, R.; Hensley, S. Incidence Angle Normalization of Radar Backscatter Data. *IEEE Trans. Geosci. Remote Sens.* **2013**, *51*, 1791–1804. [CrossRef]
67. Revill, A.; Florence, A.; Macarthur, A.; Hoad, S.; Rees, R.; Williams, M. Quantifying uncertainty and bridging the scaling gap in the retrieval of leaf area index by coupling Sentinel-2 and UAV observations. *Remote Sens.* **2020**, *12*, 1843. [CrossRef]
68. Kljun, N.; Calanca, P.; Rotach, M.W.; Schmid, H.P. A simple two-dimensional parameterisation for Flux Footprint Prediction (FFP). *Geosci. Model Dev.* **2015**, *8*, 3695–3713. [CrossRef]
69. Corbari, C.; Ravazzani, G.; Galvagno, M.; Cremonese, E.; Mancini, M. Assessing crop coefficients for natural vegetated areas using satellite data and eddy covariance stations. *Sensors* **2017**, *17*, 2664. [CrossRef]
70. Rozenstein, O.; Haymann, N.; Kaplan, G.; Tanny, J. Validation of the cotton crop coefficient estimation model based on Sentinel-2 imagery and eddy covariance measurements. *Agric. Water Manag.* **2019**, *223*, 105715. [CrossRef]
71. Huang, C.; Zheng, X.; Tait, A.; Dai, Y.; Yang, C.; Chen, Z.; Li, T.; Wang, Z. On using smoothing spline and residual correction to fuse rain gauge observations and remote sensing data. *J. Hydrol.* **2014**, *508*, 410–417. [CrossRef]
72. Wright, J.; Lillesand, T.M.; Kiefer, R.W. *Remote Sensing and Image Interpretation*, 5th ed.; Flahive, R., Powell, D., Eds.; Wiley: Hoboken, NJ, USA, 2004; ISBN 0471026093.
73. Stateczny, A.; Kazimierski, W.; Kulpa, K. Radar and Sonar Imaging and Processing. *Remote Sens.* **2020**, *12*, 1811. [CrossRef]
74. Farr, T.G.; Rosen, P.A.; Caro, E.; Crippen, R.; Duren, R.; Hensley, S.; Kobrick, M.; Paller, M.; Rodriguez, E.; Roth, L.; et al. The Shuttle Radar Topography Mission. *Rev. Geophys.* **2007**, *45*, 44. [CrossRef]
75. Inoue, Y.; Sakaiya, E.; Wang, C. Potential of X-Band Images from High-Resolution Satellite SAR Sensors to Assess Growth and Yield in Paddy Rice. *Remote Sens.* **2014**, *6*, 5995–6019. [CrossRef]
76. Ndikumana, E.; Minh, D.H.T.; Nguyen, H.T.D.; Baghdadi, N.; Courault, D.; Hossard, L.; Moussawi, I. El Estimation of rice height and biomass using multitemporal SAR Sentinel-1 for Camargue, Southern France. *Remote Sens.* **2018**, *10*, 1394. [CrossRef]
77. Flores, A.; Herndon, K.; Thapa, R.; Cherrington, E. *SAR Handbook: Comprehensive Methodologies for Forest Monitoring and Biomass Estimation*, 1st ed.; SERVIR Global: Huntsville, AL, USA, 2019.
78. Kumar, D.; Rao, S.; Sharma, J.R. Radar Vegetation Index as an Alternative to NDVI for Monitoring of Soyabean and Cotton. In Proceedings of the XXXIII INCA International Congress, Jodhpur, India, 19–21 September 2013; pp. 91–96.
79. Trudel, M.; Charbonneau, F.; Leconte, R. Using RADARSAT-2 polarimetric and ENVISAT-ASAR dual-polarization data for estimating soil moisture over agricultural fields. *Can. J. Remote Sens.* **2012**, *38*, 514–527. [CrossRef]
80. Holtgrave, A.-K.; Röder, N.; Ackermann, A.; Erasmi, S.; Kleinschmit, B. Comparing Sentinel-1 and -2 Data and Indices for Agricultural Land Use Monitoring. *Remote Sens.* **2020**, *12*, 2919. [CrossRef]

81. Richards, M. *Fundamentals of Radar Signal Processing;* McGraw-Hill: New York, NY, USA, 2005; ISBN 0-07-144474-2.
82. Mäkynen, M.P.; Manninen, A.T.; Similä, M.H.; Karvonen, J.A.; Hallikainen, M.T. Incidence angle dependence of the statistical properties of C-band HH-polarization backscattering signatures of the Baltic Sea ice. *IEEE Trans. Geosci. Remote Sens.* **2002**, *40*, 2593–2605. [CrossRef]
83. Sabel, D.; Pathe, C.; Wagner, W.; Hasenauer, S.; Bartsch, A.; Künzer, C.; Scipal, K. Using ENVISAT ScanSAR data for characterising scaling properties of scatterometer derived soil moisture information over Southern Africa. In Proceedings of the ENVISAT Symposium, Montreux, Switzerland, 23–27 April 2007.
84. Lecomte, P.; Wagner, W. ERS Wind Scatterometer Commissioning and In-Flight Calibration. Available online: https://earth.esa.int/documents/10174/1602497/WSC12.pdf (accessed on 25 June 2021).
85. Skolnik, M. *Radar Handbook*, 3rd ed.; McGraw-Hill: New York, NY, USA, 2008; ISBN 978-0-07-148547-0.
86. Rozenstein, O.; Siegal, Z.; Blumberg, D.G.; Adamowski, J. Investigating the backscatter contrast anomaly in synthetic aperture radar (SAR) imagery of the dunes along the Israel—Egypt border. *Int. J. Appl. Earth Obs. Geoinf.* **2016**, *46*, 13–21. [CrossRef]
87. Srivastava, H.S.; Patel, P.; Sharma, K.P.; Krishnamurthy, Y.V.N.; Dadhwal, V.K. A semi-empirical modelling approach to calculate two-way attenuation in radar backscatter from soil due to crop cover. *Curr. Sci.* **2011**, *100*, 1871–1874.
88. Ulaby, F. Radar response to vegetation. *IEEE Trans. Antennas Propag.* **1975**, *23*, 36–45. [CrossRef]
89. Da Silva, A.d.Q.; Paradella, W.R.; Freitas, C.C.; Oliveira, C.G. Evaluation of digital classification of polarimetric SAR data for iron-mineralized laterites mapping in the Amazon region. *Remote Sens.* **2013**, *5*, 3101–3122. [CrossRef]
90. Carver, K.R.; Elachi, C.; Ulaby, F.T. Microwave remote sensing from space. *Proc. IEEE* **1985**, *73*, 970–996. [CrossRef]
91. Richards, J.A. *Remote Sensing with Imaging Radar;* Signals and Communication Technology; Springer: Berlin/Heidelberg, Germany, 2009; ISBN 978-3-642-02019-3.
92. Moreira, A.; Prats-Iraola, P.; Younis, M.; Krieger, G.; Hajnsek, I.; Papathanassiou, K.P. A tutorial on synthetic aperture radar. *IEEE Geosci. Remote Sens. Mag.* **2013**, *1*, 6–43. [CrossRef]
93. Small, D.; Rohner, C.; Miranda, N.; Ruetschi, M.; Schaepman, M.E. Wide-Area Analysis-Ready Radar Backscatter Composites. *IEEE Trans. Geosci. Remote Sens.* **2021**, 1–14. [CrossRef]
94. Frey, O.; Santoro, M.; Werner, C.L.; Wegmüller, U. DEM-based SAR pixel-area estimation for enhanced geocoding refinement and radiometric normalization. *IEEE Geosci. Remote Sens. Lett.* **2013**, *10*, 48–52. [CrossRef]
95. Raney, R.K.; Freeman, T.; Hawkins, R.W.; Bamler, R. A plea for radar brightness. In Proceedings of the Proceedings of IGARSS '94—1994 IEEE International Geoscience and Remote Sensing Symposium, Pasadena, CA, USA, 8–12 August 1994; Volume 2, pp. 1090–1092.
96. El-Darymli, K.; Mcguire, P. Understanding the Significance of Radiometric Calibration for Synthetic Aperture Radar Imagery. In Proceedings of the 2014 IEEE 27th Canadian Conference on Electrical and Computer Engineering (CCECE), Toronto, ON, Canada, 4–7 May 2014; 2014; pp. 1–6.
97. Kreslavsky, M. Venera-15,-16 radar brightness of Venus surface: Comparison with Magellan data. *Abstr. Lunar Planet. Sci. Conf.* **1995**, *26*, 799.
98. Goldstein, R.M.; Melbourne, W.G.; Morris, G.A.; Downs, G.S.; Handley, D.A.O. Preliminary radar results of Mars. *Radio Sci.* **1970**, *5*, 475–478. [CrossRef]
99. Meyer, F.; Hinz, S. Automatic ship detection in space-borne SAR imagery. *Int. Arch. Photogram. Remote Sens. Spat. Inform. Sci.* **2009**, *38*, 1–6.
100. Maître, H. *Processing of Synthetic Aperture Radar Images;* Matre, H., Ed.; ISTE: London, UK, 2008; ISBN 9780470611111.
101. Steiger, J.H. Tests for comparing elements of a correlation matrix. *Psychol. Bull.* **1980**, *87*, 245–251. [CrossRef]
102. Fisher, R.A. On the Probable Error of a Coefficient of Correlation Deduced from a Small Sample. *Metron* **1921**, *1*, 3–32.
103. Wilcoxon, F. Individual Comparisons by Ranking Methods. *Biometrics Bull.* **1945**, *1*, 80–83. [CrossRef]
104. Button, K.S.; Ioannidis, J.P.A.; Mokrysz, C.; Nosek, B.A.; Flint, J.; Robinson, E.S.J.; Munafò, M.R. Power failure: Why small sample size undermines the reliability of neuroscience. *Nat. Rev. Neurosci.* **2013**, *14*, 365–376. [CrossRef]
105. Bréda, N.J.J. Ground-based measurements of leaf area index: A review of methods, instruments and current controversies. *J. Exp. Bot.* **2003**, *54*, 2403–2417. [CrossRef]
106. El-Shirbeny, M.A.; Abutaleb, K. Sentinel-1 Radar Data Assessment to Estimate Crop Water Stress. *World J. Eng. Technol.* **2017**, *05*, 47–55. [CrossRef]
107. Beeri, O.; Netzer, Y.; Munitz, S.; Mintz, D.F.; Pelta, R.; Shilo, T.; Horesh, A.; Mey-tal, S. Kc and LAI Estimations Using Optical and SAR Remote Sensing Imagery for Vineyards Plots. *Remote Sens.* **2020**, *12*, 3478. [CrossRef]
108. Cloude, S. *Polarisation: Applications in Remote Sensing;* Oxford University Press: Oxford, UK, 2010; ISBN 978-0-19-956973-1.
109. Verba, V.S.; Neronskiy, L.B.; Osipov, V.E.; Turuk, I.G. *Spaceborne Earth Surveillance Radar Systems;* Radiotechnika: Moscow, Russia, 2010.
110. Léonard, A.; Bériaux, E.; Defourny, P. Complementarity of Linear Polarizations in C-Band SAR Imagery to Estimate Leaf Area Index for Maize and Winter Wheat. Available online: https://ftp.space.dtu.dk/pub/Ioana/papers/s444_1leon.pdf (accessed on 25 June 2021).
111. *Naval Air Systems Command Electronic Warfare and Radar Systems Engineering Handbook*, 2nd ed.; Naval Air Warfare Center: Point Mugu, CA, USA, 1999.
112. Skolnik, M. *Introduction to Radar Systems*, 2nd ed.; McGraw-Hili Book Co.: New York, NY, USA, 1981; ISBN 9786069276969.

113. Zikidis, K. Early Warning Against Stealth Aircraft, Missiles and Unmanned Aerial Vehicles. In *Surveillance in Action. Advanced Sciences and Technologies for Security Applications*; Karampelas, P., Bourlai, T., Eds.; Springer International Publishing: Cham, Switzerland, 2018; ISBN 978-3-319-68532-8.
114. Shoshany, M.; Svoray, T.; Svoray, T.; Curran, P.J.; Foody, G.M.; Perevolotsky, A. The relationship between ERS-2 SAR backscatter and soil moisture: Generalization from a humid to semi-arid transect. *Int. J. Remote Sens.* **2000**, *21*, 2337–2343. [CrossRef]
115. Filgueiras, R.; Mantovani, E.C.; Althoff, D.; Fernandes Filho, E.I.; Cunha, F.F. da Crop NDVI Monitoring Based on Sentinel-1. *Remote Sens.* **2019**, *11*, 1441. [CrossRef]
116. Dinevich, L. Improving the accuracy of selection of bird radar echoes against a background of atomized clouds and atmospheric inhomogeneities. *Ring* **2016**, *37*, 3–18. [CrossRef]
117. NAWCWD Avionics Department. *Electronic Warfare and Radar Systems. Engineering Handbook*, 4th ed.; Naval Air Warfare Center: Point Mugu, CA, USA, 2013; ISBN 9781608077052.
118. Fenn, A.J. *Adaptive Antennas and Phased Arrays for Radar and Communications*; Artech House: Norwood, MA, USA, 2008; ISBN 9781596932739.

 land

Article

Spaceborne Estimation of Leaf Area Index in Cotton, Tomato, and Wheat Using Sentinel-2

Gregoriy Kaplan and Offer Rozenstein *

Institute of Soil, Water and Environmental Sciences, Agricultural Research Organization-Volcani Institute, HaMaccabim Road 68, P.O.B 15159, Rishon LeZion 7528809, Israel; grigorii@volcani.agri.gov.il
* Correspondence: offerr@volcani.agri.gov.il

Abstract: Satellite remote sensing is a useful tool for estimating crop variables, particularly Leaf Area Index (LAI), which plays a pivotal role in monitoring crop development. The goal of this study was to identify the optimal Sentinel-2 bands for LAI estimation and to derive Vegetation Indices (VI) that are well correlated with LAI. Linear regression models between time series of Sentinel-2 imagery and field-measured LAI showed that Sentinel-2 Band-8A—Narrow Near InfraRed (NIR) is more accurate for LAI estimation than the traditionally used Band-8 (NIR). Band-5 (Red edge-1) showed the lowest performance out of all red edge bands in tomato and cotton. A novel finding was that Band 9 (Water vapor) showed a very high correlation with LAI. Bands 1, 2, 3, 4, 5, 11, and 12 were saturated at LAI ≈ 3 in cotton and tomato. Bands 6, 7, 8, 8A, and 9 were not saturated at high LAI values in cotton and tomato. The tomato, cotton, and wheat LAI estimation performance of ReNDVI (R^2 = 0.79, 0.98, 0.83, respectively) and two new VIs (WEVI (Water vapor red Edge Vegetation Index) (R^2 = 0.81, 0.96, 0.71, respectively) and WNEVI (Water vapor narrow NIR red Edge Vegetation index) (R^2 = 0.79, 0.98, 0.79, respectively)) were higher than the LAI estimation performance of the commonly used NDVI (R^2 = 0.66, 0.83, 0.05, respectively) and other common VIs tested in this study. Consequently, reNDVI, WEVI, and WNEVI can facilitate more accurate agricultural monitoring than traditional VIs.

Keywords: Sentinel-2; spectral bands; LAI; vegetation indices

Citation: Kaplan, G.; Rozenstein, O. Spaceborne Estimation of Leaf Area Index in Cotton, Tomato, and Wheat Using Sentinel-2. *Land* **2021**, *10*, 505. https://doi.org/10.3390/land10050505

Academic Editor: Javier Cabello

Received: 26 March 2021
Accepted: 7 May 2021
Published: 9 May 2021

1. Introduction

Monitoring crop growth and performance during developmental stages is an essential aspect of agricultural management. Leaf Area Index (LAI) is a good proxy of the vegetation state [1–3] and a good yield predictor [4–6]. LAI is a dimensionless quantity that characterizes plant canopies. It is defined as the one-sided green leaf area per unit ground surface area. The LAI is an important parameter in plant ecology and a measure of the photosynthetic active area, and at the same time of the area subjected to transpiration. It is also the area that comes in contact with air pollutants. LAI is often a key biophysical variable used in biogeochemical, hydrological, and ecological models. LAI is also commonly used as a measure of crop growth and productivity at spatial scales ranging from the plot to the globe. Moreover, activities such as herbicide and fertiliser management, leaf chlorophyll content estimation, detection of crop disease, and yield prediction can be based on LAI monitoring [7].

LAI can be estimated from VIs [8–11] produced from imagery acquired by optical satellites, but this approach suffers from a low correlation between LAI and some bands that the VIs are based on. Many studies showed that LAI estimation from optical imagery suffers from saturation when LAI is greater than 3 (i.e., the LAI changes at a faster rate than the reflectance) [11–14]. Since the LAI of many crops typically exceeds this level by a large margin, optical sensors have limited use for LAI estimation. Most previous studies that defined this saturation effect were based on older sensors (e.g., Landsat, Modis, SPOT) [15–17], and accordingly, Vegetation Indices (VIs) intended for those sensors. In 2015 the first Sentinel-2 became operational, which marked the arrival of the new generation of satellites. The Multi-Spectral Instrument (MSI) onboard Sentinel-2 observes the earth at 13 spectral bands with a spatial resolution from 10 to 60 m (depending on the band) and a five-day revisit time. MSI is a spaceborne multispectral instrument that thoroughly covers the red edge spectral range, which is highly sensitive to the chlorophyll reflectance in plants [18]. The red-edge spectral range covers the wavelengths of 680–750 nm, where the change of leaf reflectance is sharp [19,20]. In order to estimate LAI from Sentinel-2, there is a need to evaluate which bands suffer from the saturation that was observed in previous generations of spaceborne sensors and explore ways to overcome this limitation.

In addition to LAI modelling based on VIs, several machine learning algorithms for LAI estimation based on Sentinel-2 bands were studied and showed mixed results [11,21–23]. Previous studies on different wavebands [24], including simulated Sentinel-2 bands, concluded that the red edge is the best spectral region for LAI estimation in several crops [2,3,25–27]. Therefore, careful selection of the bands used to derive VIs and machine learning algorithms can improve the performance and generality of the LAI estimation models based on Sentinel-2 imagery. Nevertheless, while several studies investigated the performance of MSI-based VIs and machine learning algorithms for LAI estimation of tomato, wheat, and cotton [11,28–30], very few studies investigated the performance of the real MSI bands (as opposed to synthetic data) in the LAI estimation of these crops [31].

Therefore, this study's first goal was to model LAI using real Sentinel-2 imagery and field-measured LAI to quantify the performance of individual bands and their saturation levels in cotton, tomato and wheat. The second goal of the study was to suggest well-performing VIs that employ bands not commonly used for VI derivation and facilitate better agricultural monitoring.

2. Materials and Methods

2.1. Test Sites and Field Measurements

The field data used in this study were collected during one cotton, two wheat, and three processing tomatoes experiments conducted in five locations in Israel (Figure 1).

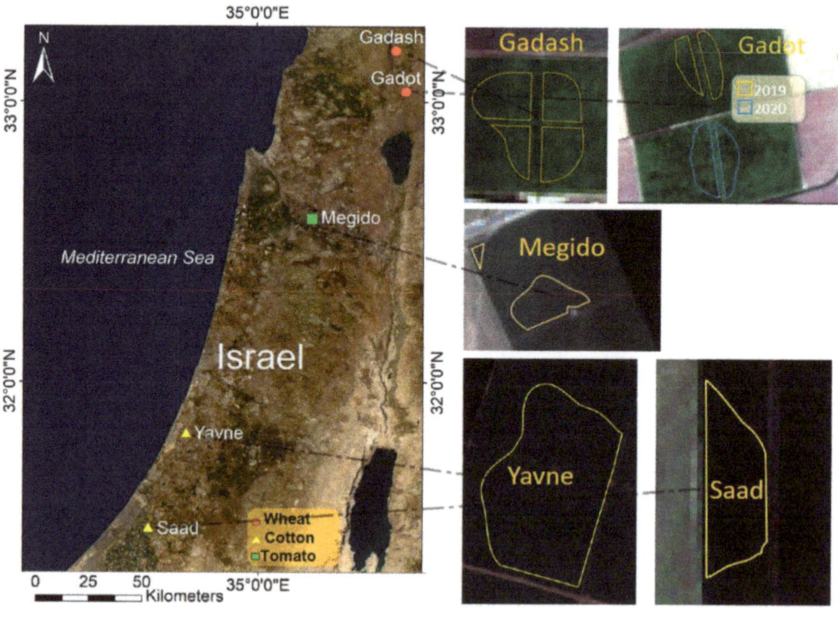

Figure 1. Locations and analysis polygons of the experiments conducted in Israel in 2018–2020.

The Sentinel-2 image inventory used for this study is presented in Table 1. Overall, 56 Sentinel-2 images were used in the study (14—wheat, 33—processing tomatoes, 9—cotton). During these experiments, LAI was measured by a SunScan Canopy Analysis System—SS1 developed by Delta-T Company (Cambridge, United Kingdom). The SunScan is a widely used, accurate, nondestructive LAI measurement system successfully employed in many previous studies [5,30,32]. The SunScan system measures LAI by calculating the difference in solar radiance received by the dome sensor installed under unobscured Sun view and the hand-held probe placed below vegetation canopy on the ground level (Figure 2).

Table 1. The Sentinel-2 bands used in the present study.

Band	Sentinel-2A		Sentinel-2B		
	Central Wavelength (nm)	**Bandwidth (nm)**	**Central Wavelength (nm)**	**Bandwidth (nm)**	**Spatial Resolution (m)**
Band 1—Coastal aerosol	442.7	21	442.2	21	60
Band 2—Blue	492.4	66	492.1	66	10
Band 3—Green	559.8	36	559.0	36	10
Band 4—Red	664.6	31	664.9	31	10
Band 5—Vegetation red edge 1	704.1	15	703.8	16	20
Band 6—Vegetation red edge 1	740.5	15	739.1	15	20
Band 7—Vegetation red edge 3	782.8	20	779.7	20	20
Band 8—NIR	832.8	106	832.9	106	10
Band 8A—Narrow NIR	864.7	21	864.0	22	20
Band 9—Water vapour	945.1	20	943.2	21	60
Band 11—SWIR	1613.7	91	1610.4	94	20
Band 12—SWIR	2202.4	175	2185.7	185	20

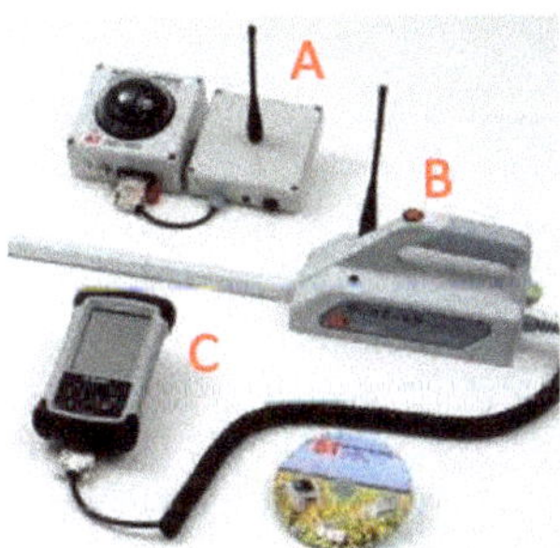

Figure 2. Main components of the SunScan system: (**A**) Dome sensor; (**B**) Probe; (**C**) Field computer.

Each LAI value used for model calibration was an average value of at least 30 field measurements. LAI was measured in the center of the fields and was correlated to average values of Sentinel-2 bands and VIs of homogenous areas in the fields' centers. In the Megido 2020 experiment, LAI was measured in two areas of the field (in the center of the field (six LAI measurements) and on the northwest corner (four LAI measurements)) where the crop developed at different rates and, thus, LAI was different. Accordingly, both time series of the field measurements were correlated with the average values of bands and VIs within defined polygons. In-field paths and their surrounding area were masked out from analysis polygons of the tomato experiments to remove bare soil areas and avoid border effects. These excluded areas consisted of approximately 20% of the overall polygon areas in the tomato fields. Therefore, each tomato polygon consisted of either two or four vegetated regions separated by the paths.

Overall, 11 averaged LAI values taken during two growing seasons were used for deriving the wheat models, nine for cotton (one season), and 23 for tomato (three seasons).

The linear regression models in this study utilised the average values derived from satellite imagery within the analysis polygons and same-date field measurements or linearly interpolated LAI values of field measurements from adjacent dates.

2.2. Satellite Imagery

Sentinel-2 is an Earth observation mission and part of the European Space Agency (ESA) Copernicus program. It includes two satellites with a payload of MSI, namely Sentinel-2A (launched 23 June 2015) and Sentinel-2B (launched 7 March 2017). Table 1 lists the spectral bands of Sentinel-2 that were used in this study. The inventory of the atmospherically and topographically corrected Level-2A Sentinel-2 images used in this study alongside the information on the LAI measurements can be found in Tables 2 and A1. Level-2A and Level-1C imagery were downloaded from the ESA Copernicus site (https://scihub.copernicus.eu/dhus/#/home, accessed on 6 April 2021) (# means "Number"). Level-1C images were processed to Level-2A using Sen2Cor algorithm [33].

Table 2. Sentinel-2 imagery and LAI measurements used in the study.

Area	Crop	Period *	# of Images	Polygon Size (Sentinel-2 Pixels)	# LAI Measurements	Range of Measured LAI
Saad	Wheat	02-March-2019 06-April-2019	6	260	4	4.8–7.1
Yavne	Wheat	11-January-2019 11-April-2019	8	550	7	3.8–7.0
Gadash	Tomato	3-May-2019 24-July-2019	8–9 **	425	6	1.4–4.7
Gadot	Tomato	25-April-2019 14-August-2019	12–13 **	249	11	0.7–9.1
Gadot	Tomato	7-May-2020 3-August-2020	11	332	6	0.9–8.6
Megido	Cotton	30-May-2020 29-July-2020	9 4	268 (Centre) 17 (NW Corner)	6 3	0.6–9.6 0.8–1.9

* Indicates the dates of the first and last images. ** A defective red edge band in a Sentinel-2 image acquired on 10 June 2019 prevented the derivation of red edge-based models for that date.

2.3. Model Calibration and Validation

Linear regression models were derived to estimate LAI for specific crops based on field measurements and Sentinel-2 bands. Similarly, regression models between LAI and VIs were derived, including NDVI [34] and NDVI based on the Narrow NIR Band-8A instead of NIR Band-8. Additionally to NDVI, models were also derived for reNDVI [35], MTCI [36], WDVI [37], EVI [38], SAVI [39], MSAVI [40], DVI [34], and two new indices: WEVI (Water vapor red Edge Vegetation Index) and WNEVI (Water vapor narrow NIR red Edge Vegetation index). For every model, the R^2 and root mean square error (RMSE) values were calculated using the Microsoft Excel software. WEVI and WNEVI were developed based on combinations of the best performing bands for LAI estimation. The following equations and Sentinel-2 bands were used for deriving the aforementioned VIs:

$$\mathrm{NDVI} = (\mathrm{B8} - \mathrm{B4})/(\mathrm{B8} + \mathrm{B4}) \quad (1)$$

$$\mathrm{NDVI8A} = (\mathrm{B8A} - \mathrm{B4})/(\mathrm{B8A} + \mathrm{B4}) \quad (2)$$

$$\mathrm{MTCI} = (\mathrm{B6} - \mathrm{B5})/(\mathrm{B5} - \mathrm{B4}) \quad (3)$$

$$\mathrm{WDVI} = \mathrm{B8} - 0.5 \times \mathrm{B4} \quad (4)$$

$$\mathrm{EVI} = (2.5 \times (\mathrm{B8} - \mathrm{B4}))/(\mathrm{B8} + 6 \times \mathrm{B4} - 7.5 \times \mathrm{B2} + 1) \quad (5)$$

$$\mathrm{SAVI} = ((\mathrm{B8} - \mathrm{B4}))/\,(\mathrm{B8} + \mathrm{B4} + 0.5)) \times 1.5 \quad (6)$$

$$\mathrm{MSAVI} = ((\mathrm{B8} - \mathrm{B4}) \times (1 + \mathrm{L}))/(\mathrm{B8} + \mathrm{B4} + \mathrm{L}) \quad (7)$$

where: L = 1 − 2 × s × NDVI × WDVI and s is the soil line slope = 0.5

$$DVI = B8 - B4 \tag{8}$$

$$reNDVI = (B8A - B6)/(B8A + B6) \tag{9}$$

$$WEVI = B9 - B6 \tag{10}$$

$$WNEVI = (B8A - B6)/(B9 + B6) \tag{11}$$

3. Results

Table 3 shows the performance of the separate Sentinel-2 bands and VIs for LAI estimation of cotton, tomato, and wheat. Overall, the bands that modelled LAI best were Band-7 (Red edge-3), Band-9 (Water vapor), and two NIR bands (8 and 8A). Notably, Band-8A (Narrow NIR) showed a higher correlation with LAI and lower RMSE in LAI estimation than Band-8 (NIR) in all three crops. Consequently, NDVI8A performed better than NDVI. Importantly, Band-4 (Red) showed average performance, and Band-5 (Red edge-1) showed weak performance relative to other bands in tomato and cotton. Therefore VIs based on the better performing bands might be beneficial for LAI estimation. One such VI, namely reNDVI, showed a very high estimation performance. Finally, the high performance in LAI prediction by the Water vapor Band-9 suggests that this band might be useful for creating VIs with good correlation to LAI. This result was confirmed by low RMSE and high R^2 values of the new WEVI and WNEVI that are based on Band-9. The two new VIs proposed in the study (WEVI and WNEVI) alongside reNDVI showed superior performance in LAI predictions compared to NDVI and NDVI8A in all three crops, with the largest difference in wheat.

Table 3. Performance of Sentinel-2 bands and VIs used in the present study. The performance of best performing bands and VIs for each crop are in bold.

	Tomato		Cotton		Wheat	
Band/VI	R^2	RMSE	R^2	RMSE	R^2	RMSE
Band 1—Coastal aerosol	0.08	2.4	0.58	2.4	0.17	1.1
Band 2—Blue	0.13	2.3	0.52	2.5	0.02	1.2
Band 3—Green	0.00	2.5	0.57	2.4	0.06	1.2
Band 4—Red	0.65	1.5	0.81	1.6	0.02	1.2
Band 5—Vegetation red edge	0.00	2.5	0.75	1.8	0.22	1.1
Band 6—Vegetation red edge	0.79	**1.1**	0.93	1.0	0.01	1.2
Band 7—Vegetation red edge	0.78	1.2	0.96	**0.7**	0.26	**1.0**
Band 8—NIR	0.78	1.2	0.96	**0.7**	0.23	1.1
Band 8A—Narrow NIR	**0.82**	**1.1**	**0.97**	**0.7**	**0.34**	**1.0**
Band 9—Water vapour	0.80	**1.1**	0.97	**0.7**	0.29	**1.0**
Band 11—SWIR	0.01	2.5	0.12	3.4	0.00	1.2
Band 12—SWIR	0.61	1.6	0.82	1.5	0.00	1.2
NDVI	0.66	1.4	0.83	1.5	0.05	1.2
NDVI8A	0.71	1.3	0.87	1.3	0.05	1.2
reNDVI	0.79	**1.1**	0.98	**0.6**	**0.83**	**0.5**
MTCI	0.16	2.3	0.95	0.8	0.53	0.8
WDVI	0.76	1.2	0.94	0.9	0.29	1.0
EVI	0.78	1.2	0.95	0.8	0.26	1.0
SAVI	0.73	1.3	0.92	1.0	0.14	1.1
MSAVI	0.75	1.2	0.93	1.0	0.15	1.5
DVI	0.77	1.2	0.94	0.9	0.19	1.1
WEVI	**0.81**	**1.1**	0.96	0.7	0.71	0.6
WNEVI	0.79	**1.1**	**0.98**	**0.6**	0.79	**0.5**

Figure 3 shows the reflectance in each band and the corresponding LAI measurements in this study's experiments. The reflectance in bands 1, 2, 3, 4, 5, 11, 12 in cotton and processing tomatoes start saturating from LAI ≈ 3 and almost no longer changing at

LAI ≈ 6. This result is especially important because bands 4 and 5 are used in many VIs. On the other hand, bands 6, 7, 8, 8A, 9 were not saturated. Insufficient satellite imagery and field measurements of LAI were acquired during the wheat experiments and hindered estimating the saturation levels of this crop.

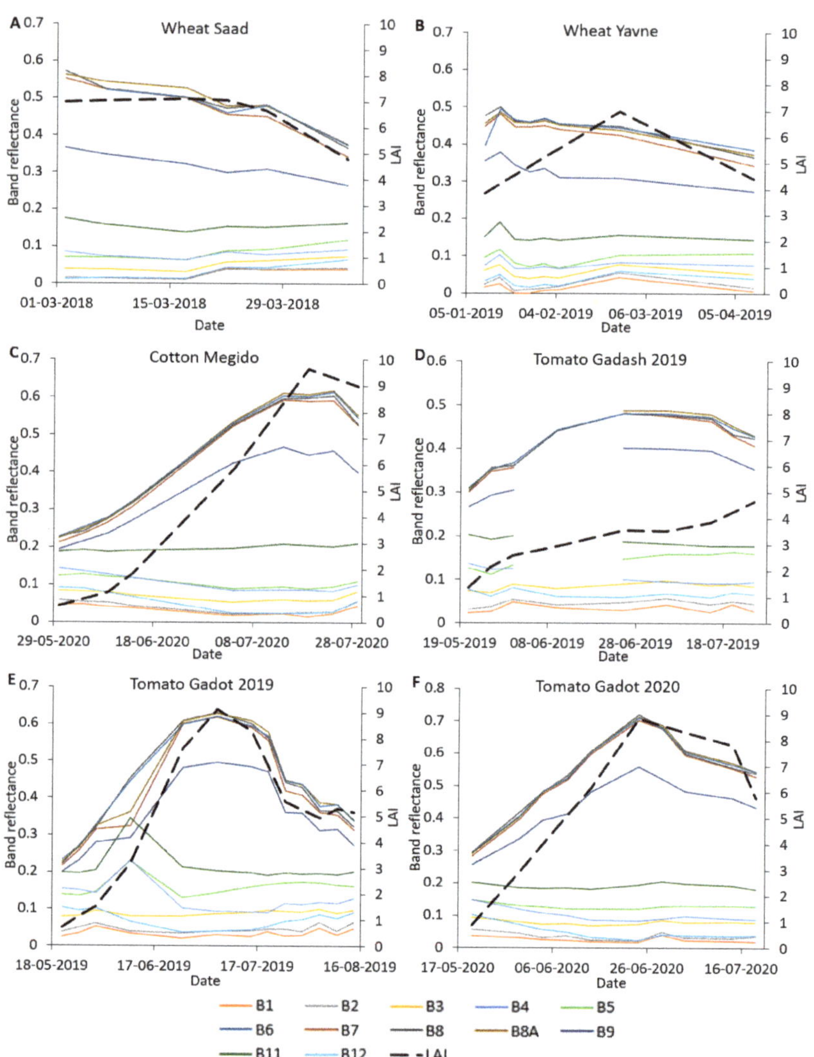

Figure 3. Band reflectance and LAI measurements in the following experiments: (**A**) Wheat Saad, (**B**) Wheat Yavne, (**C**) Cotton Megido (centre of field), (**D**) Tomato Gadash 2019, (**E**) Tomato Gadot 2019, (**F**) Tomato Gadot 2020.

Figure 4 shows the RMSE of Sentinel-2 bands LAI estimation for wheat, cotton, and tomato. While the RMSE of Sentinel-2 bands most commonly used in VIs formulae (bands 2-8A) in wheat LAI estimation is closer to each other, Band-4 and Band-5 have notably high RMSE in cotton and tomato LAI estimation, and this is especially pronounced for Band-5 in tomato.

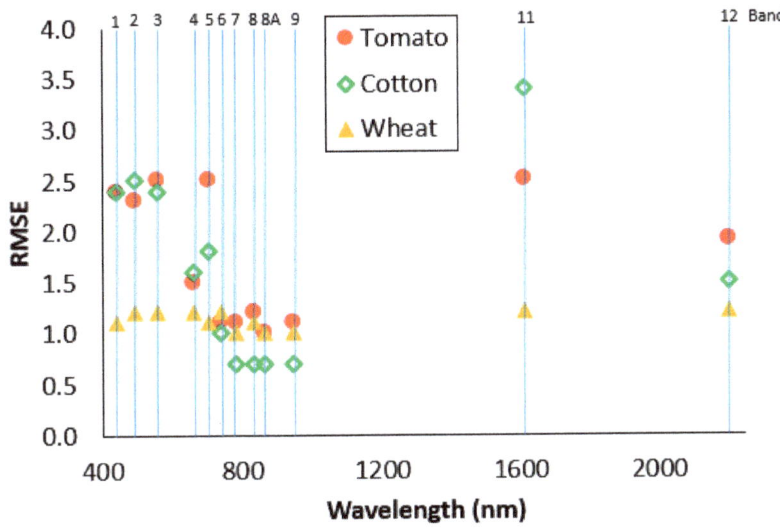

Figure 4. RMSE of Sentinel-2 bands in tomato, cotton, and wheat LAI estimation.

Figure 5 shows reNDVI, WEVI, WNEVI, NDVI, and MTCI linear regression models for tomato, cotton, and wheat.

Figure 6 shows the LAI measurements and LAI estimation based on the VIs used in this study using the models described in Table 3. While several VIs showed similar behavior in LAI estimation, MTCI, MSAVI, reNDVI, WEVI, and WNEVI were notably different. MTCI, affected by the low performance of the Band-5, did not perform well in tomato LAI estimation in Gadot 2019 and 2020. MSAVI notably underestimated wheat LAI values. Conversely, reNDVI, WEVI, and WNEVI show closer resemblance to measured LAI than all other VIs. In the present study, no difference in the spectral response of Sentinel-2A and -B satellites was observed owing to an excellent radiometric cross-calibration of the MSI on both satellites.

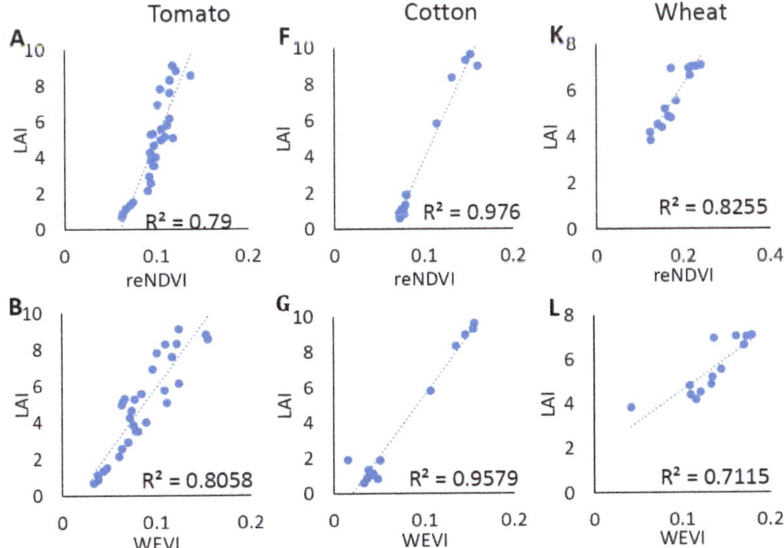

Figure 5. *Cont.*

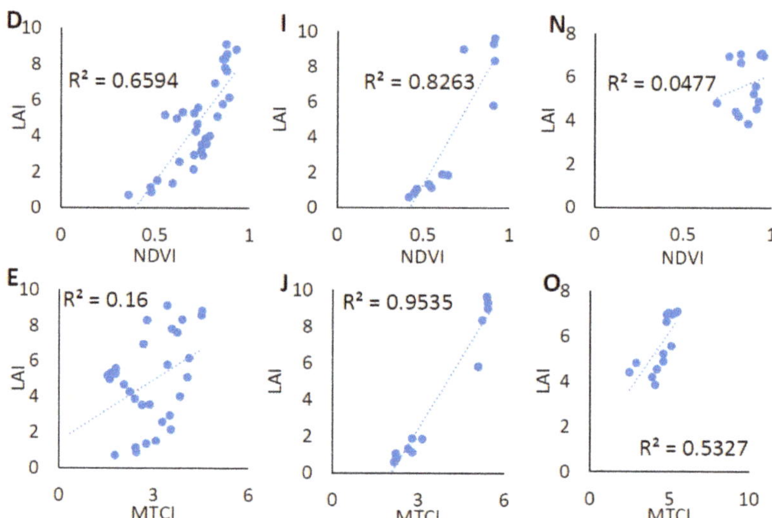

Figure 5. Tomato, cotton, and wheat LAI-VI linear regression models: (**A**) Tomato reNDVI; (**B**) Tomato WEVI; (**C**) Tomato WNEVI; (**D**) Tomato NDVI; (**E**) Tomato MTCI; (**F**) Cotton reNDVI; (**G**) Cotton WEVI; (**H**) Cotton WNEVI; (**I**) Cotton NDVI; (**J**) Cotton MTCI; (**K**) Wheat reNDVI; (**L**) Wheat WEVI; (**M**) Wheat WNEVI; (**N**) Wheat NDVI; (**O**) Wheat MTCI. The data used to derive the models is presented in Table 2, the RMSE values of the models are given in Table 3.

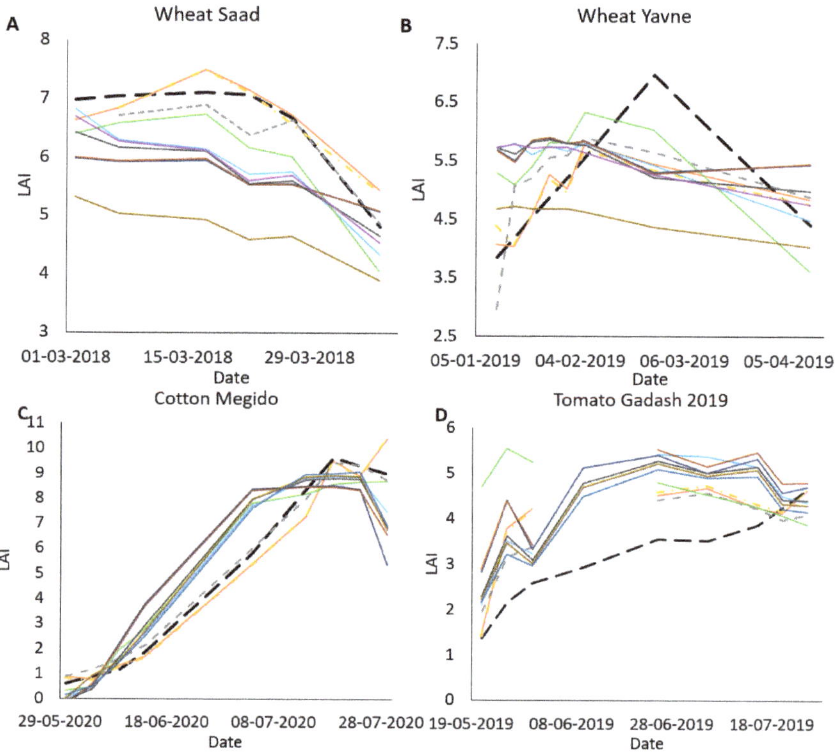

Figure 6. *Cont.*

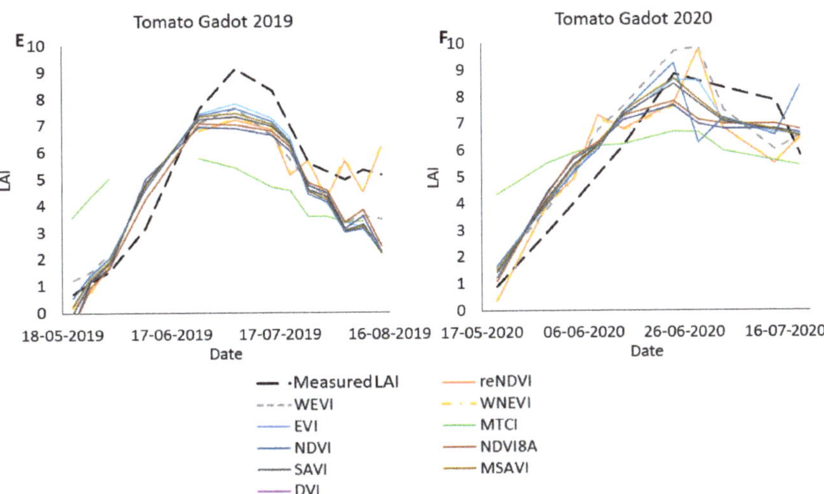

Figure 6. LAI measurements and LAI estimation based on the VIs used in the present study in the following experiments: (**A**) Wheat Saad, (**B**) Wheat Yavne, (**C**) Cotton Megido (centre of the field), (**D**) Tomato Gadash 2019, (**E**) Tomato Gadot 2019, (**F**) Tomato Gadot 2020.

4. Discussion

This study investigated the performance of the individual Sentinel-2 bands and VIs in estimating LAI of tomato, cotton, and wheat. This study's most important finding is that bands 6, 7, 8, 8A, 9 performed well in LAI estimation and did not saturate at high LAI in cotton and processing tomatoes. At the same time, the wheat data was insufficient to make this determination. Therefore, these bands can be used to create VIs for LAI monitoring. VIs such as reNDVI and two new VIs introduced in this study for the first time, WEVI and WNEVI, which are based on these bands, performed well in LAI estimation, better than the commonly used NDVI as well as all the other VIs used in the study.

Band-8A (Narrow NIR) showed better performance in LAI estimation compared to Band-8 (NIR). Therefore, NDVI derived based on Band-8A performed better than NDVI based on Band-8. Band-4 (Red) was found to have an average performance. Therefore, substituting Band-8 with Band-8A and possibly substituting Band-4 with a better-performing band (such as Band-6 used in reNDVI) is likely to improve the correlation of VIs with LAI, and facilitate more accurate agricultural monitoring. The high performance of the reNDVI achieved in the study supported this hypothesis. Unlike red edge and NIR bands, Band-9 (Water vapor) is not commonly used as a VI formulae but can be used in VIs such as WEVI and WNEVI developed in this study. The analysis of Band-9 performance, which is not commonly used for agricultural monitoring, and developing VIs based on this band that perform well in LAI estimation of the three crops, is the unique feature of the present study.

Unlike red edge bands 6 and 7 that showed high performance, Band-5 (Red edge-1), at the tail of the chlorophyll absorption peak [41], showed the lowest overall performance out of all the red edge bands. This might be explained by the negative effect of the chlorophyll content present in the leaves [10,14,42–44], which reaches maximum absorbance at about 690 nm [45]. Moreover, chlorophyll content may vary independently from LAI [46]. In this study, MTCI, based on Band-5, showed low performance in tomato LAI estimation. MTCI was previously found to have low correlation with tomato crop coefficient (K_c) and height [11]. Nevertheless, MTCI was highly correlated with LAI of cotton and wheat in the present study. MTCI was also previously found to have very high correlation with cotton K_c [47,48] as well as a very good correlation with leaf chlorophyll concentration [25,49] and LAI of many crops [3,23,50]. Consequently, despite its effective use for crop variable estimation in many cases, Band-5 and VIs based on this band (e.g., MTCI) should be used with caution to model tomato variables. Similarly, careful selection of Sentinel-2 bands might improve the performance of various machine learning algorithms, for example, the SNAP Biophysical processor [51].

The results and approach demonstrated in this study can be useful in many agricultural applications based on remote sensing data, for example Zaeen et al., [52] who developed in-season potato yield prediction models based on several VIs, and Kganyago el al., [22] that studied the performance of SNAP Biophysical processor machine learning algorithm in LAI estimation of several crops. These applications might benefit from further investigation of the correlations between Sentinel-2 bands and various vegetation variables.

In the present study, all the Sentinel-2 bands and the majority of VIs (except reNDVI, WEVI, and WNEVI) showed low performance in LAI estimation of wheat. Therefore, despite the achievements in estimating LAI using Sentinel-2 bands in tomato, cotton, and wheat, additional measurements of wheat are needed to estimate Sentinel-2 bands saturation levels in that crop. Moreover, owing to the spectral resemblance of the Sentinel-2 MSI and the VENµS sensors [2,11,53], a combination with VENµS might facilitate better agricultural monitoring, considering its high two-day temporal resolution.

Overall, the study quantified the performance of the individual Sentinel-2 bands and several VIs (including two newly developed VIs) in the LAI estimation of tomato, cotton, and wheat. Such a result facilitates deriving efficient algorithms and methods for agricultural monitoring via optical satellite imagery.

5. Conclusions

This study is a step towards improving agricultural practices such as variable rate irrigation, fertilizer and herbicide application, yield prediction, disease monitoring, and many others. This achievement is made possible because of the newly-derived VIs and models that can estimate LAI throughout the season without saturation. As a result, agricultural practices informed through remote sensing can potentially improve agricultural production.

This study found that Sentinel-2 Band-8A (Narrow NIR) is more accurate for LAI estimation than Band-8 (NIR). A very important achievement of the study is that the Band-5 (Red edge-1) showed a low correlation with LAI. Band 9 (Water vapour) showed a very high correlation with LAI alongside the red-edge bands 6 and 7 and NIR bands. Band-9 was demonstrated to be effective for LAI estimation when incorporated into new VIs suggested here for the first time, WEVI and WNEVI. Importantly, Bands 1, 2, 3, 4, 5, 11, 12 were saturated at LAI ≈ 3 and were practically not responsive to a further increase in LAI around LAI ≈ 6. Bands 6, 7, 8, 8A, 9 did not saturate at high LAI. ReNDVI, WEVI, and WNEVI were found to be the best performing VIs for LAI estimation of all three crops tested in this study.

Author Contributions: Conceptualisation G.K., O.R.; Methodology, Software, Investigation, G.K.; Writing—original draft preparation, review and editing, Visualisation, G.K., O.R.; Supervision, Project administration, Funding acquisition, O.R. All authors have read and agreed to the published version of the manuscript.

Funding: The field measurements were funded by the Chief Scientist of the Ministry of Agriculture, Israel, under grant 20-21-0006 and by the Ministry of Science and Technology, Israel, under grant numbers 3-14559, 3-15605.

Institutional Review Board Statement: Not applicable.

Informed Consent Statement: Not applicable.

Data Availability Statement: Sentinel-2 data were obtained from the ESA Copernicus Open Access Hub website (https://scihub.copernicus.eu/dhus/#/home, accessed on 26 April 2021).

Acknowledgments: We thank Bar Avni-Naor, Nitai Heymann, Lior Fine, Victor Lukyanov, and Nitzan Malachy for their contribution to data processing and fieldwork. We also thank all the growers.

Conflicts of Interest: The authors declare no conflict of interest.

Appendix A

Table A1. Sentinel-2 images inventory and LAI measurements data used in the study.

	Gadash 2019			Gadot 2019			Gadot 2020			Megido 2020			Saad 2018			Yavne 2019		
	Tomato			Tomato			Tomato			Cotton			Wheat			Wheat		
	Sentinel-2 Images	LAI Measurements	LAI Value	Sentinel-2 Images	LAI Measurements	LAI Value	Sentinel-2 Images	LAI Measurements	LAI Value	Sentinel-2 Images	LAI Measurements	LAI Value	Sentinel-2 Images	LAI Measurements	LAI Value	Sentinel-2 Images	LAI Measurements	LAI Value
1	21 May	16 May	0.6	21 May	16 May	0.3	20 May	20 May	0.8	30 May [2]	25 May [2]	0.3 [2]	2 March	1 March	7.0	11 January	6 January	3.5
2	26 May	28 May	2.5	26 May	28 May	1.3	30 May	27 May	2.3	4 June [2]	9 June [2]	1.1 [2]	7 March	14 March	7.1	16 January	20 February	6.6
3	31 May	12 June	3.0	31 May	4 June	1.8	4 June	14 June	6.2	9 June [2]	17 June [2]	2.3 [2]	17 March	25 March	7.5	21 January	7 March	7.7
4	10 June [1]	27 June	3.6	10 June [1]	12 June	3.6	9 June	24 June	8.8	14 June [2]	2 July [2]	5.3 [2]	22 March	9 April	5.0	26 January	21 March	6.8
5	25 June	10 July	3.5	25 June	27 June	8.2	14 June	14 July	7.8	4 July 20 [2]	19 July [2]	9.6 [2]	27 March			31 January	28 March	7.3
6	5 July	25 July	4.7	5 July	3 July	8.9	24 June	21 July	5.0	14 July [2]	2 August [2]	8.7 [2]	6 April			5 February	7 April	6.4
7	15 July			15 July	10 July	9.7	29 June			19 July [2]						25 February	11 April	4.4
8	20 July			20 July	25 July	5.6	4 July			24 July [2]						11 April		
9	25 July			25 July	7 August	4.8	14 July			29 July [2]								
10				30 July	11 August	5.8	19 July			30 May [3]	25 May [3]	0.6 [3]						
11				4 August	15 August	4.9				4 June [3]	9 June [3]	1.3 [3]						
12				9 August						9 June [3]	17 June [3]	2.2 [3]						
13				14 August						14 June [3]								

[1] A defective red edge band in a Sentinel-2 image acquired on 10 June 2019 prevented the derivation of red edge-based models for that date. [2] Polygon in the centre of the field. [3] NW polygon.

References

1. Ewert, F. Modelling plant responses to elevated CO2: How important is leaf area index? *Ann. Bot.* **2004**, *93*, 619–627. [CrossRef]
2. Herrmann, I.; Pimstein, A.; Karnieli, A.; Cohen, Y.; Alchanatis, V.; Bonfil, D.J. LAI assessment of wheat and potato crops by VENµS and Sentinel-2 bands. *Remote Sens. Environ.* **2011**, *115*, 2141–2151. [CrossRef]
3. Nguy-Robertson, A.L.; Peng, Y.; Gitelson, A.A.; Arkebauer, T.J.; Pimstein, A.; Herrmann, I.; Karnieli, A.; Rundquist, D.C.; Bonfil, D.J. Estimating green LAI in four crops: Potential of determining optimal spectral bands for a universal algorithm. *Agric. For. Meteorol.* **2014**, *192–193*, 140–148. [CrossRef]
4. Heuvelink, E.; Bakker, M.J.; Elings, A.; Kaarsemaker, R.; Marcelis, L.F.M. Effect of leaf area on tomato yield. *Acta Hortic.* **2005**, *691*, 43–50. [CrossRef]
5. Sadeh, Y.; Zhu, X.; Dunkerley, D.; Walker, J.P.; Zhang, Y.; Rozenstein, O.; Manivasagam, V.S.; Chenu, K. Fusion of Sentinel-2 and PlanetScope time-series data into daily 3 m surface reflectance and wheat LAI monitoring. *Int. J. Appl. Earth Obs. Geoinf.* **2021**, *96*, 102260. [CrossRef]
6. Manivasagam, V.S.; Rozenstein, O. Practices for upscaling crop simulation models from field scale to large regions. *Comput. Electron. Agric.* **2020**, *175*, 105554. [CrossRef]
7. Simic Milas, A.; Romanko, M.; Reil, P.; Abeysinghe, T.; Marambe, A. The importance of leaf area index in mapping chlorophyll content of corn under different agricultural treatments using UAV images. *Int. J. Remote Sens.* **2018**, *39*, 5415–5431. [CrossRef]
8. Kang, Y.; Özdoğan, M.; Zipper, S.C.; Román, M.O.; Walker, J.; Hong, S.Y.; Marshall, M.; Magliulo, V.; Moreno, J.; Alonso, L.; et al. How Universal Is the Relationship between Remotely Sensed Vegetation Indices and Crop Leaf Area Index? A Global Assessment. *Remote Sens.* **2016**, *8*, 597. [CrossRef]
9. Kamenova, I.; Dimitrov, P. Evaluation of Sentinel-2 vegetation indices for prediction of LAI, fAPAR and fCover of winter wheat in Bulgaria. *Eur. J. Remote Sens.* **2020**, *54*, 89–108. [CrossRef]
10. Sener, M.; Arslanoğlu, M.C. Selection of the most suitable Sentinel-2 bands and vegetation index for crop classification by using Artificial Neural Network (ANN) and Google Earth Engine (GEE). *Fresenius Environ. Bull.* **2019**, *28*, 9348–9358.
11. Kaplan, G.; Fine, L.; Lukyanov, V.; Manivasagam, V.S.; Malachy, N.; Tanny, J.; Rozenstein, O. Estimating Processing Tomato Water Consumption, Leaf Area Index, and Height Using Sentinel-2 and VENµS Imagery. *Remote Sens.* **2021**, *13*, 1046. [CrossRef]
12. Asrar, G.; Fuchs, M.; Kanemasu, E.T.; Hatfield, J.L. Estimating Absorbed Photosynthetic Radiation and Leaf Area Index from Spectral Reflectance in Wheat. *Agron. J.* **1984**, *76*, 300. [CrossRef]
13. Reichenau, T.G.; Korres, W.; Montzka, C.; Fiener, P.; Wilken, F.; Stadler, A.; Waldhoff, G.; Schneider, K. Spatial heterogeneity of Leaf Area Index (LAI) and its temporal course on arable land: Combining field measurements, remote sensing and simulation in a Comprehensive Data Analysis Approach (CDAA). *PLoS ONE* **2016**, *11*, e0158451. [CrossRef]
14. Sun, Y.; Qin, Q.; Ren, H.; Zhang, T.; Chen, S. Red-Edge Band Vegetation Indices for Leaf Area Index Estimation from Sentinel-2/MSI Imagery. *IEEE Trans. Geosci. Remote Sens.* **2020**, *58*, 826–840. [CrossRef]
15. Zheng, G.; Moskal, L.M. Retrieving Leaf Area Index (LAI) Using Remote Sensing: Theories, Methods and Sensors. *Sensors* **2009**, *9*, 2719–2745. [CrossRef] [PubMed]
16. Xavier, A.C.; Vettorazzi, C.A. Monitoring leaf area index at watershed level through NDVI from Landsat-7/ETM+ data. *Sci. Agric.* **2004**, *61*, 243–252. [CrossRef]
17. Ho, P.-G.P. (Ed.) *Geoscience and Remote Sensing*, 1st ed.; InTech: Vukovar, Croatia, 2009; ISBN 978-953-307-003-2.
18. Lemoine, G.; Defourny, P.; Gallego, J.; Davidson, A.M.; Fisette, T.; Mcnairn, H.; Daneshfar, B.; Ray, S.; Neetu; Rojas, O.; et al. *Handbook on Remote Sensing for Agricultural Statistics*; GSARS: Rome, Italy, 2017. Available online: http://www.fao.org/3/ca6394en/ca6394en.pdf (accessed on 8 May 2021).
19. Horler, D.N.H.; Dockray, M.; Barber, J. The red edge of plant leaf reflectance. *Int. J. Remote Sens.* **1983**, *4*, 273–288. [CrossRef]
20. Cui, Z.; Kerekes, J.P. Potential of red edge spectral bands in future landsat satellites on agroecosystem canopy green leaf area index retrieval. *Remote Sens.* **2018**, *10*, 1458. [CrossRef]
21. Verrelst, J.; Rivera, J.P.; Veroustraete, F.; Muñoz-Marí, J.; Clevers, J.G.P.W.; Camps-Valls, G.; Moreno, J. Experimental Sentinel-2 LAI estimation using parametric, non-parametric and physical retrieval methods-A comparison. *ISPRS J. Photogramm. Remote Sens.* **2015**, *108*, 260–272. [CrossRef]
22. Kganyago, M.; Mhangara, P.; Alexandridis, T.; Laneve, G.; Ovakoglou, G.; Mashiyi, N. Validation of Sentinel-2 leaf area index (LAI) product derived from SNAP toolbox and its comparison with global LAI products in an African semi-arid agricultural landscape. *Remote Sens. Lett.* **2020**, *11*, 883–892. [CrossRef]
23. Xie, Q.; Dash, J.; Huete, A.; Jiang, A.; Yin, G.; Ding, Y.; Peng, D.; Hall, C.C.; Brown, L.; Shi, Y.; et al. Retrieval of crop biophysical parameters from Sentinel-2 remote sensing imagery. *Int. J. Appl. Earth Obs. Geoinf.* **2019**, *80*, 187–195. [CrossRef]
24. Mutanga, O.; Skidmore, A.K. Narrow band vegetation indices overcome the saturation problem in biomass estimation. *Int. J. Remote Sens.* **2004**, *25*, 3999–4014. [CrossRef]
25. Frampton, W.J.; Dash, J.; Watmough, G.; Milton, E.J. Evaluating the capabilities of Sentinel-2 for quantitative estimation of biophysical variables in vegetation. *ISPRS J. Photogramm. Remote Sens.* **2013**, *82*, 83–92. [CrossRef]
26. Delegido, J.; Verrelst, J.; Meza, C.M.; Rivera, J.P.; Alonso, L.; Moreno, J. A red-edge spectral index for remote sensing estimation of green LAI over agroecosystems. *Eur. J. Agron.* **2013**, *46*, 42–52. [CrossRef]

27. Revill, A.; Florence, A.; MacArthur, A.; Hoad, S.P.; Rees, R.M.; Williams, M. The value of Sentinel-2 spectral bands for the assessment of winter wheat growth and development. *Remote Sens.* **2019**, *11*, 2050. [CrossRef]
28. Amin, E.; Verrelst, J.; Rivera-Caicedo, J.P.; Pipia, L.; Ruiz-Verdú, A.; Moreno, J. Prototyping Sentinel-2 green LAI and brown LAI products for cropland monitoring. *Remote Sens. Environ.* **2021**, *255*, 112168. [CrossRef]
29. Sun, Y.; Qin, Q.; Ren, H.; Zhang, Y. Decameter Cropland LAI/FPAR Estimation From Sentinel-2 Imagery Using Google Earth Engine. *IEEE Trans. Geosci. Remote Sens.* **2021**. [CrossRef]
30. Beeri, O.; Netzer, Y.; Munitz, S.; Mintz, D.F.; Pelta, R.; Shilo, T.; Horesh, A.; Mey-tal, S. Kc and LAI estimations using optical and SAR remote sensing imagery for vineyards plots. *Remote Sens.* **2020**, *12*, 3478. [CrossRef]
31. Verma, A.; Kumar, A.; Lal, K. Kharif crop characterization using combination of SAR and MSI Optical Sentinel Satellite datasets. *J. Earth Syst. Sci.* **2019**, *128*, 230. [CrossRef]
32. Revill, A.; Florence, A.; Macarthur, A.; Hoad, S.; Rees, R.; Williams, M. Quantifying uncertainty and bridging the scaling gap in the retrieval of leaf area index by coupling Sentinel-2 and UAV observations. *Remote Sens.* **2020**, *12*, 1843. [CrossRef]
33. Louis, J.; Debaecker, V.; Pflug, B.; Main-Knorn, M.; Bieniarz, J.; Mueller-Wilm, U.; Cadau, E.; Gascon, F. Sentinel-2 Sen2cor: L2a processor for users. *Proceedings Living Planet Symposium 2016*. Available online: https://elib.dlr.de/107381/1/LPS2016_sm10_3louis.pdf (accessed on 8 May 2021).
34. Tucker, C.J. Red and photographic infrared linear combinations for monitoring vegetation. *Remote Sens. Environ.* **1979**, *8*, 127–150. [CrossRef]
35. Van Beek, J.; Tits, L.; Somers, B.; Coppin, P. Stem Water Potential Monitoring in Pear Orchards through WorldView-2 Multispectral Imagery. *Remote Sens.* **2013**, *5*, 6647–6666. [CrossRef]
36. Dash, J.; Curran, P.J. Evaluation of the MERIS terrestrial chlorophyll index (MTCI). *Adv. Space Res.* **2007**, *39*, 100–104. [CrossRef]
37. Clevers, J.G.P.W. Application of a weighted infrared-red vegetation index for estimating Leaf Area Index by Correcting for Soil Moisture. *Remote Sens. Environ.* **1989**, *29*, 25–37. [CrossRef]
38. Huete, A.; Didan, K.; Miura, T.; Rodriguez, E.P.; Gao, X.; Ferreira, L.G. Overview of the radiometric and biophysical performance of the MODIS vegetation indices. *Remote Sens. Environ.* **2002**, *83*, 195–213. [CrossRef]
39. Huete, A.R. A soil-adjusted vegetation index (SAVI). *Remote Sens. Environ.* **1988**, *25*, 295–309. [CrossRef]
40. Qi, J.; Chehbouni, A.; Huete, A.R.; Kerr, Y.H.; Sorooshian, S. A modified soil adjusted vegetation index. *Remote Sens. Environ.* **1994**, *48*, 119–126. [CrossRef]
41. Pasqualotto, N.; Delegido, J.; Van Wittenberghe, S.; Rinaldi, M.; Moreno, J. Multi-crop green LAI estimation with a new simple Sentinel-2 LAI index (SeLI). *Sensors* **2019**, *19*, 904. [CrossRef]
42. Meyer, L.H.; Heurich, M.; Beudert, B.; Premier, J.; Pflugmacher, D. Comparison of Landsat-8 and Sentinel-2 data for estimation of leaf area index in temperate forests. *Remote Sens.* **2019**, *11*, 1160. [CrossRef]
43. Xie, Q.; Dash, J.; Huang, W.; Peng, D.; Qin, Q.; Mortimer, H.; Casa, R.; Pignatti, S.; Laneve, G.; Pascucci, S.; et al. Vegetation Indices Combining the Red and Red-Edge Spectral Information for Leaf Area Index Retrieval. *IEEE J. Sel. Top. Appl. Earth Obs. Remote Sens.* **2018**, *11*, 1482–1492. [CrossRef]
44. Clevers, J.; Kooistra, L.; van den Brande, M. Using Sentinel-2 Data for Retrieving LAI and Leaf and Canopy Chlorophyll Content of a Potato Crop. *Remote Sens.* **2017**, *9*, 405. [CrossRef]
45. Lanfri, S. *Vegetation Analysis Using Remote Sensing*; Cordoba National University (UNC): Cordoba, Argentina, 2010.
46. Broge, N.H.; Thomsen, A.G.; Andersen, P.B. Comparison of selected vegetation indices as indicators of crop status. *Geoinf. Eur. Integr.* **2003**, 591–596.
47. Rozenstein, O.; Haymann, N.; Kaplan, G.; Tanny, J. Estimating cotton water consumption using a time series of Sentinel 2 imagery. *Agric. Water Manag.* **2018**, *207*, 44–52. [CrossRef]
48. Rozenstein, O.; Haymann, N.; Kaplan, G.; Tanny, J. Validation of the cotton crop coefficient estimation model based on Sentinel-2 imagery and eddy covariance measurements. *Agric. Water Manag.* **2019**, *223*, 105715. [CrossRef]
49. Clevers, J.G.P.W.; Gitelson, A.A. Remote estimation of crop and grass chlorophyll and nitrogen content using red-edge bands on Sentinel-2 and-3. *Int. J. Appl. Earth Obs. Geoinf.* **2013**, *23*, 344–351. [CrossRef]
50. Viña, A.; Gitelson, A.A.; Nguy-Robertson, A.L.; Peng, Y. Comparison of different vegetation indices for the remote assessment of green leaf area index of crops. *Remote Sens. Environ.* **2011**, *115*, 3468–3478. [CrossRef]
51. Weiss, M.; Baret, F. S2ToolBox Level 2 Products: LAI, FAPAR, FCOVER. Available online: http://step.esa.int/docs/extra/ATBD_S2ToolBox_L2B_V1.1.pdf (accessed on 21 February 2021).
52. Zaeen, A.A.; Sharma, L.; Jasim, A.; Bali, S.; Buzza, A.; Alyokhin, A. In-season potato yield prediction with active optical sensors. *Agrosyst. Geosci. Environ.* **2020**, *3*, e20024. [CrossRef]
53. Manivasagam, V.S.; Kaplan, G.; Rozenstein, O. Developing Transformation Functions for VENµS and Sentinel-2 Surface Reflectance over Israel. *Remote Sens.* **2019**, *11*, 1710. [CrossRef]

 land

Review

Urban Heat Island and Its Regional Impacts Using Remotely Sensed Thermal Data—A Review of Recent Developments and Methodology

Hua Shi [1,*], George Xian [2], Roger Auch [2], Kevin Gallo [3] and Qiang Zhou [1]

1 ASRC Federal Data Solutions (AFDS), Contractor to the U.S. Geological Survey (USGS) Earth Resources Observation and Science (EROS) Center, Sioux Falls, SD 57198, USA; qzhou@contractor.usgs.gov
2 U.S. Geological Survey (USGS) Earth Resources Observation and Science (EROS) Center, Sioux Falls, SD 57198, USA; xian@usgs.gov (G.X.); auch@usgs.gov (R.A.)
3 Center for Satellite Applications and Research, National Oceanic and Atmospheric Administration (NOAA)/NESDIS, College Park, MD 20740, USA; kevin.p.gallo@noaa.gov
* Correspondence: hshi@contractor.usgs.gov; Tel.: +1-605-594-6050

Abstract: Many novel research algorithms have been developed to analyze urban heat island (UHI) and UHI regional impacts (UHIRIP) with remotely sensed thermal data tables. We present a comprehensive review of some important aspects of UHI and UHIRIP studies that use remotely sensed thermal data, including concepts, datasets, methodologies, and applications. We focus on reviewing progress on multi-sensor image selection, preprocessing, computing, gap filling, image fusion, deep learning, and developing new metrics. This literature review shows that new satellite sensors and valuable methods have been developed for calculating land surface temperature (LST) and UHI intensity, and for assessing UHIRIP. Additionally, some of the limitations of using remotely sensed data to analyze the LST, UHI, and UHI intensity are discussed. Finally, we review a variety of applications in UHI and UHIRIP analyses. The assimilation of time-series remotely sensed data with the application of data fusion, gap filling models, and deep learning using the Google Cloud platform and Google Earth Engine platform also has the potential to improve the estimation accuracy of change patterns of UHI and UHIRIP over long time periods.

Keywords: urban heat island; UHI regional impacts; non-urban areas; remote sensing; thermal band; UHI intensity

Citation: Shi, H.; Xian, G.; Auch, R.; Gallo, K.; Zhou, Q. Urban Heat Island and Its Regional Impacts Using Remotely Sensed Thermal Data—A Review of Recent Developments and Methodology. *Land* **2021**, *10*, 867. https://doi.org/10.3390/land10080867

Academic Editors: Sara Venafra, Carmine Serio and Guido Masiello

Received: 17 July 2021
Accepted: 11 August 2021
Published: 18 August 2021

1. Introduction

Urbanization is known to have substantial impacts on landscapes and ecosystems [1–4], and urban inhabitants are expected to reach 70% of the world population by 2050 [5]. Moreover, the nature of urban development has been changing from a single city model to a group of cities (urban agglomeration) worldwide. Urban heat island (UHI), urbanization, and climate change are increasingly interconnected, resulting in several environmental consequences (such as heat stress, biodiversity loss, fire risk, warming water due to run off, and diminished air quality) at both local and regional levels [2,6–9]. Such UHI related impacts are also called UHI regional impacts (UHIRIP). Generally, UHI research includes data from two major sources: air temperature data that are observed by weather or climate stations and remotely sensed data to observe UHI through land surface temperature. Before the availability of remotely sensed data, UHI was widely observed in the field, with the first scientific observation of UHI in 1833 [10]. Field observations of UHI continue to be a critical source of training and validation data [11,12]. These observations, along with modeling studies, continue to help unravel the factors that are responsible for UHI development, and are providing a basis for the development and application of sustainable adaptation strategies. Communicating scientific knowledge quickly and effectively of UHI and UHIRIP to architects, engineers, scientists, and planners could help inform urban design and decision making. Remotely sensed data have been used to observe UHI and UHIRIP on environments, ecosystems, human health, and economics in urban and non-urban areas for decades. Remote sensing offers the benefits of long data archives, repeated observations, efficiency, and multiple temporal and spatial resolutions. UHI studies using remotely sensed data have been published for hundreds of cities worldwide [6,7,13–19]. Remotely sensed data provide highly efficient, long-term, and broad-scale information for assessing UHIRIP. However, studies integrating high spatial resolution imagery (e.g., Landsat at 30 × 30 m and ECOSTRESS at 70 × 70 m) from multiple sensors to evaluate UHI and UHIRIP across a time series have been uncommon. Challenges to such studies include image frequency and calibration, cloud contamination, and the need for large storage and high-performance computing capabilities [20,21]. Early generations of broad-scale UHI assessment using remote sensing often poorly represented the spatial and temporal variance in UHI, especially at the urban and non-urban interface. As the resolution of algorithms and satellite imagery improved and interest in UHIRIP grew, researchers sought better representations of UHI. Initially, this took the form of modifications based on surface physical characteristics such as roughness length, albedo, thermal conductivity, and thermal diffusivity [22,23]. Many studies have been conducted to understand the urban thermal climate or the potential for heat island mitigation using this framework of simplified algorithms [24–26]. In more recent efforts, researchers have incorporated more sophisticated parameterization schemes that have included distributions of demography, policies, and behavior of government; ecological variables and ecosystem services; land use and land cover change (LULCC) patterns; and social and economic factors to represent the complicated impacts of UHI [27–36].

Historically, the study of UHI using remote sensing data, often Landsat data, was mainly based on comparing images at two different times using the bitemporal approach [37–39]. Although the bitemporal approach is mathematically simple and does not need large amounts of data, it is less useful than a time series approach that is able to provide a more comprehensive understanding of the complexity of UHI. Most early research [17,40–42] in UHI focused on cities or urban areas, and often ignored the urban and non-urban interface at regional scales. In recent decades, the cost of data storage has dramatically decreased, and we have witnessed an overwhelming increase in computing power and open source software that provide the foundations for time series analysis using higher resolution thermal data from satellite archives. Some studies used Landsat time series to detect historical changes [20,43–46], but few have focused on UHI and its interaction with land use and land cover (LULC) dynamics. A research team at the USGS Earth Resources Observation and Science (EROS) Center recently developed the Land Change Monitoring, Assessment, and Projection (LCMAP) project [47], which is produced with Landsat Analysis Ready Data (ARD) [48] and land surface temperature (LST) data. LCMAP data provide the potential to use Landsat LST data to analyze UHI in urban agglomerations, as well as the urban and non-urban interface at local, regional, and global scales.

This paper reviews remote sensing thermal data sources and the most up-to-date methods used for UHI and UHIRIP investigations. We start by defining UHI, UHII (UHI intensity), regional impacts, urban and non-urban interface, and remotely sensed data sources for LST. We then describe the major distinct approaches that have been used to estimate the magnitude, spatial distribution, intensity, and change pattern of UHIRIP in urban agglomerations and at different urban and non-urban interfaces. Our primary goals in this review are to describe (1) a brief historical summary in the research of UHI and UHIRIP, (2) major thermal data sources and methods used in UHI and UHIRIP research, (3) algorithms used in UHI and UHIRIP analysis, and (4) future research perspective and potential direction. Following the introduction, we discuss the development of UHI and UHIRIP in Section 2; in Section 3, we focus on the application of the remotely sensed thermal datasets in UHI and UHIRIP; we review the algorithms for UHI and UHIRIP in urban and non-urban interface studies based on remotely sensed data in Section 4; in

Section 5, we summarize UHI and UHIRIP based on remotely sensed data; and in Section 6, future research directions are discussed.

2. Development of UHI and UHIRIP Analysis

Most satellite-based investigations of UHIs can be summarized into five main objectives: (1) to examine the spatial features of urban thermal patterns and change dynamics and their relations to urban surface characteristics; (2) to study urban surface energy balances through coupling with urban climate models, including simulation and projection; (3) to study the relations between atmospheric heat islands and surface UHIs through combining coincident remote and ground-based observations; (4) to develop approaches to reduce the magnitude of the UHI and its regional impacts; and (5) to study the UHI effects on ecosystem security at a regional level. Several important reviews, bibliographies, and summaries on UHIRIP using remotely sensed data have been published (see list and descriptions in Table 1). These reviews have concentrated mostly on the various worldwide perspectives of UHI, including the definition of fundamental concepts, summary of methods, applications, exploration of output characteristics, outlines of key research findings, and potential future directions (Tables 2 and 3). The focus of this paper is on the algorithms and methods used in studies employing remote sensing thermal data for UHI and UHIRIP investigation, and future directions in this realm. We summarize (1) the disadvantages of using limited time remotely sensed data for UHI and UHIRIP analysis; (2) the limitations of data shortages due to cloud cover and satellite revisit intervals; (3) the applications of gap filling, data fusion, and deep learning; and (4) the trade-offs between high temporal frequency data (MODIS) and high spatial resolution (Landsat) time series.

Table 1. Example of main reviews, bibliographies, and summaries on UHI and UHIRIP using remotely sensed data.

Reference	Topics	Sensors	Measurements
Hall et al. [11]	Satellite remote sensing of surface energy balance success, failures, and unresolved issues in field experiment (FIFE)	Landsat, SPOT	Thermal
Gallo et al. [13]	Assessment of urban heat islands: A satellite perspective	AVHRR, Landsat MSS	Thermal
Voogt and Oke [6]	Thermal remote sensing of urban climates	Multiple, review	Thermal
Weng and Larson [49]	Satellite remote sensing of urban heat islands: current practice and prospects	Multiple, review	Thermal
Jiang et al. [50]	Land surface emissivity retrieval from combined mid-infrared and thermal infrared data of MSG-SEVIRI	Meteosat Second Generation (MSG)	Spinning Enhanced Visible and Infrared Imager (SEVIRI)
Kalma et al. [51]	Estimating land surface evaporation: A review of methods using remotely sensed surface temperature data	Multiple, review	Thermal
Racoviteanu et al. [52]	Optical remote sensing of glacier characteristics: a review focusing on the Himalaya	ASTER	Indices
Rizwan et al. [53]	A review on the generation, determination, and mitigation of urban heat island	Review	Determination of UHI
Weng [7]	Thermal infrared remote sensing for urban climate and environmental studies: Methods, applications, and trends	Multiple, review	Thermal
Bowler et al. [31]	Urban greening to cool towns and cities: A systematic review of the empirical evidence	Review	Synthesis analysis
Sailor [54]	A review of methods for estimating anthropogenic heat and moisture emissions in the urban environment	Review	Bibliometric profile
Li et al. [55]	Satellite-derived land surface temperature: current status and perspectives	Multiple, review	Thermal

Table 1. *Cont.*

Reference	Topics	Sensors	Measurements
Ngie et al. [56]	Assessment of urban heat island using satellite remotely sensed imagery: A review	Multiple, review	Thermal
Rasul et al. [16]	A review of remote sensing of urban heat and cool islands	Multiple, review	Thermal
Huang and Lu [57]	Urban heat island research from 1991 to 2015: A bibliometric analysis	Review	Bibliometric profile
Zhang et al. [58]	A bibliometric profile of the remote sensing open access journal published by MDPI between 2009 and 2018	Multiple, review	Bibliometric profile
Deilami et al. [59]	Urban heat island effect: A systematic review of spatio-temporal factors, data, methods, and mitigation measures	Multiple, review	Thermal
Zhou et al. [60]	Satellite remote sensing of surface urban heat islands: Progress, challenges, and perspectives	Multiple, review	Thermal
Becker and Zhao-Liang [61]	Surface temperature and emissivity at various scales: definition, measurement, and related problems	Multiple, review	Thermal (surface temperature and emissivity)
Dash et al. [62]	Land surface temperature and emissivity estimation from passive sensor data: theory and practice current trends	Multiple, review	Thermal (surface temperature and emissivity)

Table 2. Examples of research publications investigating UHI and UHIRIP using remotely sensed data.

UHI Applications	Example of Research
Classification with LST, index, albedo	Miles and Esau [63], Trlica et al. [64], Bonafoni [65], Wong and Nichol [66], Jin [67], Wu et al. [68], and Hu and Brunsell [69]
Regression models, geostatistical analysis	Zhang and Du [70], Wicki and Parlow [71], Dai et al. [72], Song et al. [73], Sellers et al. [74], Du et al. [75], Shahraiyni et al. [76], Chun and Guldmann [77], Ho et al. [78], and Lai et al. [79]
Multiple sensors, data fusion	Huang and Wang [80], Li et al. [81], Berger et al. [82], Liu et al. [83], Fu and Weng [84], Liang and Weng [85], and Dousset and Gourmelon [86]
Machine learning, decision support information system	Chakraborty and Lee [87], Mpakairia and Muvengwi [88], Zhang et al. [89], Tran et al. [90], Shahraiyni et al. [76], Weng and Fu [91], Mallick et al. [92], Connors et al. [93], Wentz et al. [94], Xian and Crane [95], Wilson et al. [96], and Xian et al. [97]

Table 3. The temporal frequency and spatial resolution of the main remotely sensed thermal data.

Sensor	Temporal Frequency (day)	Spatial Resolution (m)
Landsat 5 TM	16	120 (resampled to 30)
Landsat 7 ETM+	16	60 (resampled to 30)
Landsat 8 TIRS	16	100 (resampled to 30)
Terra ASTER	15	90
Terra MODIS	1	1000
Aqua MODIS	1	1000
NOAA-AVHRR	1	1000
VIIRS	1	750
ECOSTRESS	Various (randomly, 0.5)	70

2.1. Urban Heat Island

UHI studies have been conducted for over 200 years, since the first conceptualization by Luke Howard in 1818 [98]. Generally, an urban heat island (UHI) is an urban area or metropolitan area that is significantly warmer than its surrounding rural areas because of human activities. The temperature difference is usually greater at night than during the day and is most apparent when winds are weak. Some research [99,100] shows that the annual mean air temperature of a city with 1 million people or more can be 1–3 °C warmer than its surroundings. In the evening, the difference can be as high as 12 °C. Heat islands can affect communities by increasing the summertime peak energy demand (such as air conditioning costs), air pollution and greenhouse gas emissions, heat-related illness, and mortality, and decreasing water quality and ecosystem security. Higher temperature "domes" are created over an urban or industrial areas by hot air layers forming at building-top or chimney-top level. This dome is about 5 °C to 7 °C warmer than the air above it and the ground level temperature, and can trap all polluting emissions within its confines (see also temperature inversion [53,57]).

The large amount of heat generated from urban structures and pavements, as they absorb and re-radiate solar radiation, as well as the heat from other anthropogenic sources, are the main causes of UHI. These heat sources increase the temperatures of an urban area compared with its surroundings, which is known as UHI intensity (UHII). Traditionally, regardless of the methodology employed, whether it refers to (1) differences between two fixed observatories, one urban and another peripheral or non-urban; (2) mobile urban transects; or (3) remote sensing analysis, UHII provides a value of thermal differences between contrasted points, sectors, or areas, one urban and another that could be termed non-urban. Thus, the intensity of the UHI is seen in the temperature difference expressed at a given time between the hottest sector (areas) of the city and the surrounding non-urban space. The intensity of the heat island is the simplest and most quantitative indicator of the thermal modification imposed by the city upon the territory in which it is situated and of its relative warming in relation to the surrounding rural environment. The intensity could be defined for various time scales and geographical locations [101,102].

2.2. *The Study of the Spatial Structure of Urban Thermal Patterns, Change Dynamics, and Their Relation to Urban Surface Characteristics*

Based on the fractional theory of ecology [103,104], the spatial structure of urban thermal patterns and temporal change dynamics can be studied in two and three dimensions. Figure 1 shows an example of the UHI and UHIRIP profile in Sioux Falls, South Dakota, USA, and the surrounding area, derived from Landsat ARD LST over different land cover classes [97]. The study of temporal change in UHI can include multiple scales of change, including daily, day and night, monthly, seasonal, yearly, and long-term time series. The physical mechanisms driving UHI are well documented [28]. UHIRIP may be described in multiple ways with various methodological approaches to investigate each type; specifically, it can impact the ground, the surface, and various heights in the air [105,106] at a regional scale. Different pictures arise for each type of UHI when measured by different methods. Tam et al. [107] suggested that the magnitude of total change in day to day temperature variability can be used to decide a suitable urban/rural pair for any urbanization impact study. Generally, the UHI at a regional scale is best measured using remotely sensed data with one or multiple thermal bands. When explaining the character of remotely sensed UHI, Roth et al. [108] assert, "satellite-derived surface heat islands are in a separate class and it is not clear that they will match others measured by more conventional means in the urban canopy layer or the urban boundary layer". Their precautionary statement relates in part to the surface "seen" by remote sensing platforms that depend on altitude and the camera or sensor angle. Imagery collected at nadir and/or high altitude primarily consists of rooftops, streets, crop fields, and vegetation canopies. Observations from lower heights at oblique angles consist of items seen from a bird's-eye perspective plus varying degrees of vertical features in the landscape, such as the walls of buildings. As a result, angle can have a large influence on the urban surface temperatures recorded by airborne and spaceborne thermal infrared sensors [109]. Another concern regards mixed pixels (i.e., individual pixels containing surfaces having different physical properties, depending on the spatial resolution of the data), which can complicate image analysis. This is especially true for thermal sensors aboard satellites, because most have a spatial resolution that is coarser than the other spectral bands on the satellite. The typical variation of urban surface properties also complicates thermal sensors. A final consideration when using remotely sensed imagery involves correcting for atmospheric attenuation. For many applications, these issues are far outweighed by remote sensing's benefits. With high spatial resolution thermal data, these issues can typically be resolved. Additionally, from a macro research perspective, remotely sensed thermal data have the major advantage of investigating UHI and UHIRIP at a broad scale, permitting focus on environmental issues in urban agglomerations and surrounding areas, and at urban and non-urban interfaces.

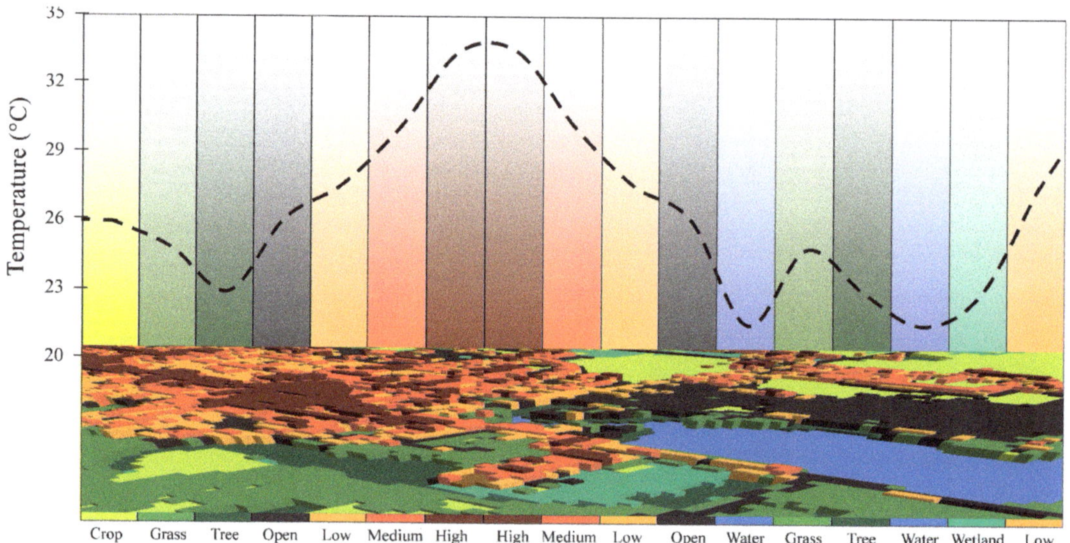

Figure 1. An example of UHI and UHIRIP in the urban and the urban and non-urban interface for part of the Sioux Falls, SD area.

2.3. Simulation and Projection of UHI and UHIRIP

Applying theories of landscape ecology [104], UHI studies focus on moving from static spatial structures of urban thermal patterns to the change dynamic of spatial patterns and processes of urban thermal characteristics. The spatial structure of UHI patterns determines the processes of UHI impacts. Li et al. [110] simulated the urban climate of various generated cities under the same weather conditions. By studying various city shapes, they generalized and proposed a reduced form to estimate UHI intensities based only on the structure of urban sites, as well as their relative distances. They concluded that in addition to the size, the UHI intensity of a city is directly related to the density and the amplifying effect that urban sites have on each other. Their approach can serve as a UHI rule of thumb for the comparison of urban development scenarios. Ramírez-Aguilar and Lucas Souza [111] present a study based on the relationship between UHI and population size (p) by considering the population density (PD) and the urban form parameters of different neighborhoods in the city of Bogotá, Colombia. They concluded that urban form, expressed by land cover and urban morphology changes caused by population density, has a great effect on temperature differences within a city. Advances in computing technology have fostered the development of new and powerful deep learning techniques that have demonstrated promising results in a wide range of applications. In particular, deep learning methods have been successfully used to classify remotely sensed data collected by Earth observation instruments [112]. Deep learning algorithms, which learn the representative and discriminative features in a hierarchical manner from the data, have recently become a hotspot in the machine-learning area, and have been introduced into the geoscience and remote sensing community for remotely sensed big data analysis [113]. With climate change, the simulation and projection of UHI and its regional impact by using computer technology (deep learning) and remotely sensed data are becoming more important for urban planning and policy makers.

2.4. Challenges for Land Surface Temperature and Emissivity Retrieval (Separation)

Land surface temperature and emissivity are two important surface parameters that can be derived from remotely sensed data after atmospheric correction [114–116]. Besides radiometric calibration and cloud detection, two main problems need to be resolved in order to obtain land surface temperature and emissivity values from various satellite sensors. These problems are often referred to as land surface temperature and emissivity separation (TES) from radiance at ground level, and as atmospheric corrections in the literature [117,118]. Reliable retrieval of urban and intra-urban thermal characteristics using satellite thermal data depends on accurate removal of the effects of atmospheric attenuations, as well as angular and land surface emissivity. In the thermal infrared of remotely sensed data, the emission of the targets is dominant when compared with the reflection, and this radiation is a function of two unknowns—the emissivity and the temperature of the target [119]. The temperature and emissivity separation is complex because of the existing non-linear relationship between temperature and radiance. The complex dynamics of these relationships within the target (atomic level) propagates in a cascade effect, reflecting variations in determining emissivity. Mohamed et al. [117] reviewed details of LST and land surface emissivity (LSE) retrieval methods and their potential for adoption in medium spatial resolution, including ASTER and Landsat. The review further comments on spatial and temporal prospects of effective intra-urban surface thermal mapping. They also suggested future development of land surface temperature and emissivity estimation for UHI assessment. Li et al. [120] described the theoretical basis of LSE measurements and reviewed the published methods. They also categorized these methods into (1) (semi-) empirical or theoretical methods, (2) multi-channel temperature emissivity separation (TES) methods, and (3) physically based methods (PBMs). Then, they discussed the validation methods that are important for verifying the uncertainty and accuracy of retrieved emissivity. Finally, the prospects for further developments are given. These studies provided a forum for assessing what had been achieved by the UHI community over four decades, and what needs to be done in the near future. It is clear that the observation, experiments, and algorithm development efforts, although completely worthwhile for scientists, need to deliver various datasets, especially from remotely sensed data to modelers working in the areas of UHI and UHIRIP at local, regional, and global levels. A lot of basic theoretical research and scientific verification work has been done on scale issues, as well as scaling issues including emissivity and temperature measurements related to remote sensing standards [121]. All of the methods described in Rolim et al. [119] represent the largest and main part of the existing methods of temperature and emissivity separation developed in the last four decades, but further research is necessary for more precise methods that are less susceptible to errors during the separation of these variables.

2.5. The Relationship between Atmospheric Heat Islands and Surface UHI through Combining Coincident Remote Sensing and Ground-Based Observations

Generally, UHI data are obtained from one of two sources—weather stations and remote sensing. Remotely sensed data have been used to observe how UHI impacts climate change in urban and non-urban areas for decades because of the multiple temporal and spatial resolutions of remotely sensed datasets. Hundreds of published studies explore UHI and its impacts by using these two data sources, but the relationship between air temperature obtained from field stations and surface temperature derived from remote sensing remains unclear. Wang et al. [122] investigated the relationship of canopy UHI (CUHI) and surface UHI (SUHI) using four observations per day, without temporal averaging, in four different cities in two different global regions, with 201 of 2232 CUHI–SUHI pairs exhibiting significant UHI differences in their spatial distributions and intensities. The results indicate that 81.09% of the UHI differences occurred during the daytime and were caused by local air advection related to wind speed $\geq$2 m/s and land surface conditions in the study areas. They concluded that a joint analysis of CUHIs and SUHIs should be conducted to characterize urban thermal environments, and that current urban planning procedures should integrate these UHI differences to develop effective mitigation strategies and adaptation measures. The combination of both types of UHI sub-components provides added value for quantifying urban thermal environments, which can assist in developing effective mitigation strategies and adaptation measures. A growing trend is to combine the two methods, both with their own advantages [59].

2.6. Develop Controlling Approaches for UHI and UHIRIP

UHIs occur when cities replace other land covers with dense concentrations of pavement, buildings, and other surfaces that absorb and retain heat. This effect increases energy costs (e.g., for air conditioning), air pollution levels, and heat-related illness and mortality. UHI results from increases in built-up surfaces in urban areas, whereas increasing vegetation cover and water surfaces within cities or urban agglomerations could improve the urban ecological function and thereby improve urban environments for humans [123]. The importance of optimizing urban LULC planning and the development of UHI mitigation methods is increasing. Progress has been made to this end [67,124,125], with the development of UHI mitigating technologies [126]. Ulpiani et al. [127] reviewed an infrared emissivity dynamic switch against overcooling, which is aimed at collecting state-of-the-art technologies and techniques to dynamically control the heat transfer to and from the radiative emitter and to ultimately modulate its cooling capacity using spacecraft thermal control, thermal camouflage, and electronics. This work discussed prominent pathways toward technically and economically effective integration in the built environment for UHI and UHIRIP.

2.7. UHI and UHIRIP on Socioeconomics and the Urban Ecosystem

2.7.1. Impacts on Human Health

Climate change, increasing urbanization, and an aging population in much of the world are likely to increase the risks to health from UHI, particularly from heat exposure. Additionally, increased urbanization has resulted in a more extensive UHI effects, causing more frequent and intense heat waves in urban regions compared with rural locales [67,128,129]. In urban and surrounding areas, heat waves will be exacerbated by the UHI effect and will have the potential to negatively influence the health and welfare of residents. Heaviside et al. [130] suggest that UHI contributed around 50% of the total heat-related mortality during the 2003 heat wave in the West Midlands of the UK. Moon [131] concluded that the mortality and morbidity risks of diabetic patients under the heat wave were mildly increased by about 18% for mortality and 10% for overall morbidity. Li et al. [132] found that high temperature significantly increases the risk of mortality in the population of Jinan, China. Most research in this topic uses both weather station and in situ measurements in order to investigate the health effects of UHI [129]. Some results [133] show that different sites (city center or surroundings) have experienced different degrees of warming as a result of increasing urbanization [131]. Johnson et al. [134] suggest that thermal remote sensing data can be utilized to improve the understanding of intra-urban variations of risk from extreme heat. The refinement of the current risk assessment systems could increase the likelihood of survival during extreme heat events and assist emergency personnel in the delivery of vital resources during such disasters. The conclusion is that UHI is directly linked to adverse health effects from exposure to extreme thermal conditions.

2.7.2. UHI and UHIRIP on LULC Differences and Change

UHI is a result of continued urbanization, urban agglomeration, and associated increases in paved areas and buildings. Mitigation strategies have been developed to increase vegetation and water surface areas within urban areas to reduce the magnitude of the temperature. One measure of UHI's ecological footprint is estimated by calculating the increase of the cooling demand caused by the heat island over the urban area, and then translating the increased energy use to environmental cost [123,125,135]. Some research shows that the UHI effect has become more prominent in areas of rapid urbanization and in urban agglomerations [136,137]. The spatial distribution of UHI has changed from a mixed pattern, where bare land, semi-bare land, and land under development were warmer than other LULC types, to extensive UHI, as contiguous urbanized blocks grew larger [38,138]. Some analyses showed that the higher temperature in the UHI had a scattered pattern and was related to certain LULC types [97]. In order to analyze the relationship between UHI and LULC changes, some studies attempted to employ a quantitative approach for exploring the relationship between surface temperature and several indices, including the Normalized Difference Vegetation Index (NDVI), Normalized Difference Water Index (NDWI), Normalized Difference Bareness Index (NDBaI), and Normalized Difference Build-up Index (NDBI). It was found that correlations between NDVI, NDWI, NDBI, and temperature are negative when NDVI and NDWI are limited in range, but there is a positive correlation between NDBI and temperature [139–142].

2.7.3. Impacts on Regional Economics

Because UHI is related to a significant increase in surface temperature and changes in precipitation patterns, it can potentially affect local economies and the social systems of cities [143]. Some studies [144,145] show that the critical sectors of services, agriculture, and tourism may be strongly affected by future UHIs. To counterbalance the consequences of the increased urban surface temperatures, important research has been carried out resulting in the development of efficient mitigation technologies. In particular, some studies [102,146] have documented the development of highly reflective materials, cool and green roofs, cool pavements, urban greens, water surface, and other mitigation technologies. UHIRIP includes economic impacts, such as increases of energy consumption for cooling purposes, as well as an increase in the peak electricity load, which is a factor for planning maximum power source capacities [147]. Scientists from Australia reported that the total economic cost to the community due to hot weather is estimated to be approximately $1.8 billion in present value terms. Approximately one-third of these impacts are due to heat waves. Of the total heat impact, the UHI effect contributes approximately $300 million (AUD) in present value terms for the city of Melbourne, Australia [9]. Estrada, Botzen and Tol [144] provided a quantitative assessment of the economic costs of the joint impacts of local and global climate change for all main cities around the world. They estimated the UHI effect for the 1692 largest cities in the world for the period 1950–2015, and predicted that the percentage of city gross domestic product (GDP) that would be lost for the median city in 2050 due to global climate change alone would be relatively small: 0.9% and 0.7% for the RCP8.5 and RCP4.5 emission scenarios, respectively [144]. At the end of the century, these impacts will increase to 3.9% and 1.2%, respectively. Cost–benefit analyses are presented of urban heat island mitigation options, including green and cool roofs and cool pavements. It has been shown that local actions can be climate risk-reduction instruments. Furthermore, limiting the urban heat island through city adaptation plans can substantially amplify the benefits of international mitigation efforts.

2.7.4. Impact on Biodiversity

Besides UHI, urban development causes wildlife habitat loss and fragmentation, threatens wildlife populations, increases fire risk, and reduces biodiversity [2,148]. These problems are of particular concern in the wildland urban interface (WUI), where homes and associated structures are built among forests, shrublands, or grasslands [1,148,149]. The WUI has received considerable attention because of recent increases in both the number of structures destroyed and the area burned annually by wildland fire. Čeplová et al. [150] studied three habitats with different disturbance regimes in 45 central European settlements of three different sizes. Their results highlight the importance of urban size as a factor shaping the biodiversity of native and alien plant communities in individual urban habitats, and the important role of habitat mosaic for maintaining high species richness in city floras. The study of Coluzzi et al. [151] represented a first step to improve the description of relevant processes to protect natural habitats and quality agriculture, therefore combating land degradation and detrimental climate change effects. Kaiser et al. [152] monitored temperature and relative air humidity in wooded sites characterized by different levels of urbanization in the surroundings, and investigated the effect of urbanization at the local and landscape scale on two key traits of biological fitness in two closely related butterfly species that differ in thermal sensitivity.

3. Remotely Sensed Thermal Datasets

Remote sensing derived LST is effective for UHI and UHIRIP studies. Satellites can quickly obtain continuous information over a large geographic area that can be maintained in long-term archives. LST for large geographic areas can be derived from surface radiation of heat measured by satellite sensors. This is particularly attractive when investigating the surface UHI in multiple cities or urban agglomerations at various spatial extents. Along with the extensive spatial coverage, many satellites record multiple wavelengths of electromagnetic energy that can be used to decipher a wealth of information, in addition to thermal information (Figure 2). Consequently, multispectral imagery allows for a comparative analysis between LST and other variables, such as land cover and vegetation indexes [50,153], specifically the interaction between UHI and LULCC [154]. Remote sensing can also be used to track the patterns of change in UHI over time through various time periods from a day, to years, and even a time series of decades [38,155–158]. Because information is desired at a high spatial resolution and dense temporal frequency, data from multiple sensors can make more accurate and reliable quantitative assessments of UHIRIP studies [60]. Table 3 includes a list of the main remotely sensed datasets that have been recently used to derive LST and analyze UHI and UHIRIP.

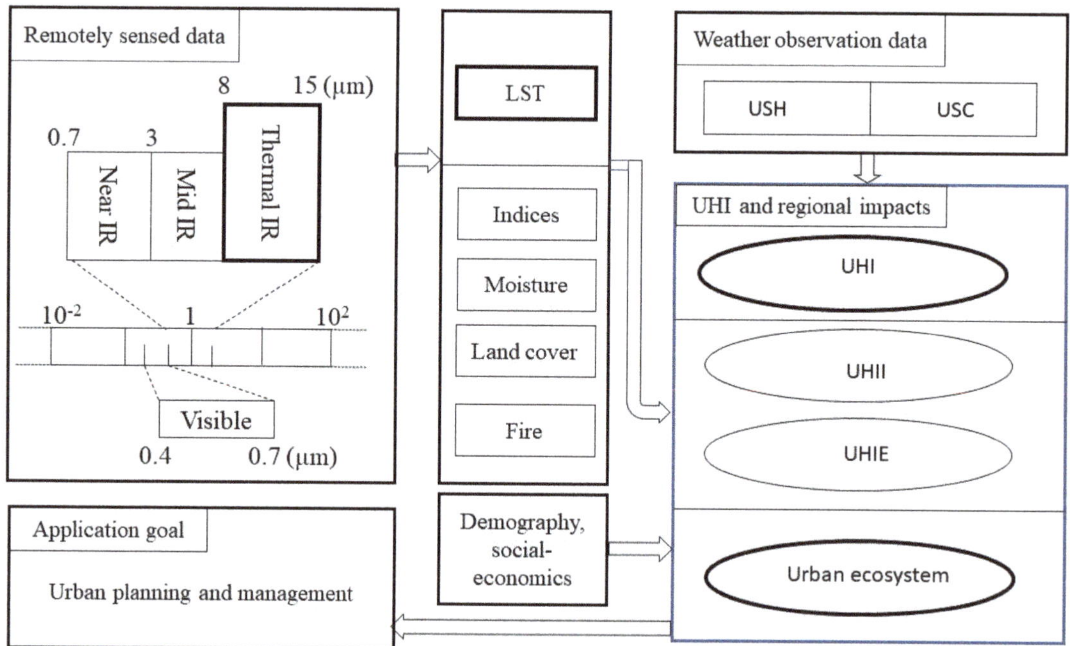

Figure 2. Schematic diagram for using remotely sensed data to evaluate UHI and UHIRIP. Bold outlines indicate high importance. USH—US historical weather data; USC—US climate data; UHI—urban heat island; UHII—UHI intensity; UHIE—UHI effects; IR—infrared band.

The rapid development of remote sensing technology offers more potential for accurate and reliable quantitative assessments of UHI (Table 3 and Figure 3). Many researchers (Table 2) have used remotely sensed LST to assess UHI over various geographic areas. However, for all of these studies, the 1 × 1 km spatial resolution of coarse datasets was found to be suitable only for broad-scale urban temperature mapping (Table 4). The higher resolution of Landsat time series is suitable for UHIRIP at various scales (Table 4).

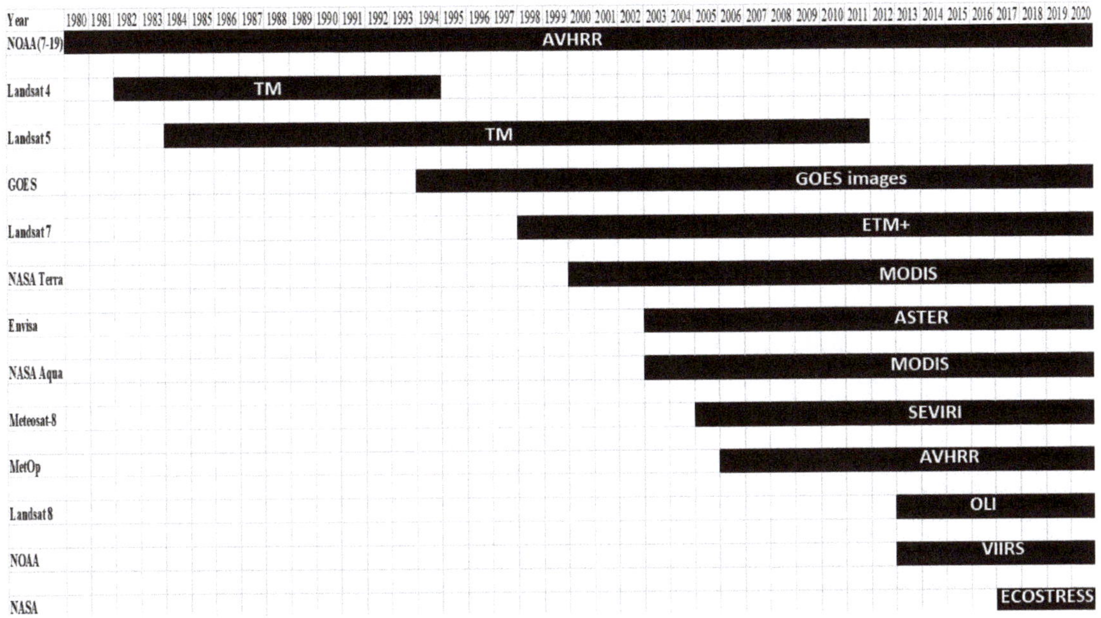

Figure 3. Timeline of satellite data availability. Data availability to 2020 indicates ongoing availability.

Table 4. Proportion of reviewed UHI, UHII, and UHIE studies using various remotely sensed data.

Sensor	%	Examples
Airborne	<1%	Liu et al. [159], and Ben-Dor and Saaroni [160]
AVHRR	4%	Stathopoulou and Cartalis [161], and Gallo and Owen [162]
MODIS	24%	French and Inamdar [115], Zhi Qiao et al. [163], and Keramitsoglou et al. [164]
ASTER	6%	Gillespie et al. [118], Ye et al. [165], Kato and Yamaguchi [166], and Lu and Weng [167]
VIIRS	<1%	Sun et al. [168], Quan et al. [169], and Gawuc and Struzewska [170]
Landsat Series	52%	Aniello et al. [171], Weng [172], Stathopoulou and Cartalis [173], and Sagris and Sepp [174]
ECOSTRESS	<1%	Hulley et al. [175] and Schultz et al. [176]
Multiple sensors	8%	Dousset and Gourmelon [86], and Elmes et al. [177]
Others	<1%	Huang and Wang [80]

Voogt and Oke [6], and others [156,178] pointed out that improved spatial and spectral resolution of sensors and advances in digital image processing techniques increase the usefulness of remote sensing for UHI and UHIRIP studies. Forster [179] also stated that satellite, radar, and airborne sensors can provide spatially continuous information pertaining to numerous variables in urban environments that complement field observations. An increasing number of studies directly relate remotely sensed data to in situ field data [180,181], and applications of remote sensing technology will expand UHI studies to various geographic extents. An exciting recent trend in UHI and UHIRIP research involves coupling remotely sensed data with ancillary and social economic datasets from multiple sources (Table 2). The typical examples include (1) fractional vegetation cover derived from satellite data to improve model simulations of UHI [182], (2) incorporation of remotely sensed data into a model that partitioned various fluxes in the surface energy balance [183], (3) integrating high-resolution multispectral data with property tax records to investigate the contribution of residential land use to UHI formation [184], (4) studying the potential application of change in urban green space as an indicator of urban environmental quality change [185], (5) using parameters from thermal satellite data and three-dimensional virtual reality models to better understand the factors controlling urban environmental quality (UEQ) [186], (6) further advancing the use of remotely sensed imagery to evaluate UEQ by estimating ground-level particulate matter (PM) concentrations using satellite-based data [187], and (7) estimating the value of U.S. urban tree cover for reducing heat-related health impacts and electricity consumption [188]. In addition, NASA's Ecosystem Spaceborne Thermal Radiometer Experiment on the International Space Station (ECOSTRESS) was launched in June 2018, and is able to image fine-scale temperatures in cities at a 70×70 m resolution throughout different times of the day, every 3–5 days on average, over most of the globe [146]. With new algorithm development, ECOSTRESS can accurately monitor UHI trends over time in vulnerable areas such as the urban and non-urban interface. With more available remotely sensed data (Figure 3), innovative studies like these hint at the potential for remote sensing to play an even more prominent role in research of urban climate, urban environment, urban ecological service, and urban planning in the future.

4. Algorithms for UHI and UHIRIP in Urban and Non-Urban Interface Studies Based on Remotely Sensed Data

Generally, the methods for evaluating UHI and UHIRIP can be summarized into four basic types: (1) historical weather station data, (2) field observation, (3) computer simulation, and (4) remote sensing technology. The limitations of the first three methods have been well documented [53,57,105,180]. In this paper, we only focus on the methods that use remote sensing technology. A number of algorithms (or methods) have been developed to estimate UHI and UHIRIP from remotely sensed data (Table 5), including simple empirical approaches to complex methods based on remotely sensed data assimilation using various models. The structure of the UHIRIP pattern centroid in three dimensions indicates the overall variation of the intensity and distribution of the UHI in space and time. The simplified relationship of thermal data and UHI has been applied from a local spatial scale using airborne very high-resolution images to a broad scale with AVHRR, MODIS, ASTER, and Landsat data at regional and continental levels. Assimilation procedures of UHI often require remotely sensed data over different spectral domains to retrieve input parameters that characterize surface properties such as thermal properties, albedo, NDVI, and other indices. A brief review of these approaches is presented in Table 5, with a discussion about the main physical bases and assumptions of various models.

Detailed knowledge of UHI and UHIRIP, especially latent and sensible heat flux components, is important for monitoring the climate change of the land surface. The main methods classically used to measure UHI are appropriate to field observations [24–26,189], but do not allow for an estimation of UHI at large spatial scales. For operational applications to ecological conservation and city planning, managers and engineers need accurate estimates of land surface temperature and UHI at broad spatial scales. New algorithms based on remotely sensed data have been developed to use the imagery of various spatial resolutions and temporal frequency to evaluate UHI [190–192]. It is often difficult to classify these methods because their complexity depends on the balance between the empirical- and physical-based modules used. Nevertheless, we summarize some algorithm (model) categories in the following subsection.

Table 5. Methods used to measure UHI and UHIRIP using remotely sensed data.

Method	Sensor	Period	Example
Calculate LST	All thermal bands	1970s–current	Avdan and Jovanovska [193], and Peng et al. [194]
Determine the UHIE	Landsat	2009	Tang et al. [195]
Determine the UHII	MODIS	2001, 2003	Tran et. al. [156]
Compare multi-temporal LST images	The normalization of the temperature based on the mean and standard deviation in high and low temperature areas.		Streutker [39]
	Common normalization of temperture based on min and max LST of the same image in the same way as for NDVI. A normalized ratio scale technique.		Chen et al. [38]
Statistical analyses of UHI	The relationship between LST, NDVI, ground vegetation (GV), and impervious surface area (ISA). Multiple linear regression. Geographically weighted regression.		Weng et al. [153], Tran et al. [156], Schwarz et al. [196], Szymanowski and Kryza [197], and Firozjaei et al. [198]
	A support vector machine regression (SVR) mode. LST	2012 (daily)	Lai et al. [79]
Data fusion	Landsat, MODIS	1988–2013,	Shen et al. [192], Wengand Fu [17], and Schmitt and Zhu [199]
Gap filling	Landsat	2020	Yan and Roy [178], Zhou et al. [60], Fu et al. [190], and Zhou et al. [200]
Time-series analysis	Landsat	1984–2015	Huang et al. [201], Peres et al. [202], Fu and Weng [203], and Xian et al. [97]
Uncertainty and accuracy assessment	MODIS, Landsat		Shen et al. [192], Lee et al. [204], Yuan and Bauer [205], and Chen et al. [206]

4.1. LST and UHI Intensity Calculation

LST calculation, including empirical direct methods where remotely sensed data are introduced directly in semi-empirical models to estimate LST, is the simplified relationship between thermal infrared remotely sensed and meteorological data [14]. This method allows for the characterization of UHI intensity both at the local scale, using ground measurements, and over large areas, using satellite data, by calculating a cumulative temperature difference [55,92]. Most current operational models [60] use remote sensing directly to estimate the input parameters and LST.

Seasonal information captures the annual profile of LST and its trend over long time periods, and is essential to the study of UHI [207]. Therefore, remote sensing has been used to accurately monitor and compare the LST difference in the same season in different years and trends over long time periods. In the last 10–15 years, thermal sensor technology has been rapidly developing (Figure 3). Three types of methods have been developed to estimate LST with remotely sensed data: the single infrared channel method; the split window method; and a new day–night MODIS LST method, which is designed to take advantage of the unique capability of the MODIS instrument [55]. Recently, Peng et al. [194] proposed a wavelet coherence approach to exploring spatial heterogeneity and the scale-dependence of the relationship between LST and multiple influencing factors. The advantages, disadvantages, and applicability of these three types of algorithms are summarized in Table 6.

Table 6. Advantages, disadvantages, and applicability of commonly used algorithms for calculating LST.

Type	Algorithm	Advantages	Disadvantages	Example
Single window	Atmosphere correction	LST for oasis in arid lands	Complicated, errors, only use for one band thermal	Landsat TM/ETM+, CBERS/IRMSS
	Qin Sing window	Accurate and applicable	Need three atmosphere parameters, only use for one band thermal	
	Universal single channel	Do not need atmosphere parameters, applicable for multiple sensors	The result impacted by standard atmosphere	
Split window	NOAA-AVHRR	Most used, accurate, applicable for most sensors, less requirement of parameters, simple models	Not accurate LST in mixed pixels	NOAA/AVHRR3 TERRA/MODIS Landsat 8/TIRS
	TERRA-MODIS			
	Landsat-TIRS		Results not stable, lower accuracy, TIRS band 11 not stable	
Other	Day and night	Accurate in MODIS	Limitations, low applicability	TERRA/MODIS TERRA/ASTER VIIRS
	Separate temperature	Accurate in ASTER	Not stable, limitations, low applicability	
	Gray matters	Good for grey matters	Sensitive in noise	

4.2. Comparing the Difference between Core Urban and Non-Urban Area

Many studies have documented the use of LST data to observe meso-scale temperature differences between urban and rural areas in cities worldwide [156,208,209]. The land surface temperature (LST) of core urban areas is generally higher than the surrounding rural areas, and has a strong correlation with land cover [153]. UHII analysis is the most common method to compute the magnitude and extent of UHI by evaluating the LST difference between urban and surrounding non-urban areas [162]. These analyses are often supported with auxiliary land surface information, such as land cover and impervious surface area (ISA). Deterministic models generally are generally based on more complex models that compute the intensity of UHIRIP in space and time. Remotely sensed data are used at different modeling levels, either as the input parameters to characterize the different surface covers, or in assimilation procedures, which aim to retrieve adequate parameters for the LST computation. Some examples of these studies are shown in Table 5. UHI intensity was typically quantified in two steps in these studies [60]. First, urban and non-urban areas were defined and delineated from land cover or ISA maps. Urban areas are usually defined as land with a relatively higher proportion of ISA [38,95], whereas non-urban areas have various definitions in different studies, but generally include non-urban land cover classes. Different sized rural and suburban zones have been used as reference areas. Other land covers, such as water bodies, cropland, forest, and low-intensity ISA, have also been used as references in the studies [101]. Second, the area-weighted mean urban-reference LST differences were calculated to reflect the UHI intensity [69,210] or magnitude. Some studies identified "hotspots" based on positive UHI intensity in certain time periods [211,212]. A positive value of UHI intensity indicated an urban heating effect, while a negative value represented a cooling effect. A few studies also quantified the UHI intensity using small numbers of representative pixels in urban and reference areas instead of the area-weighted mean value for the purpose of surface-air UHI comparison [122–124] or UHI attribution analysis [125,126]. The urban-reference difference method facilitates a comparative analysis of UHIs among cities and urban agglomerations, regions, and across the globe, but the validity of such comparisons can be limited by the large uncertainties associated with urban and reference definitions [68]. Recent research [97] performed a comprehensive and consistent analysis of surface UHI and UHIRIP using Landsat LST

ARD time series and dynamic land cover datasets in the Sioux Falls, SD, area. It shows that the use of time series of LST and land change dynamic data provided a consistent and quantitative analysis for the distribution and change of UHI intensity and UHIRIP (Figure 4). We further discuss limitations in Section 5.

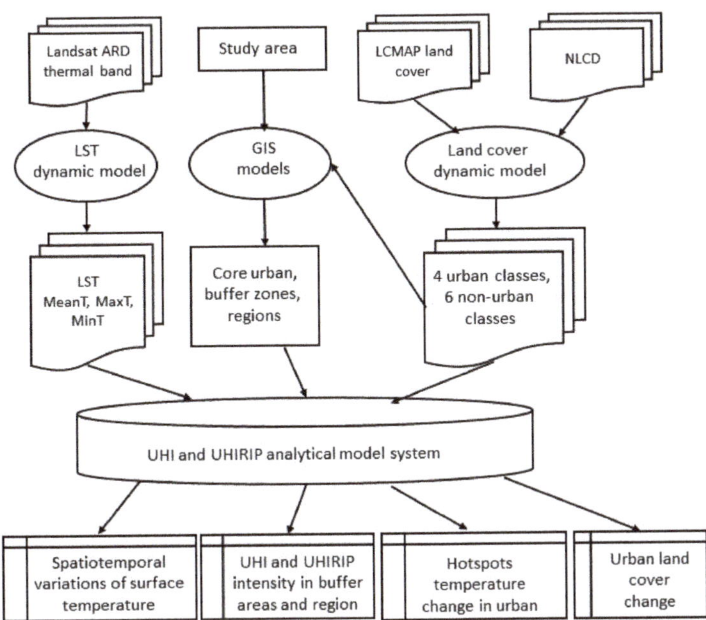

Figure 4. A general workflow chart of the use of time series of LST and land change dynamic data that provides a consistent and quantitative analysis for the distribution and change of UHI intensity and UHIRIP in Sioux Falls, SD.

4.3. UHI and UHIRIP Analysis by Using Urban Ecological Indices

Many studies have compared UHIRIP to ecological indices [103,149,213–215], vegetation fraction, and percent ISA, finding strong correlations with mean LST. Landscape metrics indicate that urban landscape configuration also influences the surface UHIRIP [216]. The latest vegetation index methods and inference methods use remote sensing to compute a reduction factor (such as Kc or Priestley Taylor-alpha parameters) for the estimation of the actual UHI [203]. Different papers deal with these approaches in the various journals, and these approaches use land cover [217,218], LST pattern [219,220], and a combination of land cover and LST pattern [221–223] as monitoring indicators of UHI.

Urban ecological status is closely related to the quality of human life and the development of urban economies. A timely and objective understanding of urban ecological status, particularly in urban and non-urban interface areas [224], has become an increasing important. Scientists have been developing a remote sensing-based ecological index for the measure of urban ecological status under UHI [213,225]. This urban ecological status index (UESI) aims to integrate four important ecological indicators that are frequently used in evaluating urban ecology. The four indicators include greenness, wetness, dryness, and heat, and can be represented by four remote sensing indices or components: NDVI, normalized difference built-up and soil index (NDBSI), wetness component of the tasseled cap transformation (Wet), and LST, respectively. Instead of a simple or weighted addition of the four indicators, a principal component analysis (PCA) can be utilized to compress the four indicators into one index in order to assess the overall urban ecological status under UHI. The calculation of the UESI can be fully automated, avoiding the need to assign threshold values or weights during the computing procedure. Therefore, the UESI can be used to easily and objectively assess urban ecological status. Combined with change detection, UESI can also be used to monitor the change of the ecological status of the core urban and surrounding non-urban areas between different years. In practice, the index was successfully applied in a multitemporal ecological status assessment [34]. Pan [221] used the G index spatial aggregation analysis to calculate the urban heat island ratio index, and the landscape metrics to quantify the changes of the spatial pattern of the UHI from the aspects of quantity, shape, and structure. Pan found that the heat island strength had a negative linear correlation with urban vegetation coverage, and a positive logarithmic correlation with urban impervious surface coverage. Bala et al. [226] developed the Urban Heat Intensity Ratio Index (UHIRI) to quantify urban heat intensity. This work analyzed the variation in LST with land cover changes in Varanasi, India, from 1989 to 2018, using Landsat images, and concluded that the replacement of vegetation with urban land cover has a severe impact on increasing UHI intensity.

4.4. Various Statistical Models

Statistical models and machine learning have also been proposed to measure UHI [227]. Among these studies, a Gaussian surface model has been utilized the most because it can provide not only the intensity, but also the spatial extent and the central location of the UHI. The kernel convolution method has also been proposed to study UHI effects because of its high efficiency in characterizing the temperature values over space in a continuous surface [227]. Chun and Guldmann [77] explored the urban determinants of UHI using two-dimensional (2D) and three-dimensional (3D) urban information as the input for spatial statistical models. The results show that solar radiations, open spaces, vegetation, building roof-top areas, and water strongly impact surface temperatures, and that spatial regressions are necessary in order to capture the neighboring effects. Recently, Li et al. [81] estimated UHI intensity by linear regression functions between LST and regionalized ISA. These statistical models could avoid the bias caused by the definitions of urban−rural areas or the choice of the representative pixels, and thus facilitate the comparison of UHI among cities. Szymanowski and Kryza [228] addressed the issue of the potential usefulness of remotely sensed data and their derivatives for UHI modeling. Statistically significant models explained 71% to 85% of the air temperature variance. It has been stressed that remotely sensed data are important sources to model urban air temperature heat islands. However, in all of these studies, such models worked less effectively in cities frequently covered by clouds, in arid landscapes, and in urban agglomerations, so they have only been applied in a few UHI studies to date. Recently, Lai et al. [79] published the statistical estimation of next-day nighttime surface urban heat islands of selected cities. Most previous studies modelled the SUHI variations for the past period, but rarely investigated the estimation for future UHIs, especially at the daily (i.e., day-to-day) scale. To address this issue, this study incorporated both meteorological and surface controls to estimate next-day nighttime UHIs using a support vector machine regression (SVR) model. Some uncertainties exist in terms of the Gaussian modelling and UHI estimators, which may limit estimation accuracy. Nevertheless, by providing a feasible yet simple approach for estimating next-day nighttime UHIs, this study fills a knowledge gap in the UHI estimation and is helpful for supporting adaptation to and mitigation of UHI and UHIRIP.

4.5. Spatial−Temporal Time-Series Algorithm

Advances in computing technology have fostered the development of new and powerful data fusion, gap filling, machine learning, and deep learning techniques that have demonstrated promising results in a wide range of applications [229]. Models to fuse data from multiple sensors and fill gaps can improve UHI monitoring using an ensemble of dense time series of thermal data with a high spatial resolution [199,230]. Zhou et al. [200] presented a new algorithm that focuses on data gap filling using clear observations from orbit overlap regions to obtain Landsat LST data. Multiple linear regression models were established for each pixel time series to estimate the stable predictions and uncertainties. Liu et al. [83] comprehensively quantified the spatial–temporal patterns of surface urban heat island by investigating the relationship between LST and the land cover types, and the associated landscape components. Such approaches have been used to generate temporally dense and high-resolution LST over long time periods by integrating the observations of Landsat, MODIS, AVHRR, VIIRS, and ECOSTRESS [191,231]. These datasets facilitate subtle analyses of monthly, seasonal, and yearly trends in UHI intensity at regional levels. Machine learning (ML) has become popular in UHI and UHIRIP, but its use has remained restricted to predicting, rather than understanding, the natural world. ML techniques may not be the solution to all the problems remotely sensed data might have. However, these techniques provide a powerful set of tools that deserve serious attention to deal with some relevant UHI and UHIRIP remotely sensed data problems [232]. Lucas [233] points out that ML differs from the broader field of statistics in two respects: (1) the estimation of parameters that relate to the real world is less emphasized, and (2) the driver of the predictions is expected to be the data rather than expert opinion and careful selection of plausible mechanistic models. The Google Earth engine (GEE) is a cloud-based platform for planetary-scale geospatial analysis that brings Google's massive computational capabilities to bear on a variety of high-impact societal and environmental issues [234]. GEE has many functions that could be used to analyze UHI and UHIRIP at local, regional, and global levels (Figure 5). Some research has generated consistent large-scale UHI and UHIRIP analysis based on optimal data and ML algorithm selection using GEE [235,236]. The advanced GEE cloud-based platform and the large number of geosciences and remote sensing datasets archived in GEE were used to analyze land the cover dynamics (236), and the results showed the advantages of using GEE to analyze the spatiotemporal dynamics of the LULCC, vegetation cover, LST, and climate for a long time series, and highlighted the importance of environmental protection. The power here lies in the way a scientist defines their questions and uses machine learning alongside other methods. Techniques for data analysis and interpretation that fully incorporate the temporal dimension remain an area of intense research and represent an important challenge for operational UHI and UHIRIP monitoring.

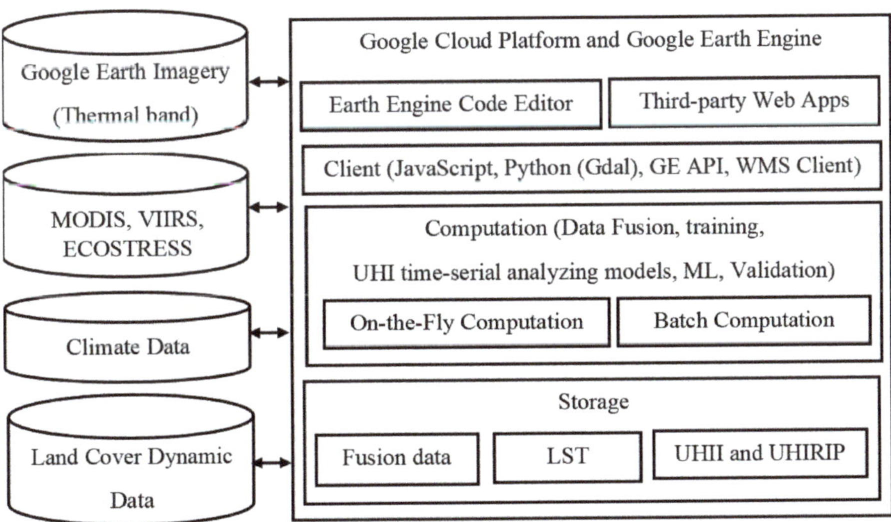

Figure 5. Simplified system diagram using the Google Cloud platform and Google Earth engine for monitoring UHI and UHIRIP.

5. Summary of UHI and UHIRIP Based on Remotely Sensed Data

This review provides an overview of research on UHI and UHIRIP based on remote sensing techniques, sensors, and algorithms, as listed in Tables 3–6, respectively. Much work has been completed on UHIRIP in the last four decades, and we have endeavored to keep updated with new methods and results. A significant research limitation still exists: the quantification of UHI and its regional impacts using high-resolution time-series remotely sensed thermal data in the urban and non-urban interface. Some of the algorithms listed in Table 5 may be the most practical approaches to assess UHI in core urban areas of cities and surrounding areas, but characterization of UHI across broad areas is necessary in order to inform monitoring, reporting, science, and policy. Being able to relate LST to UHI is especially important when such datasets are being used to inform policy decisions or to communicate outside of the scientific community. The increasing availability of remotely sensed data across a range of spatial resolutions and temporal frequencies, and technological improvements in image processing capacity and storage, have led to advances in the methods used to monitor UHI more frequently and accurately.

Assessing the uncertainty and accuracy of UHI data is important. A sensitivity analysis not only provides a framework for assessing the potential for bias and the extent of uncertainty in UHI estimates, but also reveals significant factors that determine the extent of UHIRIP in the urban and non-urban interface. Oleson et al. [237] developed an approach to evaluate the robustness of models used to simulate urban heat islands in different environments. The findings indicated that heat storage and sensible heat flux are most sensitive to uncertainties in the input parameters within the atmospheric and surface conditions considered. Sensitivity studies indicate that it is important to not only to accurately characterizing the structure of the urban area, but also to ensuring that the input data reflect the thermal admittance properties of each of the city surfaces.

Currently, a wide variety of methods are employed to characterize UHI for major cities worldwide (Table 5), although most of the applications cited were limited to small areas because of data availability and constraints of storage and computing resources. With the development of gap filling and data fusion models [238], advances in high-performance computing (HPC), and cheaper storage, applications based on high-resolution time series at larger or even regional scales will become the mainstream in the near future [199,231]. While much of the methodological variation described here will persist, future methods will evolve and adapt to greater data volumes and processing capabilities [239]. Legacy change mapping methods that rely on analyst interactions with individual scenes should decline over time given the improved ability to process and characterize time series of rich high-resolution thermal data. However, such spatial−temporal methods that are based on gap filling and data fusion should match the institutional requirements for accuracy. Near-term research objectives will require robust validation datasets in establishing which data-intensive methods are the most appropriate for quantifying UHI over large areas. Techniques for LST data analysis and interpretation that fully incorporate the temporal dimension still require intense research and represent an important challenge for operational UHI research in order to meet management needs.

Technological advances that include machining leaning and artificial intelligence in UHI and UHIRIP using remotely sensed data have led to an explosion of UHI and UHIRIP profiling data from large numbers of multiple data sources [233,240]. This rapid increase in the remotely sensed data dimension and acquisition rate is challenging conventional analysis strategies. Modern machine learning methods, such as deep learning, promise to leverage very large datasets for finding a hidden structure within them, and for making accurate predictions [232]. Deep learning methods are a powerful complement to classical machine learning tools and other analysis strategies, and have been used in a number of applications in UHII and remotely sensed image analyses [241]. The explainable artificial intelligence in UHII and UHIRIP modeling has become more and more important [242]. Interpretable machine learning methods either target a direct understanding of the model architecture (i.e., model-based interpretability) or interpret the model by analyzing the model behavior (post hoc interpretability) [242].

Currently, most of the time-series algorithms used to map UHIRIP include data from the temporal domain of AVHRR and MODIS, and the spatial domain of the data is almost entirely neglected. Although these datasets with a lower spatial resolution and higher temporal frequency can detect a change of UHI in real time, they often lack pertinent spatial detail. Even though many UHI analysis algorithms have been developed [60], most of the UHI monitoring data derived from the Landsat archive are provided in a time frame that is not near enough to real time to be relevant for specific management needs. With the advances in HPC and cheaper storage, applications based on Landsat time series at continental or even global scales will be the mainstream in the next few years.

To date, information from Landsat time-series thermal data has taken the form of statistical metrics, change metrics, pattern distribution, or trend components used in UHI impact applications [243]. Improvement of existing approaches, as well as the inclusion of novel techniques, often imported and adapted from other disciplines, are important to fully capitalize on the thermal data in order to produce monthly, seasonal, and annual LST results that meet a wide range of UHI and UHIRIP research needs. Landsat-9, which will be launched in September 2021, will continue collecting images of the Earth's surface in visible, near-infrared, and shortwave-infrared bands, as well as the thermal infrared radiation, or heat, of the Earth's surface from two thermal bands. The future European Space Agency's LSTM (Land Surface Temperature Monitoring) or Sentinel 8 mission will carry a high spatial−temporal resolution thermal infrared sensor to provide records of land-surface temperature. Land-surface temperature measurements are key variables to understand and respond to climate variability and natural hazards, such as urban heat island issues. The main objective of LSTM is to deliver global high spatial−temporal day- and night-time land surface temperature measurements. LSTM will operate from a low-Earth, polar orbit, to map both land-surface temperature and rates of evapotranspiration. It will be able to identify the temperatures of individual fields and image the Earth every three days at a 50 m resolution. Another future thermal sensor is Thermal infraRed Imaging Satellite for High-resolution Natural resource Assessment (TRISHNA), which is a future high-resolution space-time mission in the thermal infrared (TIR) led jointly by the French (CNES) and Indian (ISRO) space agencies. One of scientific objectives guiding the definition of the mission is the monitoring of the urban environment. TRISHNA will be positioned on a polar orbit and provide a revisit of three passages over 8 days with global coverage. The time of passage around 13:00 p.m. LST allows thermal data to be collected in the middle of the day, but also in the middle of the night. The instrument will offer four thermal channels (8.6 µm, 9.1 µm, 10.4 µm, and 11.6 µm) and six optical channels (485 nm, 555 nm, 650 nm, 860 nm, 1380 nm, and 1650 nm) with a spatial resolution between 50 m and 60 m for all channels. All of these observations acquired from thermal remote sensing will provide more valuable information for natural resource management, hazard monitoring, and scientific research and applications.

6. Future Research Directions

Remote sensing technology has been widely applied in the research of UHI and UHIRIP. The most important advantage of using remote sensing thermal data is the wall-to-wall coverage of UHI patterns that can meet the needs of spatial and temporal analyses. Remotely sensed data can be used to investigate the surface temperatures of cities and urban agglomerations for various ecosystems with different climate conditions, for example tropical and sub-tropical, temperate and cold temperate, coastal and inland, and arid and semi-arid land at regional scales. These studies are needed to describe surface temperature characteristics in these specific environments and how climate change may be modulating UHI patterns. UHIRIP produces an aggregate impact on weather conditions, land use, human health, biodiversity, ecosystem security, economics, and urban planning [16,244].

Land surface temperature and emissivity retrieval (separation) has always been challenging. Generally, the LSE values needed to apply the method have been estimated from a procedure that uses the visible and near-infrared bands. The algorithm was created using the brightness temperature of the thermal and emissivity of different land cover types, derived from visible and near-infrared bands of various sensors. Compared with field-based observation, remote sensing offers the advantages of a harmonized, long-term, and spatially extensive record to observe LST change. The retrieved LSTs are verified using the near surface temperature of weather station datasets, which will help to improve the accuracy of LST derived from thermal bands. The difference between retrieved LST and Automatic Weather Station (AWS) data indicates that the technique works by giving an error of ±3 °C [245]. These differences can be because of the difference between the resolutions of thermal and visible bands, and a comparison was made between the point measurement (AWS data) 2 m above the surface and surface temperature (retrieved LST). Communicating the results of time-series LST studies that are based on both field weather station observations and remote-sensing time-series data to urban planners, policymakers, and the general public could help inform urban design and decision making.

Using temporally dense time series of remotely sensed data at a high spatial resolution is a growing trend in UHI and UHIRIP research, facilitated by increasing computer capabilities to handle big datasets, machine leaning, deep learning, and Google Earth Engine applications. Landsat ARD, in particular, has great potential to derive LST. Models used to fuse data from across multiple sensors will be developed to increase data temporal density and spatial resolution. Moreover, future sensor improvement on Landsat and aircraft thermal data are possible options. On the other hand, in order to determine the temporal variation of LST using satellite data with restricted overpass times, it appears necessary to use long-time weather station observations to investigate diurnal UHI in various ecosystems, although some new sensors (e.g., ECOSTRESS) can provide this information. Future research is anticipated to improve on methods to simultaneously derive LST and land surface emission (LSE) from hyperspectral TIR, multi spectral-temporal, and TIR-microwave data; additionally, future methods will consider aerosol and cirrus effects [18]. Another viable angle of potential future studies is urban development strategies for mitigating UHI, such as increasing vegetation and water surfaces in urban development.

Climate models are the only tools that account for the complex set of processes that will determine future climate change at both a global and regional level, and assessing regional impacts of climate change begins with the development of climate projections at relevant temporal and spatial scales [246]. The most current existing climate change modeling covers large geographic areas at regional and global levels with relatively low spatial resolutions (>10 km). In the future, LST that is derived from remotely sensed data will support climate change modeling (regional climate models and statistical downscaling models) in UHI and UHIRIP analyses in urban and surrounding areas.

Our analysis indicated that determination is still a central topic of UHI research. Modeling will continue to provide vital and useful results on the spatiotemporal assessment of UHI, especially when models more effectively combine thermal data from multiple sensors. ML (DL) and AI are continuing to grow in popularity in UHI and UHIRIP research. For time series analyses with remote sensing data, a cloud computing platform such as GEE could bring about a substantial change in UHI and UHIRIP analyses, as they have the capability to process big remote sensing datasets and assess the spatiotemporal dynamics of the area quickly. A better integration of remote sensing and station measurements into models is expected. This study also suggests that direct and indirect UHIRIP, especially human health issues, heat wave impacts, air pollution, and ecological security, will receive increasing scientific attention in the future. Research on controlling and adapting to UHI impacts may warrant special attention. The interaction of UHI and UHIRIP, and their changes to LULC based on urban planning, are actively being studied.

Author Contributions: Original draft preparation, H.S.; review and editing with ideas and inputs, G.X., R.A., K.G. and Q.Z. All authors have read and agreed to the published version of the manuscript.

Funding: This research received no external funding.

Institutional Review Board Statement: Not applicable.

Informed Consent Statement: Not applicable.

Data Availability Statement: Not applicable.

Acknowledgments: We are grateful to the USGS National Land Imaging Program for supporting this research. We would like to thank Carol Deering for assistance in collecting and preparing the literary material. This manuscript was improved thanks to comments and suggestions from Norman Bliss, Matthew Rigge, and Thomas Adamson. Hua Shi and Qiang Zhou's work was performed under USGS contracts G13PC00028 and 140G0119C0001. For Kevin Gallo, the scientific results and conclusions, as well as any views or opinions expressed herein, are those of the author and do not necessarily reflect those of NOAA or the Department of Commerce. Any use of trade, firm, or product names is for descriptive purposes only and does not imply endorsement by the U.S. Government.

Conflicts of Interest: The authors declare no conflict of interest. The funding sponsors had no role in the design of the study; in the collection, analyses, or interpretation of data; in the writing of the manuscript; and in the decision to publish the results.

References

1. Radeloff, V.C.; Hammer, R.B.; Stewart, S.I.; Fried, J.S.; Holcomb, S.S.; McKeefry, J.F. The Wildland–Urban Interface in the United States. *Ecol. Appl.* **2005**, *15*, 799–805. [CrossRef]
2. Shi, H.; Singh, A.; Kant, S.; Zhu, Z.; Waller, E. Integrating habitat status, human population pressure, and protection status into biodiversity conservation priority setting. *Conserv. Biol.* **2005**, *19*, 1273–1285. [CrossRef]
3. Seto, K.C.; Shepherd, J.M. Global urban land-use trends and climate impacts. *Curr. Opin. Environ. Sustain.* **2009**, *1*, 89–95. [CrossRef]
4. Ager, A.A.; Vaillant, N.M.; Finney, M.A. A comparison of landscape fuel treatment strategies to mitigate wildland fire risk in the urban interface and preserve old forest structure. *For. Ecol. Manag.* **2010**, *259*, 1556–1570. [CrossRef]
5. Chang, Q.; Liu, X.; Wu, J.; He, P. MSPA-based urban green infrastructure planning and management approach for urban sustainability: Case study of longgang in China. *J. Urban Plan. Dev.* **2015**, *141*. [CrossRef]
6. Voogt, J.A.; Oke, T.R. Thermal remote sensing of urban climates. *Remote Sens. Environ.* **2003**, *86*, 370–384. [CrossRef]
7. Weng, Q. Thermal infrared remote sensing for urban climate and environmental studies: Methods, applications, and trends. *ISPRS J. Photogramm. Remote Sens.* **2009**, *64*, 335–344. [CrossRef]
8. Massad, R.S.; Lathière, J.; Strada, S.; Perrin, M.; Personne, E.; Stéfanon, M.; Stella, P.; Szopa, S.; de Noblet-Ducoudré, N. Reviews and syntheses: Influences of landscape structure and land uses on local to regional climate and air quality. *Biogeosciences* **2019**, *16*, 2369–2408. [CrossRef]
9. Raalte, L.V.; Nolan, M.; Thakur, P.; Xue, S.; Parker, N. *Economic Assessment of the Urban Heat Island Effect*; 60267369; AECOM Australia Pty Ltd.: Melbourne, Australia, 2012. Available online: https://www.melbourne.vic.gov.au/SiteCollectionDocuments/eco-assessment-of-urban-heat-island-effect.pdf (accessed on 20 May 2020).
10. Oke, T.R. The energetic basis of the urban heat island. *Q. J. R. Meteorol. Soc.* **1982**, *108*, 1–24. [CrossRef]
11. Hall, F.G.; Huemmrich, K.F.; Goetz, S.J.; Sellers, P.J.; Nickeson, J.E. Satellite remote sensing of surface energy balance: Success, failures, and unresolved issues in FIFE. *J. Geophys. Res.* **1992**, *97*, 19061–19089. [CrossRef]
12. Velasco, E. Go to field, look around, measure and then run models. *Urban Clim.* **2018**, *24*, 231–236. [CrossRef]
13. Gallo, K.P.; Tarpley, J.D.; McNab, A.L.; Karl, T.R. Assessment of urban heat islands: A satellite perspective. *Atmos. Res.* **1995**, *37*, 37–43. [CrossRef]
14. Courault, D.; Seguin, B.; Olioso, A. Review on estimation of evapotranspiration from remote sensing data: From empirical to numerical modeling approaches. *Irrig. Drain. Syst.* **2005**, *19*, 223–249. [CrossRef]
15. Dorigo, W.A.; Zurita-Milla, R.; de Wit, A.J.W.; Brazile, J.; Singh, R.; Schaepman, M.E. A review on reflective remote sensing and data assimilation techniques for enhanced agroecosystem modeling. *Int. J. Appl. Earth Obs. Geoinf.* **2007**, *9*, 165–193. [CrossRef]
16. Rasul, A.; Balzter, H.; Smith, C.; Remedios, J.; Adamu, B.; Sobrino, J.; Srivanit, M.; Weng, Q. A Review on Remote Sensing of Urban Heat and Cool Islands. *Land* **2017**, *6*, 38. [CrossRef]
17. Weng, Q.; Fu, P. Modeling diurnal land temperature cycles over Los Angeles using downscaled GOES imagery. *ISPRS J. Photogramm. Remote Sens.* **2014**, *97*, 78–88. [CrossRef]
18. Weng, Q.; Firozjaei, M.K.; Sedighi, A.; Kiavarz, M.; Alavipanah, S.K. Statistical analysis of surface urban heat island intensity variations: A case study of Babol city, Iran. *GIScience Remote Sens.* **2019**, *56*, 576–604. [CrossRef]
19. Ward, K.; Lauf, S.; Kleinschmit, B.; Endlicher, W. Heat waves and urban heat islands in Europe: A review of relevant drivers. *Sci. Total Environ.* **2016**, *569–570*, 527–539. [CrossRef] [PubMed]

20. Zhu, Z. Change detection using landsat time series: A review of frequencies, preprocessing, algorithms, and applications. *ISPRS J. Photogramm. Remote Sens.* **2017**, *130*, 370–384. [CrossRef]
21. Du, W.; Qin, Z.; Fan, J.; Gao, M.; Wang, F.; Abbasi, B. An efficient approach to remove thick cloud in VNIR bands of multi-temporal remote sensing images. *Remote Sens.* **2019**, *11*, 1284. [CrossRef]
22. Ling, F.; Zhang, T. A numerical model for surface energy balance and thermal regime of the active layer and permafrost containing unfrozen water. *Cold Reg. Sci. Technol.* **2004**, *38*, 1–15. [CrossRef]
23. Atkinson, B.W. Numerical modelling of urban heat-island intensity. *Bound.-Layer Meteorol.* **2003**, *109*, 285–310. [CrossRef]
24. Oke, T.R. The distinction between canopy and boundary-layer urban heat Islands. *Atmosphere* **1976**, *14*, 268–277. [CrossRef]
25. Kim, H.H. Urban heat island. *Int. J. Remote Sens.* **1992**, *13*, 2319–2336. [CrossRef]
26. Taha, H. Urban climates and heat islands: Albedo, evapotranspiration, and anthropogenic heat. *Energy Build.* **1997**, *25*, 99–103. [CrossRef]
27. Oke, T.R. *The Heat Island of the Urban Boundary Layer: Characteristics, Causes and Effects*; Springer: Dordrecht, The Netherlands, 1995; Volume 277. [CrossRef]
28. Stone, B., Jr.; Rodgers, M.O. Urban form and thermal efficiency: How the design of cities influences the urban heat island effect. *J. Am. Plan. Assoc.* **2001**, *67*, 186–198. [CrossRef]
29. Hansen, J.; Ruedy, R.; Sato, M.; Imhoff, M.; Lawrence, W.; Easterling, D.; Peterson, T.; Karl, T. A closer look at United States and global surface temperature change. *J. Geophys. Res. Atmos.* **2001**, *106*, 23947–23963. [CrossRef]
30. Golden, J.S. The Built Environment Induced Urban Heat Island Effect in Rapidly Urbanizing Arid Regions—A Sustainable Urban Engineering Complexity. *Environ. Sci.* **2004**, *1*, 321–349. [CrossRef]
31. Bowler, D.E.; Buyung-Ali, L.; Knight, T.M.; Pullin, A.S. Urban greening to cool towns and cities: A systematic review of the empirical evidence. *Landsc. Urban Plan.* **2010**, *97*, 147–155. [CrossRef]
32. Chow, W.T.L.; Brennan, D.; Brazel, A.J. Urban Heat Island Research in Phoenix, Arizona: Theoretical Contributions and Policy Applications. *Bull. Am. Meteorol. Soc.* **2011**, *93*, 517–530. [CrossRef]
33. Qin, Y. A review on the development of cool pavements to mitigate urban heat island effect. *Renew. Sustain. Energy Rev.* **2015**, *52*, 445–459. [CrossRef]
34. Xu, H.Q. A remote sensing urban ecological index and its application. *Shengtai Xuebao Acta Ecol. Sin.* **2013**, *33*, 7853–7862. [CrossRef]
35. Memon, R.A.; Leung, D.Y.C.; Liu, C.-H. An investigation of urban heat island intensity (UHII) as an indicator of urban heating. *Atmos. Res.* **2009**, *94*, 491–500. [CrossRef]
36. Sinha, P.; Coville, R.C.; Hirabayashi, S.; Lim, B.; Endreny, T.A.; Nowak, D.J. Modeling lives saved from extreme heat by urban tree cover☆. *Ecol. Model.* **2021**, *449*, 109553. [CrossRef]
37. Li, K.; Yu, Z. Comparative and combinative study of urban heat island in Wuhan City with remote sensing and CFD simulation. *Sensors* **2008**, *8*, 6692–6703. [CrossRef]
38. Chen, X.L.; Zhao, H.M.; Li, P.X.; Yin, Z.Y. Remote sensing image-based analysis of the relationship between urban heat island and land use/cover changes. *Remote Sens. Environ.* **2006**, *104*, 133–146. [CrossRef]
39. Streutker, D.R. A remote sensing study of the urban heat island of Houston, Texas. *Int. J. Remote Sens.* **2002**, *23*, 2595–2608. [CrossRef]
40. Lee, H.Y. An application of NOAA AVHRR thermal data to the study of urban heat islands. *Atmos. Environ. Part B Urban Atmos.* **1993**, *27*, 1–13. [CrossRef]
41. Gallo, K.P.; McNab, A.L.; Karl, T.R.; Brown, J.F.; Hood, J.J.; Tarpley, J.D. The use of NOAA AVHRR data for assessment of the urban heat island effect. *J. Appl. Meteorol.* **1993**, *32*, 899–908. [CrossRef]
42. Kotharkar, R.; Ramesh, A.; Bagade, A. Urban Heat Island studies in South Asia: A critical review. *Urban Clim.* **2018**, *24*, 1011–1026. [CrossRef]
43. Bullock, E.L.; Woodcock, C.E.; Holden, C.E. Improved change monitoring using an ensemble of time series algorithms. *Remote Sens. Environ.* **2019**. [CrossRef]
44. Cohen, W.B.; Yang, Z.; Healey, S.P.; Kennedy, R.E.; Gorelick, N. A LandTrendr multispectral ensemble for forest disturbance detection. *Remote Sens. Environ.* **2018**, *205*, 131–140. [CrossRef]
45. Liu, C.; Zhang, Q.; Luo, H.; Qi, S.; Tao, S.; Xu, H.; Yao, Y. An efficient approach to capture continuous impervious surface dynamics using spatial-temporal rules and dense Landsat time series stacks. *Remote Sens. Environ.* **2019**, *229*, 114–132. [CrossRef]
46. Shi, H.; Rigge, M.; Homer, C.G.; Xian, G.; Meyer, D.K.; Bunde, B. Historical Cover Trends in a Sagebrush Steppe Ecosystem from 1985 to 2013: Links with Climate, Disturbance, and Management. *Ecosystems* **2018**, *21*, 913–929. [CrossRef]
47. Zhu, Z.; Gallant, A.L.; Woodcock, C.E.; Pengra, B.; Olofsson, P.; Loveland, T.R.; Jin, S.; Dahal, D.; Yang, L.; Auch, R.F. Optimizing selection of training and auxiliary data for operational land cover classification for the LCMAP initiative. *ISPRS J. Photogramm. Remote Sens.* **2016**, *122*, 206–221. [CrossRef]
48. Dwyer, J.L.; Roy, D.P.; Sauer, B.; Jenkerson, C.B.; Zhang, H.K.; Lymburner, L. Analysis ready data: Enabling analysis of the landsat archive. *Remote Sens.* **2018**, *10*, 1363. [CrossRef]
49. Weng, Q.; Larson, R.C. Satellite Remote Sensing of Urban Heat Islands: Current Practice and Prospects. In *Geo-Spatial Technologies in Urban Environments*; 2005; pp. 91–111. Available online: https://link.springer.com/chapter/10.1007%2F3-540-26676-3_10 (accessed on 2 February 2020). [CrossRef]

50. Jiang, G.M.; Li, Z.L.; Nerry, F. Land surface emissivity retrieval from combined mid-infrared and thermal infrared data of MSG-SEVIRI. *Remote Sens. Environ.* **2006**, *105*, 326–340. [CrossRef]
51. Kalma, J.D.; McVicar, T.R.; McCabe, M.F. Estimating land surface evaporation: A review of methods using remotely sensed surface temperature data. *Surv. Geophys.* **2008**, *29*, 421–469. [CrossRef]
52. Racoviteanu, A.E.; Williams, M.W.; Barry, R.G. Optical remote sensing of glacier characteristics: A review with focus on the Himalaya. *Sensors* **2008**, *8*, 3355–3383. [CrossRef]
53. Rizwan, A.M.; Dennis, L.Y.C.; Liu, C. A review on the generation, determination and mitigation of Urban Heat Island. *J. Environ. Sci.* **2008**, *20*, 120–128. [CrossRef]
54. Sailor, D.J. A review of methods for estimating anthropogenic heat and moisture emissions in the urban environment. *Int. J. Climatol.* **2011**, *31*, 189–199. [CrossRef]
55. Li, Z.L.; Tang, B.H.; Wu, H.; Ren, H.; Yan, G.; Wan, Z.; Trigo, I.F.; Sobrino, J.A. Satellite-derived land surface temperature: Current status and perspectives. *Remote Sens. Environ.* **2013**, *131*, 14–37. [CrossRef]
56. Ngie, A.; Abutaleb, K.; Ahmed, F.; Darwish, A.; Ahmed, M. Assessment of urban heat island using satellite remotely sensed imagery: A review. *S. Afr. Geogr. J.* **2014**, *96*, 198–214. [CrossRef]
57. Huang, Q.; Lu, Y. Urban heat island research from 1991 to 2015: A bibliometric analysis. *Theor. Appl. Climatol.* **2018**, *131*, 1055–1067. [CrossRef]
58. Zhang, Y.; Thenkabail, P.S.; Wang, P. A bibliometric profile of the Remote Sensing Open Access Journal published by MDPI between 2009 and 2018. *Remote Sens.* **2019**, *11*, 91. [CrossRef]
59. Deilami, K.; Kamruzzaman, M.; Liu, Y. Urban heat island effect: A systematic review of spatio-temporal factors, data, methods, and mitigation measures. *Int. J. Appl. Earth Obs. Geoinf.* **2018**, *67*, 30–42. [CrossRef]
60. Zhou, D.; Xiao, J.; Bonafoni, S.; Berger, C.; Deilami, K.; Zhou, Y.; Frolking, S.; Yao, R.; Qiao, Z.; Sobrino, J.A. Satellite remote sensing of surface urban heat islands: Progress, challenges, and perspectives. *Remote Sens.* **2019**, *11*, 48. [CrossRef]
61. Becker, F.; Li, Z.L. Surface temperature and emissivity at various scales: Definition, measurement and related problems. *Remote Sens. Rev.* **1995**, *12*, 225–253. [CrossRef]
62. Dash, P.; Göttsche, F.M.; Olesen, F.S.; Fischer, H. Land surface temperature and emissivity estimation from passive sensor data: Theory and practice-current trends. *Int. J. Remote Sens.* **2002**, *23*, 2563–2594. [CrossRef]
63. Miles, V.; Esau, I. Seasonal and spatial characteristics of Urban Heat Islands (UHIs) in northern West Siberian cities. *Remote Sens.* **2017**, *9*, 989. [CrossRef]
64. Trlica, A.; Hutyra, L.R.; Schaaf, C.L.; Erb, A.; Wang, J.A. Albedo, Land Cover, and Daytime Surface Temperature Variation Across an Urbanized Landscape. *Earths Future* **2017**, *5*, 1084–1101. [CrossRef]
65. Bonafoni, S. Spectral index utility for summer urban heating analysis. *J. Appl. Remote Sens.* **2015**, *9*. [CrossRef]
66. Wong, M.S.; Nichol, J.E. Spatial variability of frontal area index and its relationship with urban heat island intensity. *Int. J. Remote Sens.* **2013**, *34*, 885–896. [CrossRef]
67. Jin, M.S. Developing an Index to Measure Urban Heat Island Effect Using Satellite Land Skin Temperature and Land Cover Observations. *J. Clim.* **2012**, *25*, 6193–6201. [CrossRef]
68. Wu, C.D.; Lung, S.C.C.; Jan, J.F. Development of a 3-D urbanization index using digital terrain models for surface urban heat island effects. *Isprs J. Photogramm. Remote Sens.* **2013**, *81*, 1–11. [CrossRef]
69. Hu, L.Q.; Brunsell, N.A. The impact of temporal aggregation of land surface temperature data for surface urban heat island (SUHI) monitoring. *Remote Sens. Environ.* **2013**, *134*, 162–174. [CrossRef]
70. Zhang, Z.W.; Du, Q.Y. A Bayesian Kriging Regression Method to Estimate Air Temperature Using Remote Sensing Data. *Remote Sens.* **2019**, *11*, 767. [CrossRef]
71. Wicki, A.; Parlow, E. Multiple Regression Analysis for Unmixing of Surface Temperature Data in an Urban Environment. *Remote Sens.* **2017**, *9*, 684. [CrossRef]
72. Dai, Z.; Guldmann, J.M.; Hu, Y. Spatial regression models of park and land-use impacts on the urban heat island in central Beijing. *Sci. Total Environ.* **2018**, *626*, 1136–1147. [CrossRef]
73. Song, J.; Du, S.; Feng, X.; Guo, L. The relationships between landscape compositions and land surface temperature: Quantifying their resolution sensitivity with spatial regression models. *Landsc. Urban Plan.* **2014**, *123*, 145–157. [CrossRef]
74. Sellers, P.J.; Meeson, B.W.; Hall, F.G.; Asrar, G.; Murphy, R.E.; Schiffer, R.A.; Bretherton, F.P.; Dickinson, R.E.; Ellingson, R.G.; Field, C.B.; et al. Remote sensing of the land surface for studies of global change: Models—Algorithms—Experiments. *Remote Sens. Environ.* **1995**, *51*, 3–26. [CrossRef]
75. Du, S.H.; Xiong, Z.Q.; Wang, Y.C.; Guo, L. Quantifying the multilevel effects of landscape composition and configuration on land surface temperature. *Remote Sens. Environ.* **2016**, *178*, 84–92. [CrossRef]
76. Shahraiyni, H.T.; Sodoudi, S.; El-Zafarany, A.; Abou El Seoud, T.; Ashraf, H.; Krone, K. A Comprehensive Statistical Study on Daytime Surface Urban Heat Island during Summer in Urban Areas, Case Study: Cairo and Its New Towns. *Remote Sens.* **2016**, *8*, 643. [CrossRef]
77. Chun, B.; Guldmann, J.M. Spatial statistical analysis and simulation of the urban heat island in high-density central cities. *Landsc. Urban Plan.* **2014**, *125*, 76–88. [CrossRef]
78. Ho, H.C.; Knudby, A.; Sirovyak, P.; Xu, Y.M.; Hodul, M.; Henderson, S.B. Mapping maximum urban air temperature on hot summer days. *Remote Sens. Environ.* **2014**, *154*, 38–45. [CrossRef]

79. Lai, J.; Zhan, W.; Quan, J.; Bechtel, B.; Wang, K.; Zhou, J.; Huang, F.; Chakraborty, T.; Liu, Z.; Lee, X. Statistical estimation of next-day nighttime surface urban heat islands. *ISPRS J. Photogramm. Remote Sens.* **2021**, *176*, 182–195. [CrossRef]
80. Huang, X.; Wang, Y. Investigating the effects of 3D urban morphology on the surface urban heat island effect in urban functional zones by using high-resolution remote sensing data: A case study of Wuhan, Central China. *ISPRS J. Photogramm. Remote Sens.* **2019**, *152*, 119–131. [CrossRef]
81. Li, H.; Zhou, Y.; Li, X.; Meng, L.; Wang, X.; Wu, S.; Sodoudi, S. A new method to quantify surface urban heat island intensity. *Sci. Total Environ.* **2018**, *624*, 262–272. [CrossRef] [PubMed]
82. Berger, C.; Rosentreter, J.; Voltersen, M.; Baumgart, C.; Schmullius, C.; Hese, S. Spatio-temporal analysis of the relationship between 2D/3D urban site characteristics and land surface temperature. *Remote Sens. Environ.* **2017**, *193*, 225–243. [CrossRef]
83. Liu, K.; Su, H.B.; Li, X.K.; Wang, W.M.; Yang, L.J.; Liang, H. Quantifying Spatial-Temporal Pattern of Urban Heat Island in Beijing: An Improved Assessment Using Land Surface Temperature (LST) Time Series Observations From LANDSAT, MODIS, and Chinese New Satellite GaoFen-1. *IEEE J. Sel. Top. Appl. Earth Obs. Remote Sens.* **2016**, *9*, 2028–2042. [CrossRef]
84. Fu, P.; Weng, Q.H. Consistent land surface temperature data generation from irregularly spaced Landsat imagery. *Remote Sens. Environ.* **2016**, *184*, 175–187. [CrossRef]
85. Liang, B.Q.; Weng, Q.H. Assessing Urban Environmental Quality Change of Indianapolis, United States, by the Remote Sensing and GIS Integration. *IEEE J. Sel. Top. Appl. Earth Obs. Remote Sens.* **2011**, *4*, 43–55. [CrossRef]
86. Dousset, B.; Gourmelon, F. Satellite multi-sensor data analysis of urban surface temperatures and landcover. *ISPRS J. Photogramm. Remote Sens.* **2003**, *58*, 43–54. [CrossRef]
87. Chakraborty, T.; Lee, X. A simplified urban-extent algorithm to characterize surface urban heat islands on a global scale and examine vegetation control on their spatiotemporal variability. *Int. J. Appl. Earth Obs. Geoinf.* **2019**, *74*, 269–280. [CrossRef]
88. Mpakairia, K.S.; Muvengwi, J. Night-time lights and their influence on summer night land surface temperature in two urban cities of Zimbabwe: A geospatial perspective. *Urban Clim.* **2019**. [CrossRef]
89. Zhang, Y.; Jiang, P.; Zhang, H.; Cheng, P. Study on urban heat island intensity level identification based on an improved restricted Boltzmann machine. *Int. J. Environ. Res. Public Health* **2018**, *15*, 186. [CrossRef]
90. Tran, D.X.; Pla, F.; Latorre-Carmona, P.; Myint, S.W.; Gaetano, M.; Kieu, H.V. Characterizing the relationship between land use land cover change and land surface temperature. *ISPRS J. Photogramm. Remote Sens.* **2017**, *124*, 119–132. [CrossRef]
91. Weng, Q.H.; Fu, P. Modeling annual parameters of clear-sky land surface temperature variations and evaluating the impact of cloud cover using time series of Landsat TIR data. *Remote Sens. Environ.* **2014**, *140*, 267–278. [CrossRef]
92. Mallick, J.; Rahman, A.; Singh, C.K. Modeling urban heat islands in heterogeneous land surface and its correlation with impervious surface area by using night-time ASTER satellite data in highly urbanizing city, Delhi-India. *Adv. Space Res.* **2013**, *52*, 639–655. [CrossRef]
93. Connors, J.P.; Galletti, C.S.; Chow, W.T.L. Landscape configuration and urban heat island effects: Assessing the relationship between landscape characteristics and land surface temperature in Phoenix, Arizona. *Landsc. Ecol.* **2013**, *28*, 271–283. [CrossRef]
94. Wentz, E.A.; Quattrochi, D.A.; Netzband, M.; Myint, S.W. Synthesizing urban remote sensing through application, scale, data and case studies. *Geocarto Int.* **2012**, *27*, 425–442. [CrossRef]
95. Xian, G.; Crane, M. An analysis of urban thermal characteristics and associated land cover in Tampa Bay and Las Vegas using Landsat satellite data. *Remote Sens. Environ.* **2006**, *104*, 147–156. [CrossRef]
96. Wilson, J.S.; Clay, M.; Martin, E.; Stuckey, D.; Vedder-Risch, K. Evaluating environmental influences of zoning in urban ecosystems with remote sensing. *Remote Sens. Environ.* **2003**, *86*, 303–321. [CrossRef]
97. Xian, G.Z.; Shi, H.; Auch, R.; Gallo, K.P.; Zhou, Q.; Wu, Z.; Kolian, M. The effects of urban land cover dynamics on urban heat island intensity and temporal trends. *GIScience Remote Sens.* **2021**. [CrossRef]
98. Howard, L. *The Climate of London, Deduced From Meteorological Observations, Made at Different Places in the Neighbourhood of the Metropolis*; Phillips, W., Ed.; George Yard: London, UK, 1818.
99. Stewart, I.D.; Oke, T.R.; Krayenhoff, E.S. Evaluation of the 'local climate zone' scheme using temperature observations and model simulations. *Int. J. Climatol.* **2014**, *34*, 1062–1080. [CrossRef]
100. US EPA. *Reducing Urban Heat Islands: Compendium of Strategies*; US EPA: Washington, DC, USA, 2008.
101. Martin-Vide, J.; Sarricolea, P.; Moreno-García, M.C. On the definition of urban heat island intensity: The "rural" reference. *Front. Earth Sci.* **2015**, *3*. [CrossRef]
102. Giridharan, R.; Emmanuel, R. The impact of urban compactness, comfort strategies and energy consumption on tropical urban heat island intensity: A review. *Sustain. Cities Soc.* **2018**, *40*, 677–687. [CrossRef]
103. Zhang, F.S.; Liu, Z.X. Fractal theory and its application in the analysis of soil spatial variability: A review. *J. Appl. Ecol.* **2011**, *22*, 1351–1358.
104. Gustafson, E.J. How has the state-of-the-art for quantification of landscape pattern advanced in the twenty-first century? *Landsc. Ecol.* **2019**, *34*, 2065–2072. [CrossRef]
105. Hu, L.Q.; Brunsell, N.A. A new perspective to assess the urban heat island through remotely sensed atmospheric profiles. *Remote Sens. Environ.* **2015**, *158*, 393–406. [CrossRef]
106. Hu, L.Q.; Monaghan, A.J.; Brunsell, N.A. Investigation of Urban Air Temperature and Humidity Patterns during Extreme Heat Conditions Using Satellite-Derived Data. *J. Appl. Meteorol. Climatol.* **2015**, *54*, 2245–2259. [CrossRef]

107. Tam, B.Y.; Gough, W.A.; Mohsin, T. The impact of urbanization and the urban heat island effect on day to day temperature variation. *Urban Clim.* **2015**, *12*, 1–10. [CrossRef]
108. Roth, M.; Oke, T.R.; Emery, W.J. Satellite-derived urban heat islands from three coastal cities and the utilization of such data in urban climatology. *Int. J. Remote Sens.* **1989**, *10*, 1699–1720. [CrossRef]
109. Lagouarde, J.P.; Moreau, P.; Irvine, M.; Bonnefond, J.M.; Voogt, J.A.; Solliec, F. Airborne experimental measurements of the angular variations in surface temperature over urban areas: Case study of Marseille (France). *Remote Sens. Environ.* **2004**, *93*, 443–462. [CrossRef]
110. Li, Y.; Schubert, S.; Kropp, J.P.; Rybski, D. On the influence of density and morphology on the Urban Heat Island intensity. *Nat. Commun.* **2020**, *11*, 2647. [CrossRef] [PubMed]
111. Ramírez-Aguilar, E.A.; Lucas Souza, L.C. Urban form and population density: Influences on Urban Heat Island intensities in Bogotá, Colombia. *Urban Clim.* **2019**, *29*. [CrossRef]
112. Paoletti, M.E.; Haut, J.M.; Plaza, J.; Plaza, A. Deep learning classifiers for hyperspectral imaging: A review. *ISPRS J. Photogramm. Remote Sens.* **2019**, *158*, 279–317. [CrossRef]
113. Zhang, L.; Zhang, L.; Du, B. Deep learning for remote sensing data: A technical tutorial on the state of the art. *IEEE Geosci. Remote Sens. Mag.* **2016**, *4*, 22–40. [CrossRef]
114. Chen, M.; Jiang, X.; Wu, H.; Wang, N.; Tang, R. An in-Scene Atmospheric Compensation Algorithm for Aster Thermal Band. In Proceedings of the International Geoscience and Remote Sensing Symposium (IGARSS), Yokohama, Japan, 28 July–2 August 2019; pp. 1876–1879. [CrossRef]
115. French, A.N.; Inamdar, A. Land cover characterization for hydrological modelling using thermal infrared emissivities. *Int. J. Remote Sens.* **2010**, *31*, 3867–3883. [CrossRef]
116. Sobrino, J.A.; Jiménez-Muñoz, J.C.; Sòria, G.; Romaguera, M.; Guanter, L.; Moreno, J.; Plaza, A.; Martínez, P. Land surface emissivity retrieval from different VNIR and TIR sensors. *IEEE Trans. Geosci. Remote Sens.* **2008**, *46*, 316–327. [CrossRef]
117. Mohamed, A.A.; Odindi, J.; Mutanga, O. Land surface temperature and emissivity estimation for Urban Heat Island assessment using medium- and low-resolution space-borne sensors: A review. *Geocarto Int.* **2017**, *32*, 455–470. [CrossRef]
118. Gillespie, A.; Rokugawa, S.; Matsunaga, T.; Steven Cothern, J.; Hook, S.; Kahle, A.B. A temperature and emissivity separation algorithm for advanced spaceborne thermal emission and reflection radiometer (ASTER) images. *IEEE Trans. Geosci. Remote Sens.* **1998**, *36*, 1113–1126. [CrossRef]
119. Rolim, S.B.A.; Grondona, A.; Hackmann, C.L.; Rocha, C. A Review of Temperature and Emissivity Retrieval Methods: Applications and Restrictions. *Am. J. Environ. Eng.* **2016**, *6*, 119–128. [CrossRef]
120. Li, Z.L.; Wu, H.; Wang, N.; Qiu, S.; Sobrino, J.A.; Wan, Z.; Tang, B.H.; Yan, G. Land surface emissivity retrieval from satellite data. *Int. J. Remote Sens.* **2013**, *34*, 3084–3127. [CrossRef]
121. Ma, M.; Chen, S.B.; Lu, T.Q.; Lu, P.; Xiao, Y. Study on scale problems based on the diviner thermal infrared emissivity of LRO satellite. *Hongwai Yu Haomibo Xuebao J. Infrared Millim. Waves* **2018**, *37*, 315–324. [CrossRef]
122. Wang, W.; Yao, X.; Shu, J. Air advection induced differences between canopy and surface heat islands. *Sci. Total Environ.* **2020**, *725*. [CrossRef]
123. Oláh, A.B. The possibilities of decreasing the urban heat Island. *Appl. Ecol. Environ. Res.* **2012**, *10*, 173–183. [CrossRef]
124. Gaur, A.; Eichenbaum, M.K.; Simonovic, S.P. Analysis and modelling of surface Urban Heat Island in 20 Canadian cities under climate and land-cover change. *J. Environ. Manag.* **2018**, *206*, 145–157. [CrossRef]
125. Cai, Z.; Han, G.; Chen, M. Do water bodies play an important role in the relationship between urban form and land surface temperature? *Sustain. Cities Soc.* **2018**, *39*, 487–498. [CrossRef]
126. Akbari, H.; Cartalis, C.; Kolokotsa, D.; Muscio, A.; Pisello, A.L.; Rossi, F.; Santamouris, M.; Synnefa, A.; Wong, N.H.; Zinzi, M. Local climate change and urban heat island mitigation techniques—The state of the art. *J. Civ. Eng. Manag.* **2016**, *22*, 1–16. [CrossRef]
127. Ulpiani, G.; Ranzi, G.; Shah, K.W.; Feng, J.; Santamouris, M. On the energy modulation of daytime radiative coolers: A review on infrared emissivity dynamic switch against overcooling. *Sol. Energy* **2020**, *209*, 278–301. [CrossRef]
128. Pickett, S.T.A.; Cadenasso, M.L.; Grove, J.M.; Boone, C.G.; Groffman, P.M.; Irwin, E.; Kaushal, S.S.; Marshall, V.; McGrath, B.P.; Nilon, C.H.; et al. Urban ecological systems: Scientific foundations and a decade of progress. *J. Environ. Manag.* **2011**, *92*, 331–362. [CrossRef] [PubMed]
129. Smargiassi, A.; Goldberg, M.S.; Plante, C.; Fournier, M.; Baudouin, Y.; Kosatsky, T. Variation of daily warm season mortality as a function of micro-urban heat islands. *J. Epidemiol. Community Health* **2009**, *63*, 659–664. [CrossRef] [PubMed]
130. Heaviside, C.; Vardoulakis, S.; Cai, X.M. Attribution of mortality to the urban heat island during heatwaves in the West Midlands, UK. *Environ. Health A Glob. Access Sci. Source* **2016**, *15*. [CrossRef] [PubMed]
131. Moon, J. The effect of the heatwave on the morbidity and mortality of diabetes patients; a meta-analysis for the era of the climate crisis. *Environ. Res.* **2021**, *195*. [CrossRef]
132. Li, J.; Xu, X.; Yang, J.; Liu, Z.; Xu, L.; Gao, J.; Liu, X.; Wu, H.; Wang, J.; Yu, J.; et al. Ambient high temperature and mortality in Jinan, China: A study of heat thresholds and vulnerable populations. *Environ. Res.* **2017**, *156*, 657–664. [CrossRef]
133. Tan, J.; Zheng, Y.; Tang, X.; Guo, C.; Li, L.; Song, G.; Zhen, X.; Yuan, D.; Kalkstein, A.J.; Li, F.; et al. The urban heat island and its impact on heat waves and human health in Shanghai. *Int. J. Biometeorol.* **2010**, *54*, 75–84. [CrossRef]

134. Johnson, D.P.; Wilson, J.S.; Luber, G.C. Socioeconomic indicators of heat-related health risk supplemented with remotely sensed data. *Int. J. Health Geogr.* **2009**, *8*, 57. [CrossRef] [PubMed]
135. Chen, L.; Jiang, R.; Xiang, W.N. Surface Heat Island in Shanghai and Its Relationship with Urban Development from 1989 to 2013. *Adv. Meteorol.* **2016**. [CrossRef]
136. Zhou, D.; Bonafoni, S.; Zhang, L.; Wang, R. Remote sensing of the urban heat island effect in a highly populated urban agglomeration area in East China. *Sci. Total Environ.* **2018**, *628–629*, 415–429. [CrossRef]
137. Zhou, B.; Rybski, D.; Kropp, J.P. The role of city size and urban form in the surface urban heat island. *Sci. Rep.* **2017**, *7*, 4791. [CrossRef]
138. Zhou, D.; Zhang, L.; Hao, L.; Sun, G.; Liu, Y.; Zhu, C. Spatiotemporal trends of urban heat island effect along the urban development intensity gradient in China. *Sci. Total Environ.* **2016**, *544*, 617–626. [CrossRef] [PubMed]
139. Adeyeri, O.E.; Akinsanola, A.A.; Ishola, K.A. Investigating surface urban heat island characteristics over Abuja, Nigeria: Relationship between land surface temperature and multiple vegetation indices. *Remote Sens. Appl. Soc. Environ.* **2017**, *7*, 57–68. [CrossRef]
140. Wu, X.; Cheng, Q. Coupling Relationship of Land Surface Temperature, Impervious Surface Area and Normalized Difference Vegetation Index for Urban Heat Island Using Remote Sensing. In Proceedings of the SPIE—The International Society for Optical Engineering 2007. Available online: https://www.spiedigitallibrary.org/conference-proceedings-of-spie/6749/1/Coupling-relationship-of-land-surface-temperature-impervious-surface-area-and/10.1117/12.737550.full?SSO=1 (accessed on 10 October 2020). [CrossRef]
141. Alexander, C. Normalised difference spectral indices and urban land cover as indicators of land surface temperature (LST). *Int. J. Appl. Earth Obs. Geoinf.* **2020**, *86*, 102013. [CrossRef]
142. Diaz-Pacheco, J.; Gutiérrez, J. Exploring the limitations of CORINE Land Cover for monitoring urban land-use dynamics in metropolitan areas. *J. Land Use Sci.* **2014**, *9*, 243–259. [CrossRef]
143. Giannakopoulos, C.; Kostopoulou, E.; Varotsos, K.V.; Tziotziou, K.; Plitharas, A. An integrated assessment of climate change impacts for Greece in the near future. *Reg. Environ. Chang.* **2011**, *11*, 829–843. [CrossRef]
144. Estrada, F.; Botzen, W.J.W.; Tol, R.S.J. A global economic assessment of city policies to reduce climate change impacts. *Nat. Clim. Chang.* **2017**, *7*, 403–406. [CrossRef]
145. Li, X.; Li, W.; Middel, A.; Harlan, S.L.; Brazel, A.J.; Turner, B.L. Remote sensing of the surface urban heat island and land architecture in Phoenix, Arizona: Combined effects of land composition and configuration and cadastral-demographic-economic factors. *Remote Sens. Environ.* **2016**, *174*, 233–243. [CrossRef]
146. Cai, G.Y.; Liu, Y.; Du, M.Y. Impact of the 2008 Olympic Games on urban thermal environment in Beijing, China from satellite images. *Sustain. Cities Soc.* **2017**, *32*, 212–225. [CrossRef]
147. Bonafoni, S.; Baldinelli, G.; Verducci, P. Sustainable strategies for smart cities: Analysis of the town development effect on surface urban heat island through remote sensing methodologies. *Sustain. Cities Soc.* **2017**, *29*, 211–218. [CrossRef]
148. Pickett, S.T.A.; Cadenasso, M.L.; Grove, J.M.; Nilon, C.H.; Pouyat, R.V.; Zipperer, W.C.; Costanza, R. Urban Ecological Systems: Linking Terrestrial Ecological, Physical, and Socioeconomic Components of Metropolitan Areas. *Annu. Rev. Ecol. Syst.* **2001**, *32*, 127–157. [CrossRef]
149. Ciardini, V.; Caporaso, L.; Sozzi, R.; Petenko, I.; Bolignano, A.; Morelli, M.; Melas, D.; Argentini, S. Interconnections of the urban heat island with the spatial and temporal micrometeorological variability in Rome. *Urban Clim.* **2019**, *29*. [CrossRef]
150. Čeplová, N.; Kalusová, V.; Lososová, Z. Effects of settlement size, urban heat island and habitat type on urban plant biodiversity. *Landsc. Urban Plan.* **2017**, *159*, 15–22. [CrossRef]
151. Coluzzi, R.; D'Emilio, M.; Imbrenda, V.; Giorgio, G.A.; Lanfredi, M.; Macchiato, M.; Ragosta, M.; Simoniello, T.; Telesca, V. Investigating climate variability and long-term vegetation activity across heterogeneous basilicata agroecosystems. *Geomat. Nat. Hazards Risk* **2019**, *10*, 168–180. [CrossRef]
152. Kaiser, A.; Merckx, T.; van Dyck, H. The Urban Heat Island and its spatial scale dependent impact on survival and development in butterflies of different thermal sensitivity. *Ecol. Evol.* **2016**, *6*, 4129–4140. [CrossRef]
153. Weng, Q.; Lu, D.; Schubring, J. Estimation of land surface temperature-vegetation abundance relationship for urban heat island studies. *Remote Sens. Environ.* **2004**, *89*, 467–483. [CrossRef]
154. Wang, H.T.; Zhang, Y.Z.; Tsou, J.Y.; Li, Y. Surface Urban Heat Island Analysis of Shanghai (China) Based on the Change of Land Use and Land Cover. *Sustainability* **2017**, *9*, 1538. [CrossRef]
155. Stathopoulou, M.; Cartalis, C.; Keramitsoglou, I. Mapping micro-urban heat islands using NOAA/AVHRR images and CORINE Land Cover: An application to coastal cities of Greece. *Int. J. Remote Sens.* **2004**, *25*, 2301–2316. [CrossRef]
156. Tran, H.; Uchihama, D.; Ochi, S.; Yasuoka, Y. Assessment with satellite data of the urban heat island effects in Asian mega cities. *Int. J. Appl. Earth Obs. Geoinf.* **2006**, *8*, 34–48. [CrossRef]
157. Zhang, Y.; Zhan, Y.; Yu, T.; Ren, X. Urban green effects on land surface temperature caused by surface characteristics: A case study of summer Beijing metropolitan region. *Infrared Phys. Technol.* **2017**, *86*, 35–43. [CrossRef]
158. Marina, S.; Constantinos, C. Study of the Urban Heat Island of Athens, Greece during Daytime and Night-Time. In Proceedings of the 2007 Urban Remote Sensing Joint Event, URS. Available online: https://www.researchgate.net/publication/4254319_Study_of_the_urban_heat_island_of_Athens_Greece_during_daytime_and_night-time (accessed on 5 May 2020). [CrossRef]

159. Liu, K.; Su, H.B.; Zhang, L.F.; Yang, H.; Zhang, R.H.; Li, X.K. Analysis of the Urban Heat Island Effect in Shijiazhuang, China Using Satellite and Airborne Data. *Remote Sens.* **2015**, *7*, 4804–4833. [CrossRef]
160. Ben-Dor, E.; Saaroni, H. Airborne video thermal radiometry as a tool for monitoring microscale structures of the urban heat island. *Int. J. Remote Sens.* **1997**, *18*, 3039–3053. [CrossRef]
161. Stathopoulou, M.; Cartalis, C. Downscaling AVHRR land surface temperatures for improved surface urban heat island intensity estimation. *Remote Sens. Environ.* **2009**, *113*, 2592–2605. [CrossRef]
162. Gallo, K.P.; Owen, T.W. Assessment of urban heat islands: A multi-sensor perspective for the Dallas-Ft. Worth, USA region. *Geocarto Int.* **1998**, *13*, 35–41. [CrossRef]
163. Zhi Qiao, Z.; Chen Wu, C.; Dongqi Zhao, D.; Xinliang Xu, X.; Jilin Yang, J.; Li Feng, L.; Sun, Z.; Liu, L. Determining the Boundary and Probability of Surface Urban Heat Island Footprint Based on a Logistic Model. *Remote Sens.* **2019**, *11*, 1368. [CrossRef]
164. Keramitsoglou, I.; Kiranoudis, C.T.; Ceriola, G.; Weng, Q.; Rajasekar, U. Identification and analysis of urban surface temperature patterns in Greater Athens, Greece, using MODIS imagery. *Remote Sens. Environ.* **2011**, *115*, 3080–3090. [CrossRef]
165. Ye, C.; Wang, M.J.; Li, J. Derivation of the characteristics of the surface urban heat island in the greater toronto area using thermal infrared remote sensing. *Remote Sens. Lett.* **2017**, *8*, 637–646. [CrossRef]
166. Kato, S.; Yamaguchi, Y. Estimation of storage heat flux in an urban area using ASTER data. *Remote Sens. Environ.* **2007**, *110*, 1–17. [CrossRef]
167. Lu, D.S.; Weng, Q.H. Spectral mixture analysis of ASTER images for examining the relationship between urban thermal features and biophysical descriptors in Indianapolis, Indiana, USA. *Remote Sens. Environ.* **2006**, *104*, 157–167. [CrossRef]
168. Sun, Y.; Gao, C.; Li, J.; Li, W.; Ma, R. Examining urban thermal environment dynamics and relations to biophysical composition and configuration and socio-economic factors: A case study of the Shanghai metropolitan region. *Sustain. Cities Soc.* **2018**, *40*, 284–295. [CrossRef]
169. Quan, J.L.; Zhan, W.F.; Chen, Y.H.; Wang, M.J.; Wang, J.F. Time series decomposition of remotely sensed land surface temperature and investigation of trends and seasonal variations in surface urban heat islands. *J. Geophys. Res. Atmos.* **2016**, *121*, 2638–2657. [CrossRef]
170. Gawuc, L.; Struzewska, J. Impact of MODIS Quality Control on Temporally Aggregated Urban Surface Temperature and Long-Term Surface Urban Heat Island Intensity. *Remote Sens.* **2016**, *8*, 374. [CrossRef]
171. Aniello, C.; Morgan, K.; Busbey, A.; Newland, L. Mapping micro-urban heat islands using LANDSAT TM and a GIS. *Comput. Geosci.* **1995**, *21*, 965–967, 969. [CrossRef]
172. Weng, Q.H. Fractal analysis of satellite-detected urban heat island effect. *Photogramm. Eng. Remote Sens.* **2003**, *69*, 555–566. [CrossRef]
173. Stathopoulou, M.; Cartalis, C. Daytime urban heat islands from Landsat ETM+ and Corine land cover data: An application to major cities in Greece. *Sol. Energy* **2007**, *81*, 358–368. [CrossRef]
174. Sagris, V.; Sepp, M. Landsat-8 TIRS Data for Assessing Urban Heat Island Effect and Its Impact on Human Health. *IEEE Geosci. Remote Sens. Lett.* **2017**, *14*, 2385–2389. [CrossRef]
175. Hulley, G.; Hook, S.; Fisher, J.; Lee, C. ECOSTRESS, A NASA Earth-Ventures Instrument for studying links between the water cycle and plant health over the diurnal cycle. In Proceedings of the International Geoscience and Remote Sensing Symposium (IGARSS), Fort Worth, TX, USA, 23–28 July 2017; pp. 5494–5496. [CrossRef]
176. Schultz, J.A.; Hartmann, M.; Heinemann, S.; Janke, J.; Jürgens, C.; Oertel, D.; Rücker, G.; Thonfeld, F.; Rienow, A. DIEGO: A Multispectral Thermal Mission for Earth Observation on the International Space Station. *Eur. J. Remote Sens.* **2019**. [CrossRef]
177. Elmes, A.; Rogan, J.; Williams, C.; Ratick, S.; Nowak, D.; Martin, D. Effects of urban tree canopy loss on land surface temperature magnitude and timing. *ISPRS J. Photogramm. Remote Sens.* **2017**, *128*, 338–353. [CrossRef]
178. Yan, L.; Roy, D.P. Spatially and temporally complete Landsat reflectance time series modelling: The fill-and-fit approach. *Remote Sens. Environ.* **2020**, *241*. [CrossRef]
179. Forster, B. Some urban measurements from Landsat data (Sydney, Australia). *Photogramm. Eng. Remote Sens.* **1983**, *49*, 1693–1707.
180. Zhao, W.; Li, A. A Review on Land Surface Processes Modelling over Complex Terrain. *Adv. Meteorol.* **2015**, *2015*. [CrossRef]
181. Nichol, J.E.; Fung, W.Y.; Lam, K.s.; Wong, M.S. Urban heat island diagnosis using ASTER satellite images and 'in situ' air temperature. *Atmos. Res.* **2009**, *94*, 276–284. [CrossRef]
182. Hirano, Y.; Yasuoka, Y.; Ichinose, T. Urban climate simulation by incorporating satellite-derived vegetation cover distribution into a mesoscale meteorological model. *Theor. Appl. Climatol.* **2004**, *79*, 175–184. [CrossRef]
183. Liou, Y.A.; Kar, S.K. Evapotranspiration estimation with remote sensing and various surface energy balance algorithms-a review. *Energies* **2014**, *7*, 2821–2849. [CrossRef]
184. Stone, B.; Norman, J.M. Land use planning and surface heat island formation: A parcel-based radiation flux approach. *Atmos. Environ.* **2006**, *40*, 3561–3573. [CrossRef]
185. Zubair, O.A.; Weilert, T. Potential Application of Change in Urban Green Space as an Indicator of Urban Environmental Quality Change. *Univers. J. Geosci.* **2014**, *2*, 222–228. [CrossRef]
186. Nichol, J.; Wong, M.S. Modeling urban environmental quality in a tropical city. *Landsc. Urban Plan.* **2005**, *73*, 49–58. [CrossRef]
187. Shin, M.; Kang, Y.; Park, S.; Im, J.; Yoo, C.; Quackenbush, L.J. Estimating ground-level particulate matter concentrations using satellite-based data: A review. *GISci. Remote Sens.* **2020**, *57*, 174–189. [CrossRef]

188. McDonald, R.I.; Kroeger, T.; Zhang, P.; Hamel, P. The Value of US Urban Tree Cover for Reducing Heat-Related Health Impacts and Electricity Consumption. *Ecosystems* **2019**. [CrossRef]
189. Arnfield, A.J. Two decades of urban climate research: A review of turbulence, exchanges of energy and water, and the urban heat island. *Int. J. Climatol.* **2003**, *23*, 1–26. [CrossRef]
190. Fu, P.; Xie, Y.; Weng, Q.; Myint, S.; Meacham-Hensold, K.; Bernacchi, C. A physical model-based method for retrieving urban land surface temperatures under cloudy conditions. *Remote Sens. Environ.* **2019**, *230*. [CrossRef]
191. Hulley, G.; Shivers, S.; Wetherley, E.; Cudd, R. New ECOSTRESS and MODIS land surface temperature data reveal fine-scale heat vulnerability in cities: A case study for Los Angeles County, California. *Remote Sens.* **2019**, *11*, 2136. [CrossRef]
192. Shen, H.F.; Huang, L.W.; Zhang, L.P.; Wu, P.H.; Zeng, C. Long-term and fine-scale satellite monitoring of the urban heat island effect by the fusion of multi-temporal and multi-sensor remote sensed data: A 26-year case study of the city of Wuhan in China. *Remote Sens. Environ.* **2016**, *172*, 109–125. [CrossRef]
193. Avdan, U.; Jovanovska, G. Algorithm for automated mapping of land surface temperature using LANDSAT 8 satellite data. *J. Sens.* **2016**, *2016*. [CrossRef]
194. Peng, J.; Qiao, R.; Liu, Y.; Blaschke, T.; Li, S.; Wu, J.; Xu, Z.; Liu, Q. A wavelet coherence approach to prioritizing influencing factors of land surface temperature and associated research scales. *Remote Sens. Environ.* **2020**, *246*. [CrossRef]
195. Tang, Y.Q.; Lan, C.Y.; Feng, H.H. Effect analysis of land-use pattern with landscape metrics on an urban heat island. *J. Appl. Remote Sens.* **2018**, *12*. [CrossRef]
196. Schwarz, N.; Schlink, U.; Franck, U.; Großmann, K. Relationship of land surface and air temperatures and its implications for quantifying urban heat island indicators—An application for the city of Leipzig (Germany). *Ecol. Indic.* **2012**, *18*, 693–704. [CrossRef]
197. Szymanowski, M.; Kryza, M. GIS-based techniques for urban heat island spatialization. *Clim. Res.* **2009**, *38*, 171–187. [CrossRef]
198. Firozjaei, M.K.; Weng, Q.; Zhao, C.; Kiavarz, M.; Lu, L.; Alavipanah, S.K. Surface anthropogenic heat islands in six megacities: An assessment based on a triple-source surface energy balance model. *Remote Sens. Environ.* **2020**, *242*. [CrossRef]
199. Schmitt, M.; Zhu, X.X. Data Fusion and Remote Sensing: An ever-growing relationship. *IEEE Geosci. Remote Sens. Mag.* **2016**, *4*, 6–23. [CrossRef]
200. Zhou, Q.; Xian, G.; Shi, H. Gap fill of land surface temperature and reflectance products in landsat analysis ready data. *Remote Sens.* **2020**, *12*, 1192. [CrossRef]
201. Huang, F.; Zhan, W.; Voogt, J.; Hu, L.; Wang, Z.; Quan, J.; Ju, W.; Guo, Z. Temporal upscaling of surface urban heat island by incorporating an annual temperature cycle model: A tale of two cities. *Remote Sens. Environ.* **2016**, *186*, 1–12. [CrossRef]
202. Peres, L.D.F.; Lucena, A.J.D.; Rotunno Filho, O.C.; França, J.R.D.A. The urban heat island in Rio de Janeiro, Brazil, in the last 30 years using remote sensing data. *Int. J. Appl. Earth Obs. Geoinf.* **2018**, *64*, 104–116. [CrossRef]
203. Fu, P.; Weng, Q. A time series analysis of urbanization induced land use and land cover change and its impact on land surface temperature with Landsat imagery. *Remote Sens. Environ.* **2016**, *175*, 205–214. [CrossRef]
204. Lee, S.J.; Balling, R.; Gober, P. Bayesian maximum entropy mapping and the soft data problem in urban climate research. *Ann. Assoc. Am. Geogr.* **2008**, *98*, 309–322. [CrossRef]
205. Yuan, F.; Bauer, M.E. Comparison of impervious surface area and normalized difference vegetation index as indicators of surface urban heat island effects in Landsat imagery. *Remote Sens. Environ.* **2007**, *106*, 375–386. [CrossRef]
206. Chen, F.; Yang, S.; Yin, K.; Chan, P. Challenges to quantitative applications of Landsat observations for the urban thermal environment. *J. Environ. Sci.* **2017**, *59*, 80–88. [CrossRef]
207. Pielke Sr, R.A.; Davey, C.A.; Niyogi, D.; Fall, S.; Steinweg-Woods, J.; Hubbard, K.; Lin, X.; Cai, M.; Kim, Y.K.; Li, H.; et al. Unresolved issues with the assessment of multidecadal global land surface temperature trends. *J. Geophys. Res. Atmos.* **2007**, *112*. [CrossRef]
208. Wong, K.V.; Chaudhry, S. Use of Satellite Images for Observational and Quantitative Analysis of Urban Heat Islands around the World. *J. Energy Resour. Technol. Trans. ASME* **2012**, *134*. [CrossRef]
209. Yan, Z.-W.; Wang, J.; Xia, J.-J.; Feng, J.-M. Review of recent studies of the climatic effects of urbanization in China. *Adv. Clim. Chang. Res.* **2016**, *7*, 154–168. [CrossRef]
210. Imhoff, M.L.; Zhang, P.; Wolfe, R.E.; Bounoua, L. Remote sensing of the urban heat island effect across biomes in the continental USA. *Remote Sens. Environ.* **2010**, *114*, 504–513. [CrossRef]
211. Coutts, A.M.; Harris, R.J.; Phan, T.; Livesley, S.J.; Williams, N.S.G.; Tapper, N.J. Thermal infrared remote sensing of urban heat: Hotspots, vegetation, and an assessment of techniques for use in urban planning. *Remote Sens. Environ.* **2016**, *186*, 637–651. [CrossRef]
212. Ruan, Y.L.; Zou, Y.H. Monitoring the Spatio-Temporal Trajectory of Urban Area Hotspots in Wuhan, China Using Time-Series Nighttime Light Images. *ISPRS Int. Arch. Photogramm. Remote Sens. Spat. Inf. Sci.* **2019**, 71–76. [CrossRef]
213. Chen, L.; Sun, R.; Lu, Y. A conceptual model for a process-oriented landscape pattern analysis. *Sci. China Earth Sci.* **2019**, *62*, 2050–2057. [CrossRef]
214. Jorgensen, S.E. *Integration of Ecosystem Theories: A Pattern*; 1992. Available online: https://link.springer.com/book/10.1007/978-94-011-2682-3 (accessed on 22 July 2021). [CrossRef]
215. Li, L.; Zha, Y. Satellite-based spatiotemporal trends of canopy urban heat islands and associated drivers in China's 32 major cities. *Remote Sens.* **2019**, *11*, 102. [CrossRef]

216. Li, J.X.; Song, C.H.; Cao, L.; Zhu, F.G.; Meng, X.L.; Wu, J.G. Impacts of landscape structure on surface urban heat islands: A case study of Shanghai, China. *Remote Sens. Environ.* **2011**, *115*, 3249–3263. [CrossRef]
217. Zhang, P.; Imhoff, M.L.; Wolfe, R.E.; Bounoua, L. Characterizing urban heat islands of global settlements using MODIS and nighttime lights products. *Can. J. Remote Sens.* **2010**, *36*, 185–196. [CrossRef]
218. Chen, W.; Zhang, Y.; Pengwang, C.Y.; Gao, W.J. Evaluation of Urbanization Dynamics and its Impacts on Surface Heat Islands: A Case Study of Beijing, China. *Remote Sens.* **2017**, *9*, 453. [CrossRef]
219. Schwarz, N.; Lautenbach, S.; Seppelt, R. Exploring indicators for quantifying surface urban heat islands of European cities with MODIS land surface temperatures. *Remote Sens. Environ.* **2011**, *115*, 3175–3186. [CrossRef]
220. Cheval, S.; Dumitrescu, A. The July urban heat island of Bucharest as derived from modis images. *Theor. Appl. Climatol.* **2009**, *96*, 145–153. [CrossRef]
221. Pan, J.H. Area Delineation and Spatial-Temporal Dynamics of Urban Heat Island in Lanzhou City, China Using Remote Sensing Imagery. *J. Indian Soc. Remote Sens.* **2016**, *44*, 111–127. [CrossRef]
222. Xu, Y.M.; Qin, Z.H.; Wan, H.X. Spatial and Temporal Dynamics of Urban Heat Island and Their Relationship with Land Cover Changes in Urbanization Process: A Case Study in Suzhou, China. *J. Indian Soc. Remote Sens.* **2010**, *38*, 654–663. [CrossRef]
223. Wang, J.; Kuffer, M.; Sliuzas, R.; Kohli, D. The exposure of slums to high temperature: Morphology-based local scale thermal patterns. *Sci. Total Environ.* **2019**, *650*, 1805–1817. [CrossRef]
224. De Noronha Vaz, E.; Cabral, P.; Caetano, M.; Nijkamp, P.; Painho, M. Urban heritage endangerment at the interface of future cities and past heritage: A spatial vulnerability assessment. *Habitat Int.* **2012**, *36*, 287–294. [CrossRef]
225. Larsen, L.G.; Eppinga, M.B.; Passalacqua, P.; Getz, W.M.; Rose, K.A.; Liang, M. Appropriate complexity landscape modeling. *Earth Sci. Rev.* **2016**, *160*, 111–130. [CrossRef]
226. Bala, R.; Prasad, R.; Yadav, V.P. Quantification of urban heat intensity with land use/land cover changes using Landsat satellite data over urban landscapes. *Theor. Appl. Climatol.* **2021**. [CrossRef]
227. Rajasekar, U.; Weng, Q.H. Urban heat island monitoring and analysis using a non-parametric model: A case study of Indianapolis. *ISPRS J. Photogramm. Remote Sens.* **2009**, *64*, 86–96. [CrossRef]
228. Szymanowski, M.; Kryza, M. Application of remotely sensed data for spatial approximation of urban heat island in the city of Wrocław, Poland. In Proceedings of the 2011 Joint Urban Remote Sensing Event, JURSE 2011—Proceedings, Munich, Germany, 11–13 April 2011; pp. 353–356. [CrossRef]
229. Audebert, N.; Le Saux, B.; Lefevre, S. Deep learning for classification of hyperspectral data: A comparative review. *IEEE Geosci. Remote Sens. Mag.* **2019**, *7*, 159–173. [CrossRef]
230. Bioucas-Dias, J.M.; Plaza, A.; Camps-Valls, G.; Scheunders, P.; Nasrabadi, N.M.; Chanussot, J. Hyperspectral remote sensing data analysis and future challenges. *IEEE Geosci. Remote Sens. Mag.* **2013**, *1*, 6–36. [CrossRef]
231. Sun, W.; Du, Q. Hyperspectral band selection: A review. *IEEE Geosci. Remote Sens. Mag.* **2019**, *7*, 118–139. [CrossRef]
232. Crisci, C.; Ghattas, B.; Perera, G. A review of supervised machine learning algorithms and their applications to ecological data. *Ecol. Model.* **2012**, *240*, 113–122. [CrossRef]
233. Lucas, T.C.D. A translucent box: Interpretable machine learning in ecology. *Ecol. Monogr.* **2020**, *90*. [CrossRef]
234. Gorelick, N.; Hancher, M.; Dixon, M.; Ilyushchenko, S.; Thau, D.; Moore, R. Google Earth Engine: Planetary-scale geospatial analysis for everyone. *Remote Sens. Environ.* **2017**, *202*, 18–27. [CrossRef]
235. Chu, L.; Oloo, F.; Bergstedt, H.; Blaschke, T. Assessing the link between human modification and changes in land surface temperature in hainan, china using image archives from google earth engine. *Remote Sens.* **2020**, *12*, 888. [CrossRef]
236. Hao, B.; Ma, M.; Li, S.; Li, Q.; Hao, D.; Huang, J.; Ge, Z.; Yang, H.; Han, X. Land use change and climate variation in the three gorges reservoir catchment from 2000 to 2015 based on the google earth engine. *Sensors* **2019**, *19*, 2118. [CrossRef]
237. Oleson, K.W.; Bonan, G.B.; Feddema, J.; Vertenstein, M. An urban parameterization for a global climate model. Part II: Sensitivity to input parameters and the simulated urban heat island in offline simulations. *J. Appl. Meteorol. Climatol.* **2008**, *47*, 1061–1076. [CrossRef]
238. Ghamisi, P.; Rasti, B.; Yokoya, N.; Wang, Q.; Hofle, B.; Bruzzone, L.; Bovolo, F.; Chi, M.; Anders, K.; Gloaguen, R.; et al. Multisource and multitemporal data fusion in remote sensing: A comprehensive review of the state of the art. *IEEE Geosci. Remote Sens. Mag.* **2019**, *7*, 6–39. [CrossRef]
239. Ghamisi, P.; Yokoya, N.; Li, J.; Liao, W.; Liu, S.; Plaza, J.; Rasti, B.; Plaza, A. Advances in Hyperspectral Image and Signal Processing: A Comprehensive Overview of the State of the Art. *IEEE Geosci. Remote Sens. Mag.* **2017**, *5*, 37–78. [CrossRef]
240. Sherafati, S.A.; Saradjian, M.R.; Niazmardi, S. Urban Heat Island growth modeling using Artificial Neural Networks and Support Vector Regression: A case study of Tehran, Iran. *Int. Arch. Photogramm. Remote Sens. Spatial Inf. Sci.* **2013**, 399–403. [CrossRef]
241. Brook, A.; de Micco, V.; Battipaglia, G.; Erbaggio, A.; Ludeno, G.; Catapano, I.; Bonfante, A. A smart multiple spatial and temporal resolution system to support precision agriculture from satellite images: Proof of concept on Aglianico vineyard. *Remote Sens. Environ.* **2020**, *240*. [CrossRef]
242. Ryo, M.; Angelov, B.; Mammola, S.; Kass, J.M.; Benito, B.M.; Hartig, F. Explainable artificial intelligence enhances the ecological interpretability of black-box species distribution models. *Ecography* **2021**, *44*, 199–205. [CrossRef]
243. Wulder, M.A.; Loveland, T.R.; Roy, D.P.; Crawford, C.J.; Masek, J.G.; Woodcock, C.E.; Allen, R.G.; Anderson, M.C.; Belward, A.S.; Cohen, W.B.; et al. Current status of Landsat program, science, and applications. *Remote Sens. Environ.* **2019**, *225*, 127–147. [CrossRef]

244. Shepherd, J.M.A.; Strother, C.T.; Horst, A.; Bounoua, L.; Mitra, C. *Urban Climate Archipelagos: A New Framework for Urban Impacts on Climate*; Earthzine: Washington, DC, USA, 2013. Available online: https://earthzine.org/urban-climate-archipelagos-a-new-framework-for-urban-impacts-on-climate/ (accessed on 13 August 2020).
245. Jeevalakshmi, D.; Narayana Reddy, S.; Manikiam, B. Land surface temperature retrieval from LANDSAT data using emissivity estimation. *Int. J. Appl. Eng. Res.* **2017**, *12*, 9679–9687.
246. Hayhoe, K.; VanDorn, J.; Croley, T.; Schlegal, N.; Wuebbles, D. Regional climate change projections for Chicago and the US Great Lakes. *J. Great Lakes Res.* **2010**, *36*, 7–21. [CrossRef]

www.ingramcontent.com/pod-product-compliance
Lightning Source LLC
LaVergne TN
LVHW071304170726
843515LV00006BA/1479

MDPI
St. Alban-Anlage 66
4052 Basel
Switzerland
Tel. +41 61 683 77 34
Fax +41 61 302 89 18
www.mdpi.com

Land Editorial Office
E-mail: land@mdpi.com
www.mdpi.com/journal/land